super SAFE mark

TRAINER'S KIT

RETAIL BEST PRACTICES TO FOOD SAFETY AND SANITATION

David McSane, H.S.D. • David E. Morrow • Anna Graf Williams, Ph.D.
Richard Linton, Ph.D. • Nancy R. Rue, Ph.D.

Production Services and Development by Learnovation ®, LLC

FMI
FOOD MARKETING INSTITUTE

Editor-in-Chief: Stephen Helba
Acquisitions Editor: Vernon R. Anthony
Developmental Editors: Anna Graf Williams & David E. Morrow
Desktop Publishing: Karen J. Hall & George D. Williams
Copy Editor: Cheryl Pontius
Illustrations: John Wise

Pearson Education LTD.
Pearson Education Singapore, Pte. Ltd
Pearson Education, Canada, Ltd
Pearson Education–Japan
Pearson Education Australia PTY, Limited
Pearson Education North Asia Ltd
Pearson Educación de Mexico, S.A. de C.V.
Pearson Education Malaysia, Pte. Ltd

PEARSON
Prentice
Hall

Printed in the United States of America

10 9 8 7 6 5 4 3 2

ISBN 0-13-095038-6

WARNING: ©COPYRIGHTED MATERIAL

We are pleased that you have purchased a copy of this Trainer's Kit and hope you find it to be a highly affective training tool. Please be advised that this is a copyright protected publication, and you must obtain permission for any re-use or duplication, in writing, from the publisher.

The publisher is available to customize this Trainer's Kit to fit your specific needs and will happily work with you to make this product one you can succeed with. However, failure to abide by the copyrighted status of this publication is against the law and violators will be prosecuted.

Thank you for your cooperation!

PEARSON
Prentice Hall

FOOD MARKETING INSTITUTE

Retail Best Practices and Guide to Food Safety and Sanitation Trainers Kit

Table of Contents

Note from FMI

Food safety has always been a paramount priority for the food industry. Today it is even more so due to improved methods of detecting safety hazards and to increased consumer concerns about both food safety and security. As the last link in the global supply chain that delivers food to consumers, retailers play a special role in protecting public health.

In recent years, the industry, government and scientific community have designed effective programs to safeguard the food supply. Some are required by law, and others are embodied in good retail practices, standard operating procedures and training that teaches lasting lessons. The programs may vary, but all feature the same basic principles, including sanitation, employee health and hygiene - all aimed to prevent contamination before customers purchase food products.

The guidelines in the following pages are one of several educational tools developed through the cooperative efforts of the Food Marketing Institute and Prentice Hall. The authors of this manual are among the world's leading food safety experts and educators. They bring decades of experience training retail food handlers and managers how to systematically prevent contamination.

This training manual identifies best practices that prevent foodborne illnesses at the retail level. Knowing these is only the beginning in fully understanding food safety. Apply this knowledge daily each time you prepare, handle, display or offer food. These practices will help ensure that your customers become loyal patrons, confident your food is safe and wholesome.

Tim Hammonds
President and CEO
Food Marketing Institute

Session 1

How to Use the Trainer's Kit

The *Retail Best Practices and Trainer's Guide to Food Safety and Sanitation* consists of the trainer's kit manual and four compact disks consisting of:

1. CD1: 16 Hour PowerPoint Slideshow
2. CD2: 8 Hour PowerPoint Slideshow
3. CD3: Posters
4. CD4: Image Bank

Trainer's Kit Manual:

The Trainer's Kit manual consists of 11 training sessions. The information in each training session corresponds directly to chapters in the *Retail Best Practices and Guide to Food Safety and Sanitation* and to the *Retail Best Practices and Supervisor's Guide to Food Safety and Sanitation.*

Each training session is comprised of lesson pages. The lesson pages consist of:

- ➢ The slide that is being seen by the learner.
- ➢ The essential information from which you will teach. The slide shows directly correspond to the lesson pages so your learners may follow your lesson.

Lesson page description:

- The key information to be delivered is displayed on the slide.
- Additional information to be delivered has been bulleted for easy reference during instruction and reduced preparation time. Much of this information is fundamental to the success of the learner.
- You will encounter certain words in boldface, underlined print. These words are the **Essential Terms** from the *Retail Best Practices and Guide to Food Safety and Sanitation* and are considered to be important words the learners need to fully understand.
- Slide mapping – Page numbers appear at the lower right hand portion of each slide. These numbers correspond to a page, or set of pages in both the *Retail Best Practices and Guide to Food Safety and Sanitation* and the *Retail Best Practices and Supervisor's Guide to Food Safety and Sanitation* where the information on the slide can be found.

Types of Slides

The Microsoft® PowerPoint© 2000 slide shows consist of six different types of slides you will use to deliver and review the necessary information. These slides include:

1) Session Introduction slide
    - Session setup.
2) Points to Ponder slide(s)
    - Learning objectives.
3) Lesson slides.
    - Core information
4) Safety Stop! slide(s).
    - Summary breaks
5) Concepts to Keep slide(s).
    - Session review.
6) Take This Back To Your Team! slide(s)
    - Information to take back into operations.

1) Session Introduction Slide

The first lesson page in every session contains an introduction slide and information you may use to set up your session. The information on the first lesson page is designed to help you customize the content of the training program to effectively meet the needs of your learners. The introductory information includes:

- An overview of the topics to be discussed in the session.
- A brief description of the ***Safety Stop!*** slides (see page 4) found within the session, including the number of:
    - ***Safety Stop!*** slides included in the session
    - ***Sample Test Questions*** included in the session
    - ***Real-life*** scenarios included in the session
    - ***Making it Real*** practical exercises included in the session.
- Timelines for the 16-hour and 8-hour training courses. A suggested sample agenda for both the 8-hour and 16-hour trainings can be found in Appendix B.
    - In the 16-hour course, it is suggested you review all the slides and further develop the training with the **Sample Test Questions, Real-life**, and **Making it Real** review exercises provided at each **Safety Stop!** The **Sample Test Questions** and **Making it Real** exercises allow the learner to become more involved. The **Real-life** examples will drive it all home.
    - The 8-hour course allows enough time to review the information contained in the 8-hour slide show. Specific slides have been "hidden" in the slide show to allow for a shortened version of the training that is still effective. A full list of the hidden slides in the 8-hour set can be found in Appendix A. The 8-hour slide set is a suggested set of slides for the 8-hour training; however, you can customize them to fit your specific training needs. If time allows, you may choose to incorporate some of the training exercises found on the **Safety Stop!** slides in the 16-hour session.
- Suggestions for videos, posters, and additional references have been provided which can be used to enhance your training session.

Example: Session Introduction Slide

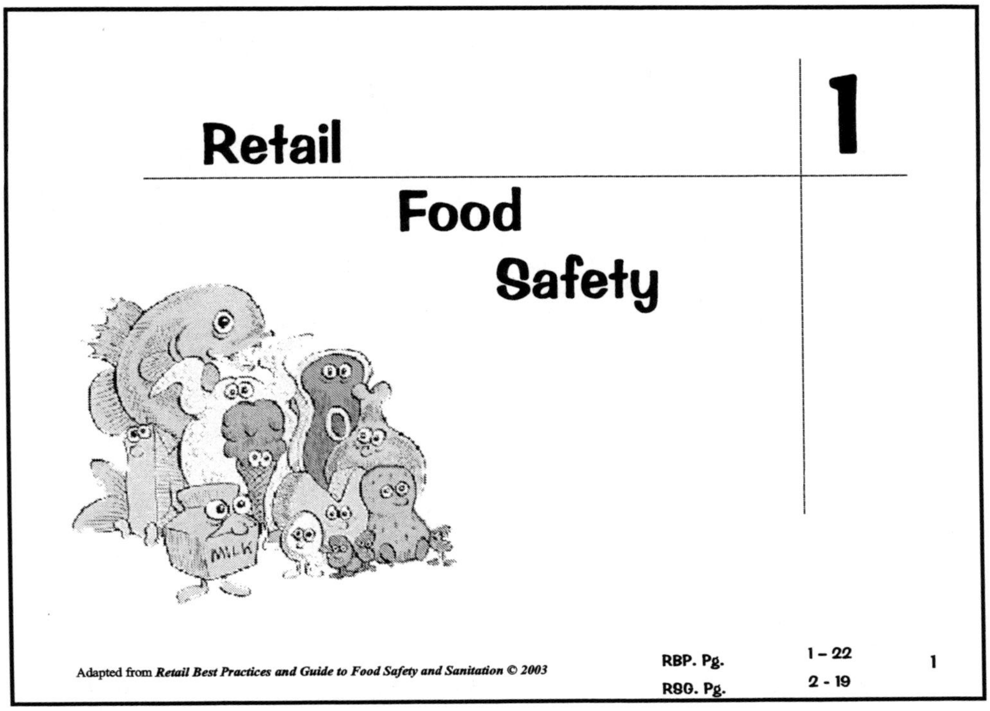

2) Points to Ponder Slides

The next slide(s) you encounter will be the session's *Points to Ponder*. The *Points to Ponder* stem from the *Learning Objectives* in the *Retail Best Practices and Guide to Food Safety and Sanitation*, and they complement the *Learn How To's* in the *Retail Best Practices and Supervisor's Guide to Food Safety and Sanitation* as well. They have been written in question form to allow you to break the ice with your learners and to better understand how much time you will need to spend on the session. Space has been left between each point allowing you to make notes on the answers you are seeking for each question. The answers to the questions are found in the information throughout the session.

If your learners can answer the questions with ease, you may choose to spend less time on the session; however, if they struggle with the questions, you may choose to spend more time on the session. These questions are written as an introduction to each session; however, they are equally as effective as a review of each session.

Example: Points to Ponder Slide

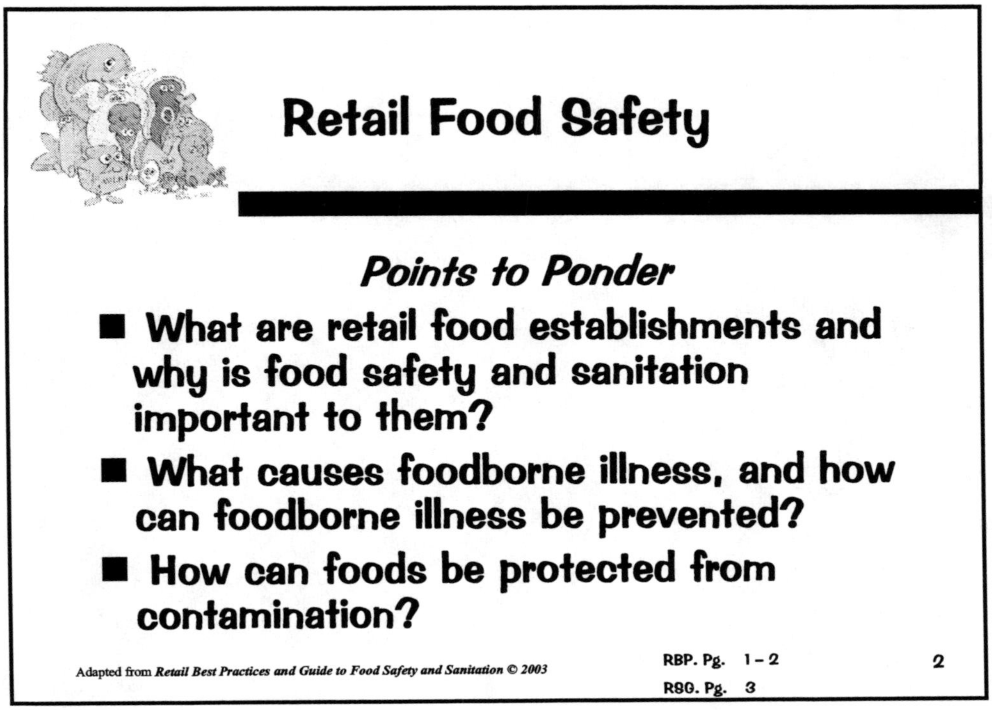

3) Lesson Slides

Lesson slides contain the core information from which you will teach (see page 1 of "How to Use the Trainer's Kit" for information on "Lesson page description"). The information has been broken down into short phrases or bulleted lists allowing you easy access to the key points. Your learners will only see the information contained on the slide. It is up to you to get the additional information contained on the lesson page across to your learner.

Example: Lesson Slide

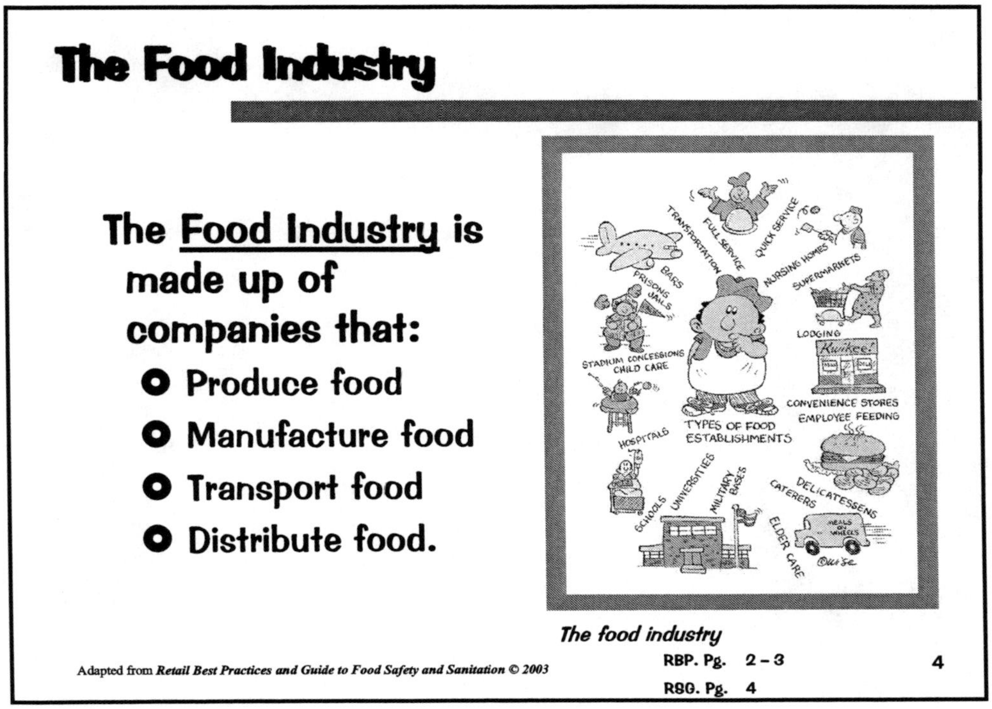

4) Safety Stop! Slides

Throughout each session, you will encounter ***Safety Stop!*** slides. The number of ***Safety Stop!*** slides varies in each session. These slides provide you with logical break points during longer sessions, as well as the opportunity to summarize and review the important information that was just covered.

For 8-hour training courses, all the ***Safety Stop!*** slides have been "hidden." However, you may choose to utilize some of the practical exercises at each ***Safety Stop!*** It is suggested all the practical exercises be used for the 16-hour training courses. This information is vital to first-time learners and will help reinforce key concepts from the session to even the most experienced food safety employees.

Each ***Safety Stop!*** slide includes three different types of practical exercises:

- **Sample Test Questions**
  - The **Sample Test Questions** offer sample test questions on the lesson.
  - Utilize **Sample Test Questions** to facilitate class discussions on individual topics or to quiz knowledge.
- **Real-life Exercises**
  - Short stories that describe a potential real-life occurrence in the work environment.
  - Directly links the topic being discussed to an everyday experience of your learners.
  - **Real-life** helps your learners relate more closely to the topic.
  - **Real-life** can enhance the important information.
  - Ask learners to share similar real-life experiences related to the topic to encourage further discussions.

- **Making it Real** (Classroom Activities)
  - These require a little pre-planning. Many of these require you to bring supplies with you into class.
  - These exercises provide learners the opportunity to learn through the sensory evaluation techniques of sight, sound, smell, and touch.
  - Many exercises allow the learners to physically perform an activity (such as calibrating the dial-faced bi-metallic thermometer).
  - Some exercises involve allowing the learner to see, touch, and sometimes experiment with important food safety utensils.

Example: Safety Stop! Slide

5) Concepts to Keep Slides

The ***Concepts to Keep*** slides highlight some of the most important concepts taught in the lesson. These should be used to review the entire session. Some larger lessons may have two to three ***Concepts to Keep*** slides. This information is extremely important when it comes to keeping food safe in retail food establishments.

Example: Concepts to Keep Slide

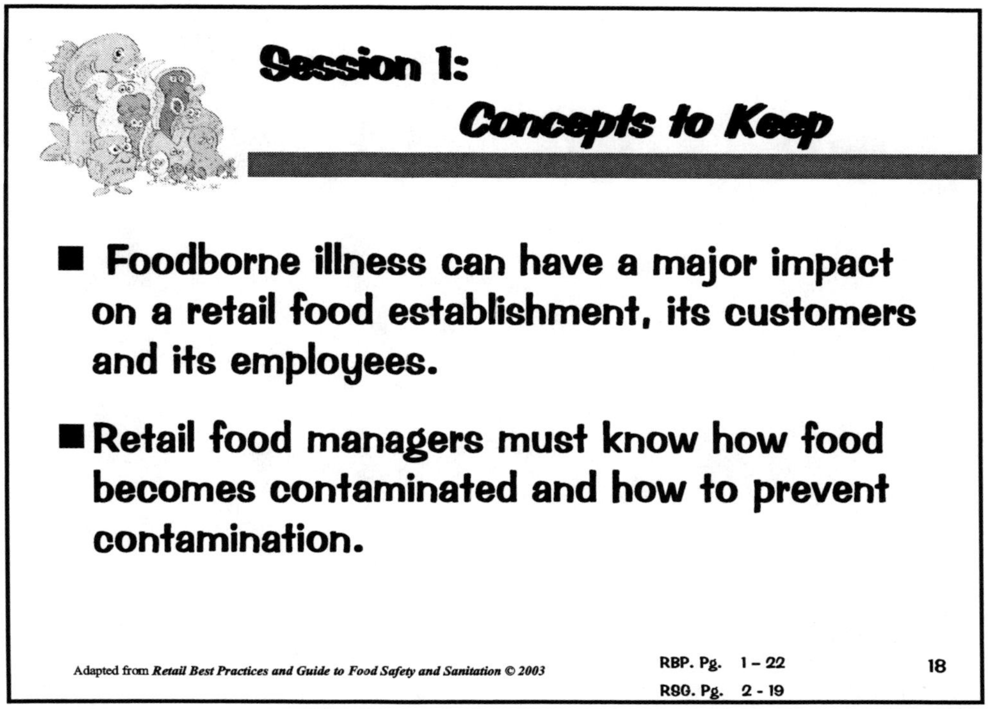

6) Take This Back To Your Team! Slides

This slide contains any information reviewed in the session that would be helpful for the learner to take back to their operation. The information contained here should be shared with employees who may not be certified in food safety in order to communicate and emphasize good food safety and sanitation practices.

Example: Take This Back To Your Team! Slide

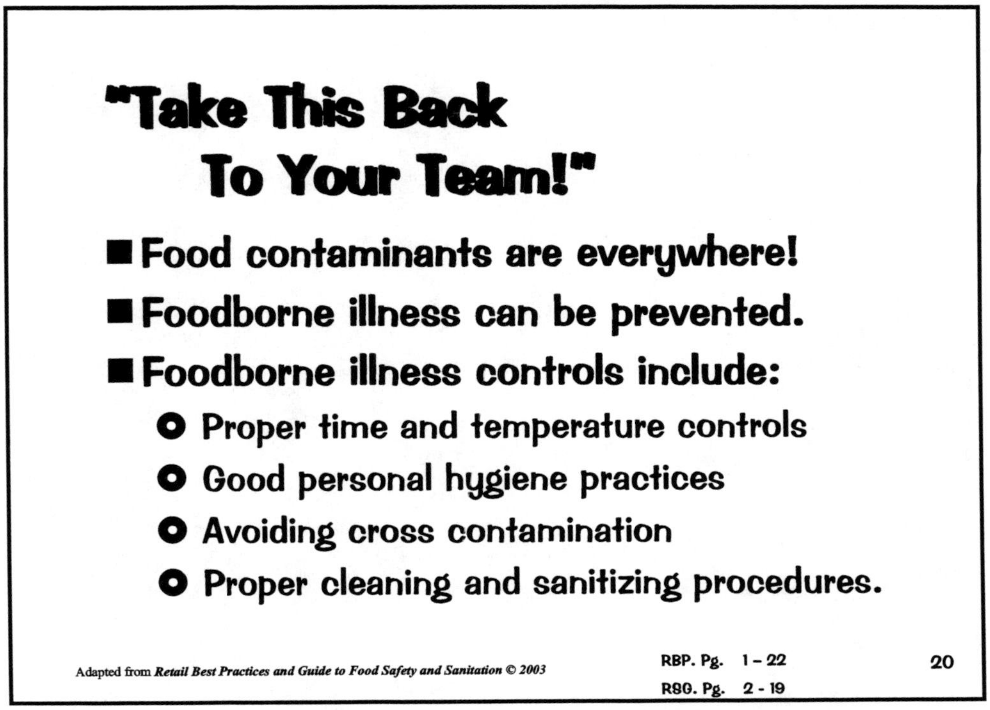

Compact Disks

Slide Show CDs

CD1: 16 Hour PowerPoint Slideshow
CD2: 8 Hour PowerPoint Slideshow

The first two CDs contain the 16-hour and 8-hour Microsoft® PowerPoint© 2000 slide shows you will be showing to your learners. The CDs include a self-executing PowerPoint© slide show viewer, which will allow you to run and view the PowerPoint© slides on your computer without having Microsoft® PowerPoint© 2000 installed on the computer. *In order to customize the slide shows, you must own a copy of Microsoft® PowerPoint© and have it installed on your computer.*

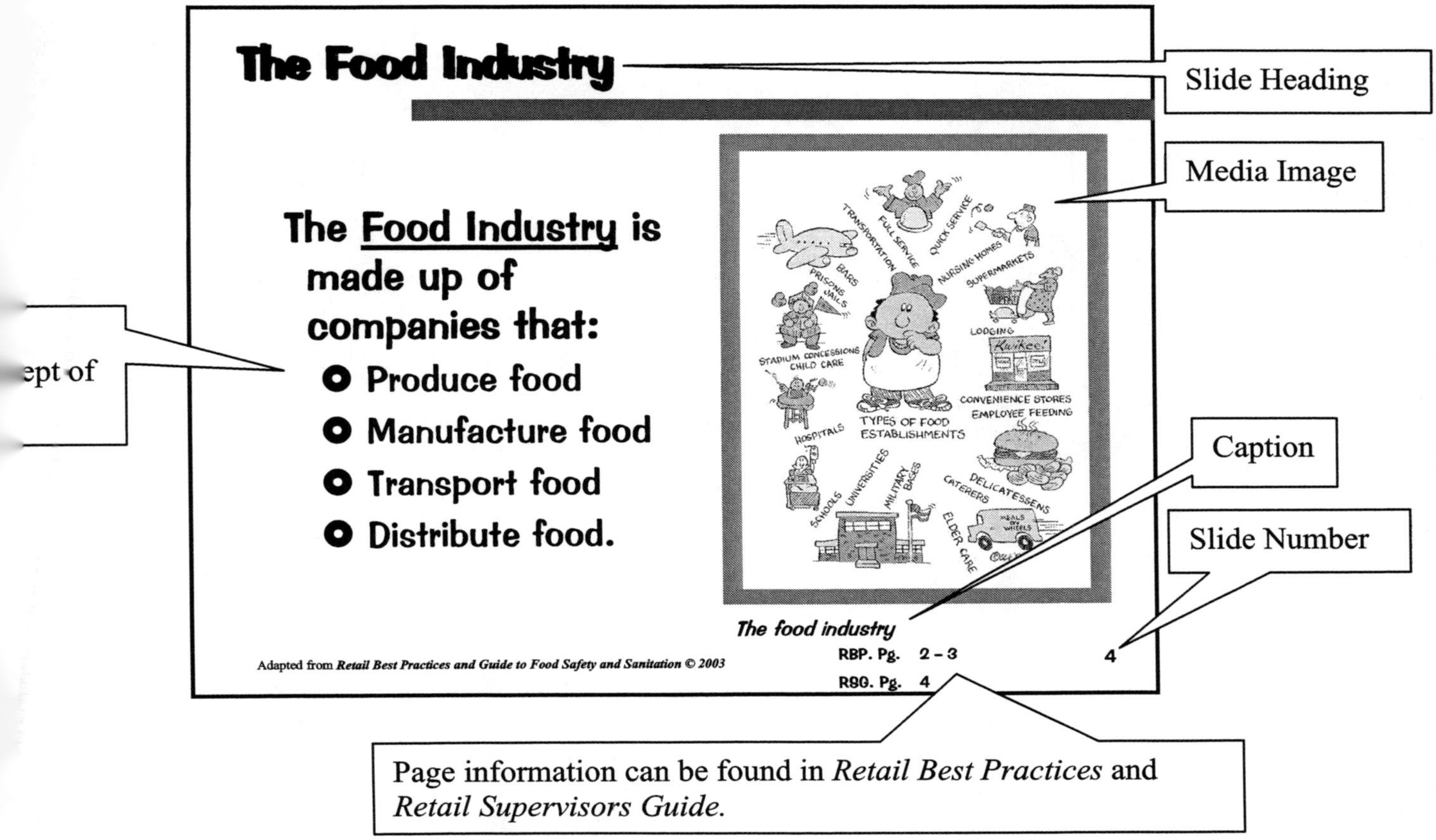

- The Microsoft® PowerPoint© 2000 slide show is organized by sessions. Each session has a variety of slide types including:
  - **Introduction Slide** (1 slide) – This slide allows you to do any classroom management or setup for the upcoming session.
  - **Points to Ponder** (1 to 3 slides) – These are your learning objectives. Some larger lessons may have two to three Points to Ponder slides.
  - **Lesson Pages** (multiple slides) – These are the slides that contain the core of the information for the session.
  - **Safety Stop!** (Varies depending upon session) – Filled with sample test questions and practical exercises you can use to review key points and to customize your training.
  - **Concepts to Keep** (1 to 3 slides) – Here is your review of the entire session. Easy to read, bulleted highlights of the needed information. Some larger lessons may have two to three Concepts to Keep slides.
  - **Take This Back To Your Team!** (1 to 2 slides) – This slide contains any information reviewed in the session that would be helpful for the learner to review with his or her staff members. The information contained here will assist employees that may not have to be certified to utilize good food safety and sanitation practices.
- The slides have illustrations and photos, as well as large-print key points.
- The bottom right-hand corner of each slide offers a page reference to both the *Retail Best Practices and Guide to Food Safety and Sanitation* and

Retail Best Practices and Supervisor's Guide to Food Safety and Sanitation textbooks so you, or the learner, can easily find more detailed information on the topic you are reviewing.

Microsoft PowerPoint© Slideshows: Viewing and Installation

(See Appendix F)

The PowerPoint© Training Sessions can be run from the enclosed CD (CD1: 16 Hour PowerPoint Slideshow, or CD2: 8 Hour PowerPoint Slideshow) or downloaded onto your computer's hard drive and run from there. In order to run the Training Sessions on your computer you need to have Microsoft PowerPoint© installed on your computer.

Running PowerPoint© Training Sessions from the enclosed CD

1) Insert Compact Disk (16-hour or 8-hour) into your computer's CDROM drive.
2) Click on the "Start" button in the bottom left corner of the screen.
3) Click on "Run"
4) Type the name of the drive you are using (usually the "D" drive) and a colon, i.e. "D:"
5) A screen will appear that shows all of the folders on the CD
   - Session 1
   - Session 2
   - Session 3
   - Session 4
   - Session 5
   - Session 6
   - Session 7
   - Session 8
   - Session 9
   - Session 10
   - Session 11
   - A "Read Me" file
6) Double-click on the training session you want to open.
7) To run the training session as a slideshow:
   - Click on View
   - Click on Slideshow
   - Or press F5

Installing the Training Sessions onto Your Computer

Before beginning installation, it is suggested all other programs be closed. Before you attempt to install the Training Sessions on your computer, it is suggested you carefully read all of the instructions first. Then follow the instructions as you install your new PowerPoint© Training Sessions onto your computer.

1) Insert Compact Disk (16-hour or 8-hour) into your computer's CDROM drive.
2) Click on the "Start" button in the bottom left corner of the screen.
3) Click on "Run"
4) Type the name of the drive you are using (usually the "D" drive) and a colon, i.e. "D:"

5) A screen will appear that shows all of the folders on the CD
   - Session 1
   - Session 2
   - Session 3
   - Session 4
   - Session 5
   - Session 6
   - Session 7
   - Session 8
   - Session 9
   - Session 10
   - Session 11
   - A "Read Me" file
6) <u>Double-click on the training session</u> you want to save to your hard drive. This will open the file.
7) <u>Click on File</u>
8) <u>Click on Save As</u>
9) <u>Open the directory</u> (or create a directory) where you want to save the file
10) <u>Click Save.</u>

You may want to create a directory to house you new PowerPoint© Training Sessions on. To do this, follow these instructions:

<u>Creating a Directory (Optional)</u>

It is suggested you create a directory on your computer to house your new PowerPoint© Training Sessions. To do this, follow these instructions:

1) Follow the above instructions (PowerPoint© Training Session Installation) to <u>Step 9</u>
2) Click on the down arrow to the left of the box called "Save In"
3) <u>Click on the "C"</u> drive (or the drive where you want to house your training sessions).
4) <u>Click on the "Create New Folder"</u> icon to the right of the "Save In" box (looks like a file folder with an asterisk [*] above it)
5) <u>Type in a name for the folder.</u> The folder name should be no longer than 8 letters and not contain spaces. We suggest "FoodSafe"
6) <u>Click OK</u>
7) <u>Click Save.</u>

<u>Poster CD</u>

CD3: Posters

The third CD contains 11 easily printable food safety posters. The posters can be viewed through the "Explore" menu (found by "right clicking" over the "Start" button) on your Microsoft Windows-based computer.

- You can print the posters in a couple of different ways:

  - Print posters from the "Explore" menu.
  - Bring the CD to a local printer or copy store (such as Kinko's) and ask them to print the posters in color at any size.
- The posters focus on the five key areas in food safety and sanitation:
  - Proper personal hygiene
  - Time and temperature management
  - Preventing cross contamination
  - Cleaning and sanitizing
  - Management of food safety.
- A thumbnail printout of each poster can be found in Appendix E.
- The posters are great tools to assist in reinforcing key topics during training sessions.

Image Bank CD

CD4: Image Bank

The fourth CD contains photos and illustrations that can be used to customize additional slides you may choose to add to the presentation, or to change the message being delivered in a current slide to better fit your operation. Simply open your installed Trainer Session file and use the following steps to insert a photo.

1. Open the presentation you wish to edit.
2. Scroll to the slide you wish to edit.
3. Click on "Insert."
4. Place your mouse over "Picture."
5. Click on "From File."
6. From the "Insert Picture" box, choose your compact disk drive.
7. Double click on the picture you want to insert.

To delete a photo, illustration, or text:

1. Click on the image you want to delete.
2. Press delete.

Resizing an Image

All of the images are saved at a specific size, and if they are enlarged too much, especially those with text or numbers, they will get blurry. If you want the photo to be larger, the best solution is to open the photo in a photo software (i.e., Photoshop©) and save the photo to a larger size.

They can also be resized in PowerPoint©, but you run the risk of blurring the image this way. To resize a photo in PowerPoint©, click and drag the corner of the photo, not the edges. Better still

1. Click on the photo
2. Click on Format
3. Click on Object
4. Click on Size
5. Adjust the height/width from there.

Use all of these customizable tools to make your training sessions fun, educational, and effective. Only you know the education level, the timelines, and the individual needs of your learners. Utilize only the tools you need to best impact food safety and sanitation in your employees' environment.

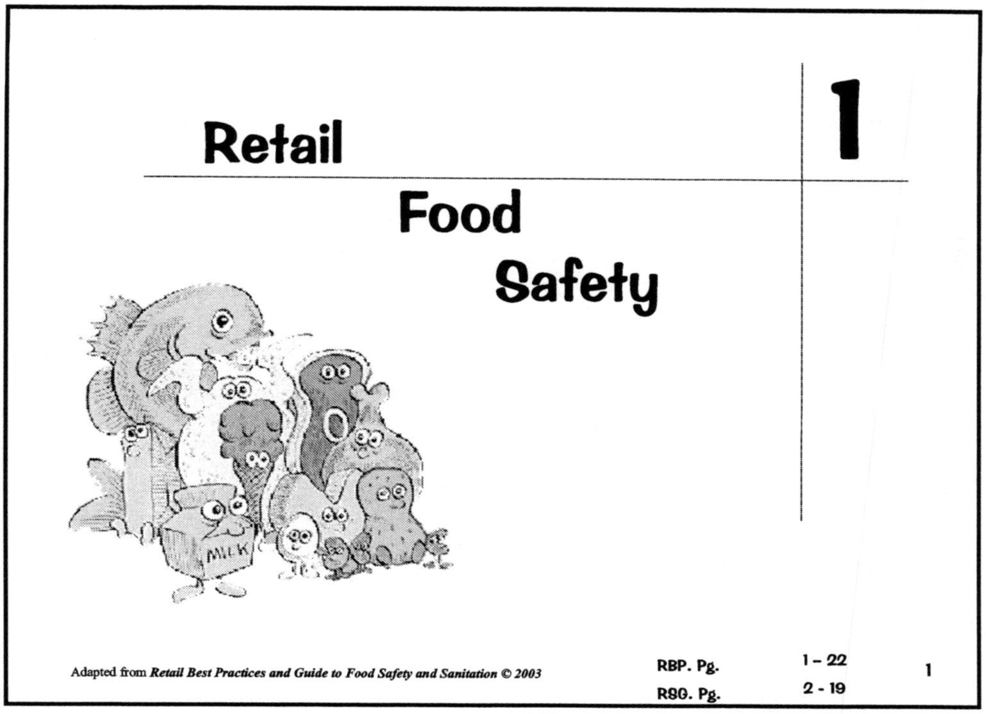

Retail Food Safety – Session 1 Key Points

Slide 1

Training Tip:
The content of each slide provides the most essential information in the session. The written text below each slide is provided to help you understand and better teach the essential information. The written text may also be used to answer questions that may be posed during the session. Depending on the length of your training session, you may not be able to address all the information contained beneath each slide.

Overview:
Session 1 briefly discusses the topic of foodborne illness and overviews important food safety and sanitation concepts from the remaining 10 sessions. The learner should leave this session with an understanding of what foodborne illness is, how it affects the retail food industry, and some basic preventive measures for foodborne illness.

Suggested Timeline:
8-hour course: 20 minutes using the PowerPoint© slides only

- 15 ***Slides.***

16-hour course: 30 minutes using the 20 PowerPoint© slides and the 1 ***Safety Stop*** slide containing:

- 10 ***Sample Test Questions***
- 6 ***Real-life*** scenarios
- 3 ***Making it Real*** interactive exercises.

Video suggestions and supplemental resources:

Video

The Invisible Challenge, Food Marketing Institute, 1-301-843-3084

Retail Food Safety

Points to Ponder

- What are retail food establishments and why is food safety and sanitation important to them?
- What causes foodborne illness, and how can foodborne illness be prevented?
- How can foods be protected from contamination?

Adapted from *Retail Best Practices and Guide to Food Safety and Sanitation* © 2003

RBP. Pg. 1 – 2

RSG. Pg. 3

2

Slide 2

Points to Ponder

- What are retail food establishments and why is food safety and sanitation important to them?

- What causes foodborne illness, and how can foodborne illness be prevented?

- How can foods be protected from contamination?

Retail Food Safety

More Points to Ponder

- What is the role of the retail food manager in food safety and sanitation?
- Why is food protection manager certification important?

Adapted from *Retail Best Practices and Guide to Food Safety and Sanitation* © 2003

RBP. Pg. 1 – 2
RSG. Pg. 3

3

Slide 3

More Points to Ponder

- What is the role of the retail food manager in food safety and sanitation?

- Why is food protection manager certification important?

The Food Industry

The Food Industry is made up of companies that:

- Produce food
- Manufacture food
- Transport food
- Distribute food.

The food industry

Adapted from *Retail Best Practices and Guide to Food Safety and Sanitation* © 2003

RBP. Pg. 2 – 3
RSG. Pg. 4

4

Slide 4

The **food industry** is made up of companies that:

- Produce food
- Manufacture food
- Transport food
- Distribute food.

Food establishments are facilities involved in food distribution, such as:

- Restaurants
- Supermarkets
- Institutions (hospitals, schools, hotels, etc.)
- Convenience stores
- Caterers
- Temporary food services.

Retail Food Establishments

- **Facts and figures according to the Food Marketing Institute:**
  - There are more than 127,000 retail food establishments averaging more than 49,225 items in inventory
  - The retail food establishment industry employs more than 3.5 million people.

Retail food establishment

Adapted from *Retail Best Practices and Guide to Food Safety and Sanitation* © 2003 RBP. Pg. 3 - 4 RSG. Pg. 4 - 5 5

Slide 5

The term **<u>retail food establishment</u>** includes:

- Any store selling a line of dry retail food, canned goods or non-food items plus perishable items.

There are 15 different store formats currently operating in the retail food industry exceeding $570 billion in sales in the year 2000. These include:

- Conventional supermarket – the original format offering groceries, meats and produce
- Superstore – have at least 40,000 sq. ft. of selling area offering an expanding selection of non-food items (at least 10% GM (General Merchandise)/HBC (Health, Beauty, and Cosmetics)
- Food/Drug combo – combination of superstore and drugstore
- Warehouse store – low margin grocery store with a reduced variety of products
- Limited assortment store – barebones, low-priced grocery with limited services and perishables
- Other – small corner grocery stores with limited selections of staples and convenience goods
- **<u>Convenience Store</u>**
  - Traditional – small, higher margin store offering staples, non-food, and convenience foods
  - Petroleum-based – gas stations with a traditional convenience store attached
- Nontraditional grocery:
  - Hypermarket
  - Wholesale club
  - Mini-club
  - Supercenters
  - Deep discount drug stores
  - Internet.

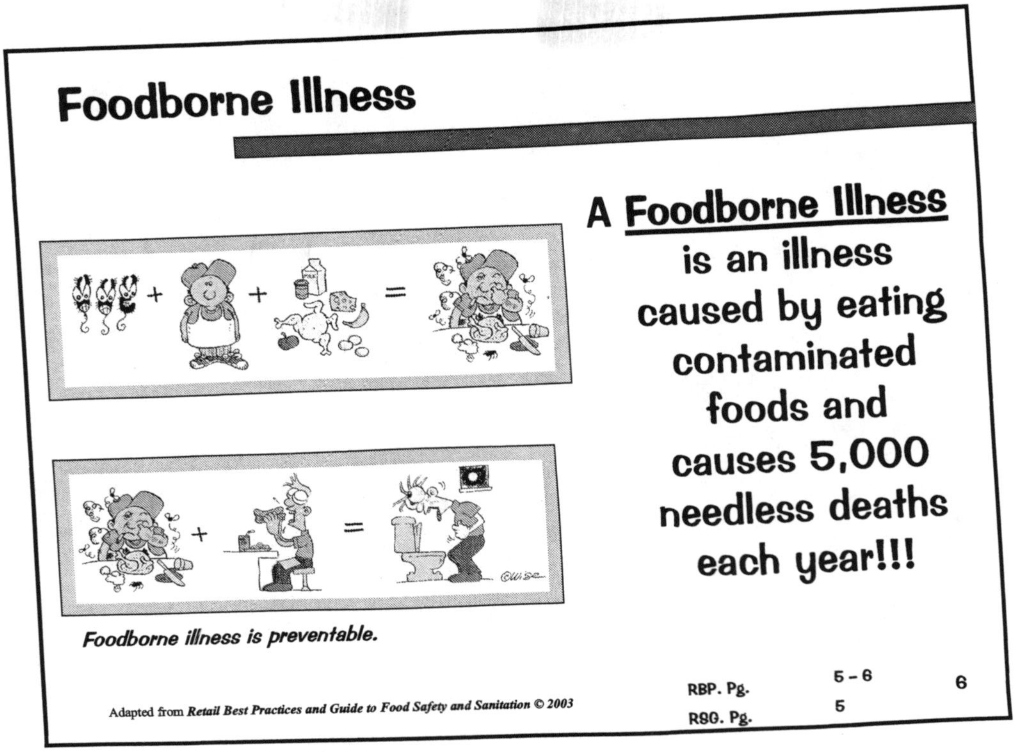

Slide 6

A **foodborne illness** is:
An illness caused by the consumption of a contaminated food.
According to the Centers for Disease Control, foodborne illness:

- Causes 76 million illnesses each year
- Causes 325,000 hospitalizations each year
- Causes 5,000 deaths each year.

Foodborne illness:

- Affects millions of people each year
- Costs the retail food industry billions of dollars each year
- Can result in legal action and damaged reputation
- May cost you your job.

Most cases of foodborne illness in retail food establishments are caused by foods that have been:

- Improperly cooked and/or held at improper temperatures
- Handled by infected employees who practice poor personal hygiene
- Exposed to disease-causing agents by cross contamination
- In contact with contaminated equipment not properly cleaned and sanitized
- Obtained from unsafe sources.

The Food Marketing Institute reports:

- Nearly 75% of shoppers are completely satisfied or mostly confident the food the establishments is safe
- Seven out of 10 shoppers say product safety is "very important" when they sho

y buy in retail food

p for food.

Preventing Foodborne Illness

- Control Time and Temperature
- Practice proper Personal Hygiene techniques
- Avoid Cross Contamination
- Properly Clean and Sanitize food-contact surfaces.

Retail food managers can prevent foodborne illness.

Adapted from *Retail Best Practices and Guide to Food Safety and Sanitation* © 2003

RBP. Pg. 7 – 8
RSG. Pg. 5 - 7

7

Slide 7

As a retail food manager, you must understand foodborne illness can be prevented if the basic rules of food safety are routinely followed.

Foodborne illness can be prevented in many ways including:

- Controlling the time and temperature of food
- Cooking foods to proper temperatures
- Holding potentially hazardous foods at the proper temperatures
- Following proper personal health and hygiene guidelines
- Proper daily health habits, including hand washing
- Preventing cross contamination
- Preventing the transfer of harmful microorganisms or chemicals between foods
- Following proper cleaning and sanitizing procedures.

Trends in food consumption are changing and retail food establishments are changing to meet consumer needs:

- Consumption of fresh produce is on the rise
- Convenience foods are more popular
- In-store food preparation is increasing.

Key food safety issues include:

- *Listeria monocytogenes* in ready-to-eat processed foods
- Shiga toxin-producing *Escherichia coli* in raw meat and unprocessed fruit juices
- Hepatitis A virus in deli sandwich operations and shellfish
- Parasites in fresh produce.

More than half the food eaten by Americans is prepared outside the home.

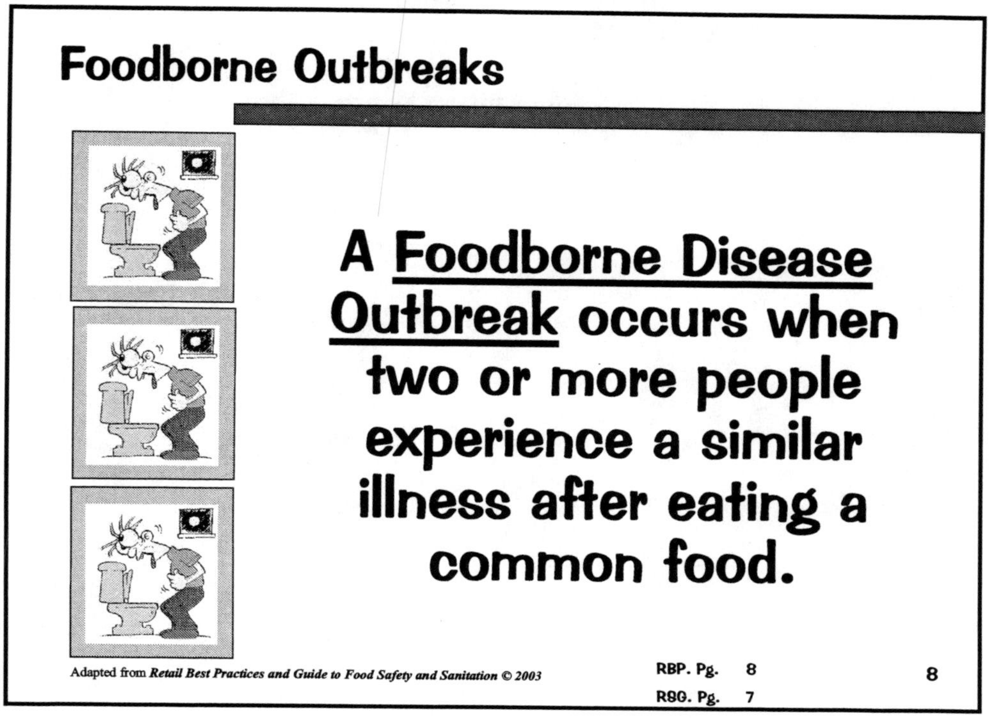

Slide 8

Foodborne disease outbreak:

- An incident in which two or more people experience a similar illness after eating a common food.

Recent foodborne disease outbreaks include:

- Shiga toxin-producing *Escherichia coli* bacteria in lettuce, unpasteurized apple cider, and radish sprouts
- *Salmonella spp*. in cut melons, alfalfa sprouts, ice cream, and dry cereal
- Hepatitis A virus in raw and lightly cooked oysters
- *Listeria monocytogenes* in hot dogs and luncheon meats.

Outbreaks of foodborne illness receive widespread media attention. As a result, public awareness about the problem, and how it occurs, has increased dramatically in the last few years.

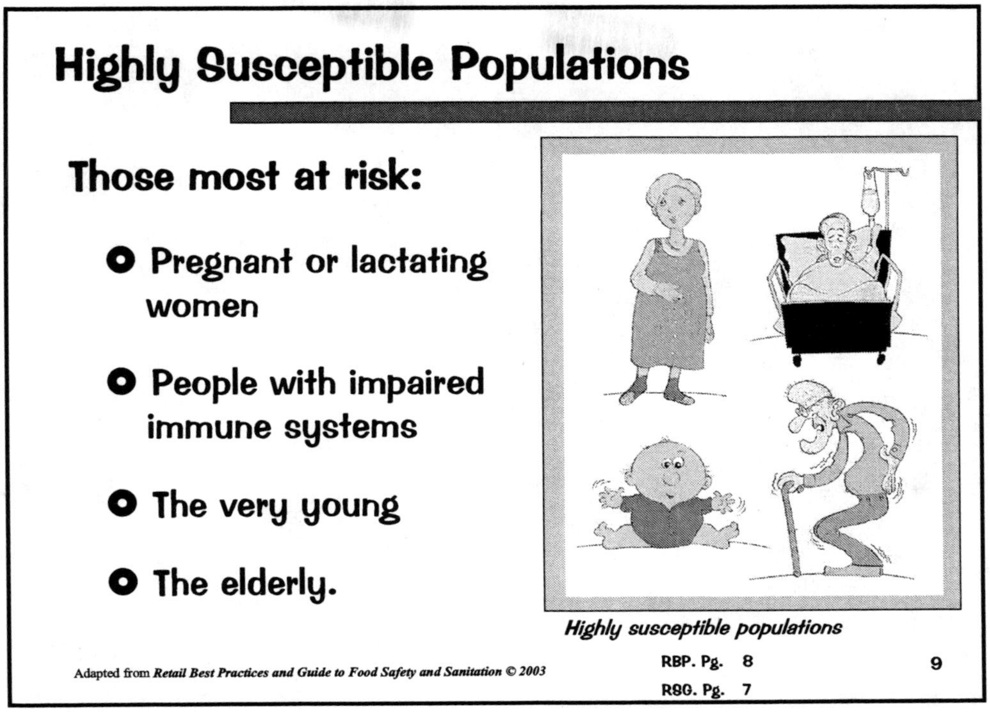

Slide 9

A growing number of people are especially susceptible to foodborne illness. Examples of highly susceptible populations include:

- Pregnant or lactating women
- People with impaired immune systems due to cancer, AIDS, HIV, diabetes, or medications that suppress response to infection
- Very young children
- The elderly.

Foodborne illness can cause severe reactions, even death, for individuals in these highly susceptible categories.

The availability of a safe food supply is critical to these people.

Sources of Contamination

Contaminants can come from anywhere and can include:

- Soil
- Water
- Humans
- Contact surfaces
- And MORE!!!

Sources of contamination

Adapted from *Retail Best Practices and Guide to Food Safety and Sanitation* © 2003

RBP. Pg. 9 – 11
RSG. Pg. 8

10

Slide 10

Contamination is:

- The presence of substances or conditions in food that can be harmful to humans.

Contamination can occur at a variety of points as food flows from the farm to the table. Common sources of contamination include:

- Soil
- Water
- Humans
- Contact surfaces
- Air
- Plants
- Animals
- Packing materials
- Ingredients.

Many types of food contamination can cause illness without changing the appearance, odor, or taste of food.

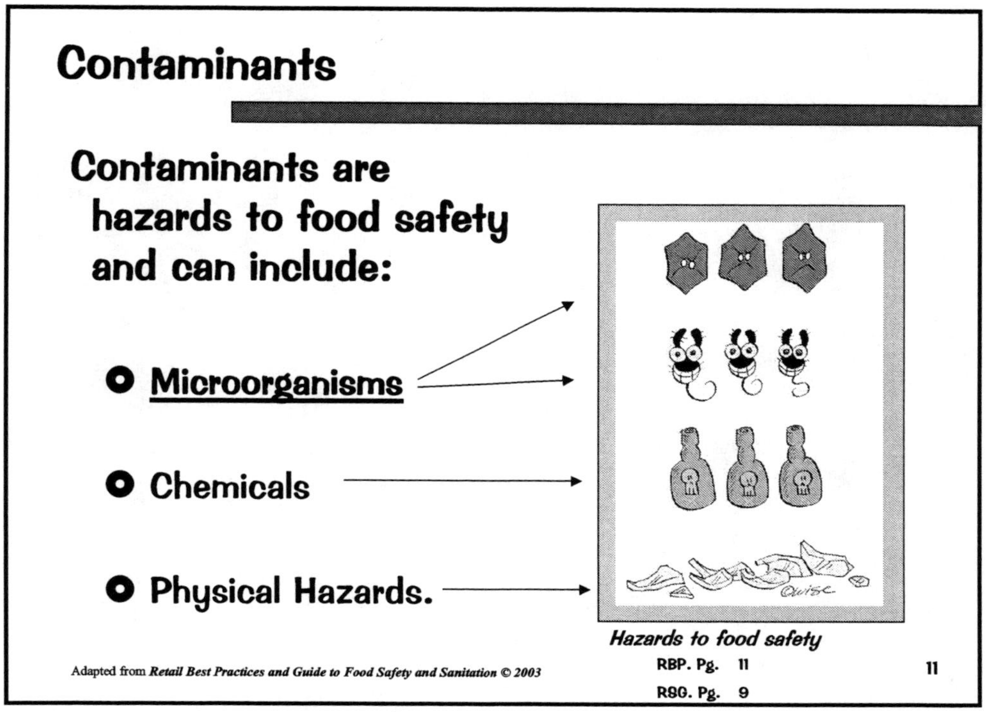

Slide 11

Common contaminants include:

- **Microorganisms** are bacteria, viruses, molds, and other tiny organisms that are too small to be seen with the naked eye. The organisms are also referred to as microbes because they cannot be seen without the aid of a microscope.
- Chemicals are toxic substances that may occur naturally or may be added during the processing of food.
- Physical hazards are hard or soft foreign objects in food that can cause illness and injury.

Microorganisms or **microbes** are:

- Living creatures that can only be seen with a microscope
  - They present an “invisible challenge” because they cannot be seen with the naked eye
  - **Germ** is another term for microorganism.

There are four types of microorganisms:

- Bacteria
- Viruses
- Parasites
- Fungi.

Bacteria and viruses cause most foodborne illnesses.

Keep foods away from these potential contaminants to help keep it safe.

Food Flow

Food can be contaminated at any point throughout the flow of food within retail food establishments!

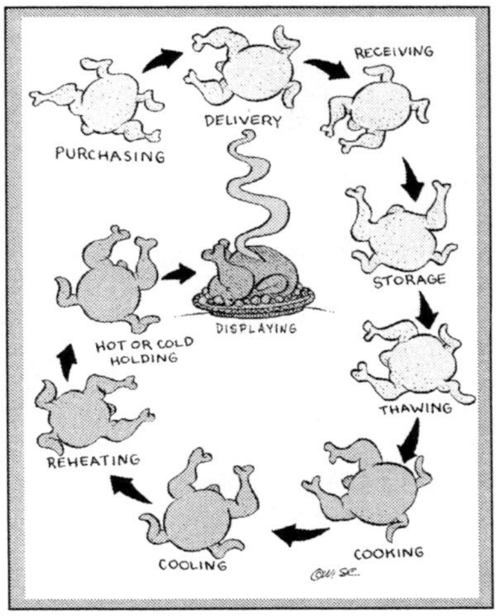

The flow of food

Adapted from *Retail Best Practices and Guide to Food Safety and Sanitation* © 2003

RBP. Pg. 11 – 12
RSG. Pg. 9 - 10

12

Slide 12

Food flow consists of food products and the ingredients used to make them.

The flow of food outside retail food establishments includes:

- Harvesting
- Production and processing
- Manufacturing
- Transportation of foods.

Within retail food establishments, the flow of food begins at the time food is purchased and ends at the time it is sold.

Food flow in retail food establishments typically consists of:

- Purchasing
- Delivery
- Receiving
- Storing
- Preparation
    - Thawing
    - Cooking
    - Cooling
    - Reheating
- Display
    - Hot- or Cold-holding
- Checkout and bagging.

Food must be protected against contaminants throughout the food flow cycle.

ryone that comes in contact with food has a responsibility to prevent contamination. This includes:

- Retail food establishments
- Operators
- Managers
- Food handlers and employees.

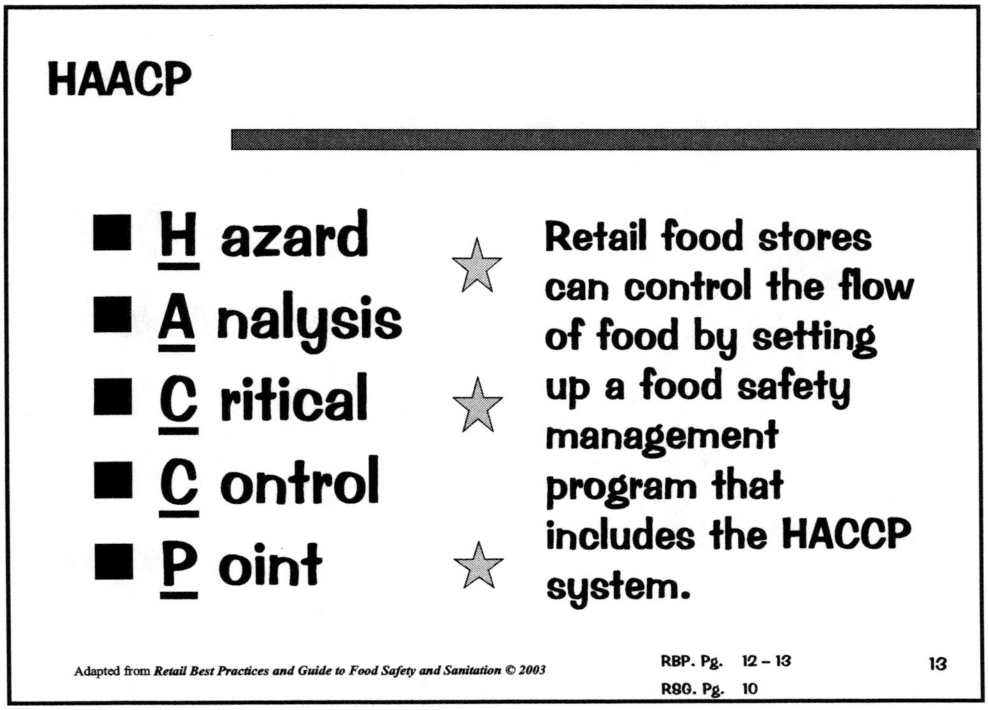

Slide 13

The *FDA Food Code* recommends using the Hazard Analysis Critical Control Point (HACCP) system to help protect people from foodborne illness.

HACCP:

- Is a system designed by NASA in the 1960s to keep food safe for astronauts
- Follows the food flow throughout the retail food establishment
- Identifies each step in the process where contamination might cause food to become unsafe.

Controlling food safety:

- The size of the retail food industry makes it impossible for regulatory inspectors to manage all aspects of food safety; therefore, governmental agencies and food industry must work together to create effective total food safety programs
- It is important for individual retail food establishments to perform self-audits
- Protect your customers by following food safety and sanitation best practices within your establishment.

Keep Facilities and Equipment Clean and Sanitary

Just because it looks clean does not mean it's sanitary!!!

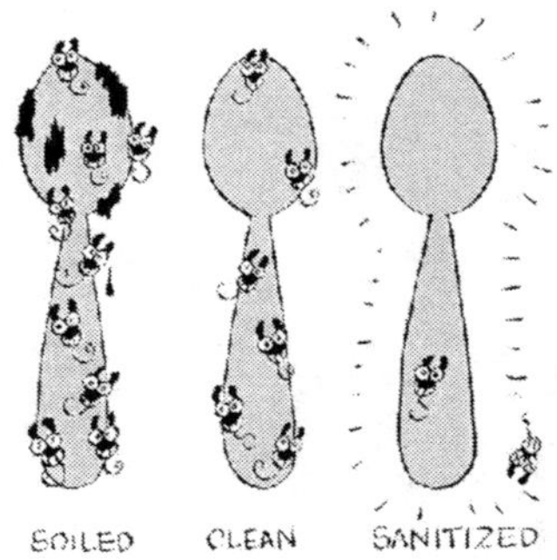

Clean vs. sanitary

Adapted from *Retail Best Practices and Guide to Food Safety and Sanitation* © 2003

RBP. Pg. 13 – 14
RSG. Pg. 11

14

Slide 14

Customers' impressions about cleanliness of an operation are frequently influenced by what they see both inside and outside your establishment.

Proper design and planning of your facility and equipment will improve shopper satisfaction as well as food safety.

There is a difference between cleaning and sanitary.

Cleaning:

- Involves removal of visible dirt, soil and food particles from the surfaces of equipment and utensils

Sanitizing:

- Involves reducing the number of disease-causing microorganisms to acceptable public health levels.

Keeping your retail food establishment clean and sanitary is important because it:

- Will increase customer satisfaction and help develop a regular clientele
- Reduces the possibility of food contamination
- Minimizes the attraction of pests
- Increases the life of equipment
- Improves employee morale.

Accident Prevention and Training

Ensuring a safe environment for employees and customers requires ongoing training and continuous monitoring.

Adapted from *Retail Best Practices and Guide to Food Safety and Sanitation* © 2003

RBP. Pg. 14 – 15
RGG. Pg. 12

15

Slide 15

Accident prevention programs are necessary in every retail food establishment.

Many accidents can be avoided through proper training, however, human error can always play an unexpected role. For instance:

- Oven doors left open
- Poorly stacked boxes
- Spills.

Disruption of water, electricity or sewer systems can also create havoc.

- The *FDA Food Code* contains specific instructions on how to operate when basic services are lost
- A good disaster plan is invaluable.

Proper education and training of food employees can also prevent foodborne illness. Train employees on:

- Where contaminants come from
- How they get into food
- What can be done to eliminate or control them.

Role of the Government and Benefits of Food Safety Certification

Why Get Certified?

- Certifications may be required
- Demonstrate knowledge of food safety guidelines which can help prevent foodborne illness
- Keep current on industry Best Practices and the ***FDA Food Code.***

Adapted from *Retail Best Practices and Guide to Food Safety and Sanitation* © 2003

RBP. Pg. 16 – 19
RSG. Pg. 13 - 16

16

Slide 16

The purpose of government regulation in food safety is to oversee the food-producing system and protect food intended for human consumption.

The U.S. Food and Drug Administration (FDA) produces the ***FDA Food Code*** which is a set of recommendations intended to be a model for state and local food safety regulations.

Food safety certification classes should provide you with the knowledge, skills and abilities to:

- Identify the causes of foodborne illness
- Understand the relationship between time and temperature and the growth of microorganisms
- Understand the importance of proper personal hygiene techniques
- Prevent cross contamination
- Describe proper cleaning and sanitizing procedures and the difference between them
- Set up proper management systems for monitoring food safety.

Slide 17

Safety Stop! 1.1

Sample Test Questions

1. The food industry consists of companies that:
   - A. Produce food.
   - B. Distribute food.
   - C. Manufacture food.
   - **D. All of the above.**

2. Foodborne illness is an illness caused by:
   - **A. Consuming contaminated food.**
   - B. Poor cooking procedures.
   - C. Consuming food you dislike.
   - D. Eating in dirty restaurants.

3. Foodborne illness can be prevented by
   - **A. Controlling the time and temperature of food products.**
   - B. Washing your hands after work.
   - C. Handling knives properly.
   - D. Wearing your uniform to work.

4. How many victims must become ill after eating a common food in order to qualify as an outbreak?
   - **A. 2.**
   - B. 3.
   - C. 5.
   - D. 1.

People who are highly susceptible to foodborne illness include:

A. People that eat out often.
B. Middle-aged men.
C. Young women.
D. Children.

Contamination is:

A. The presence of germs in food that can cause illness.
B. The presence of harmful substances or conditions in food that can cause illness or injury.
C. The presence of chemicals in food that can cause illness.
D. All of the above.

In retail food establishments, the flow of food begins where food is purchased and ends after food is:

A. Cooked.
B. Received.
C. Sold at checkout and bagging.
D. Displayed.

8. Which of the following is not a microorganism that can cause foodborne illness?

A. Bacteria.
B. Virus.
C. Yeast.
D. Parasites.

9. HACCP stands for:

A. Hazard Analysis Crucial Control Point.
B. Hazard Assessment Critical Control Point.
C. Hazard Analysis Critical Control Point.
D. Hazard Analysis Critical Contact Point.

10. Sanitary is:

A. When the number of microorganisms has been reduced to 100 ppm.
B. When all microorganisms have been eliminated.
C. When the number of microorganisms has been reduced to acceptable public health levels.
D. When soil has been removed from the surface of equipment and utensils.

Real-life

Slide 7
Seafood Counter

Patty was working at the seafood counter. A customer ordered a live lobster from the live tank. She removed it, priced and bagged it, and gave it to the customer. The next customer ordered a pound of cooked and peeled, ready-to-eat shrimp. Patty, without washing her hands and without changing her gloves, put the ready-to-eat shrimp in a bag for that customer. What was her mistake?

Patty touched a live animal, the lobster, and without washing her hands and changing her gloves, she handled ready-to-eat shrimp.

Slide 9
Campylobacter jejuni
The local grocery store provided a catering service to an assisted living center for a function. They provided several salads including a chicken salad. That salad contained chicken that was not fully cooked. The salad was contaminated with *Campylobacter jejuni*. What issues compounded the seriousness of the problem?

The residents of assisted living centers are frequently the elderly who are part of the highly susceptible populations for foodborne illness.

Slide 10
Hand washing
Joanne was working the bakery department putting the finishing touches to the birthday cake orders for the day. She stopped to use the restroom. When she returned, she neglected to wash her hands before she put on new gloves. She continued icing the cakes. The next day, after several birthday parties, several children were hospitalized with cases of *Salmonella spp.* What was Joanne's mistake?

Joanne created a foodborne outbreak because she is a carrier of *Salmonella spp.* and neglected to wash her hands after using the restroom.

Slide 12
Checkout
Mark was working hard bagging groceries. He grabbed a package of raw chicken that was dripping in juices. He wrapped it in one bag, and placed it in another. He then placed some apples in the bag with the chicken. What was his mistake?

Mark made two mistakes. He should have washed his hands after handling the raw chicken, and he should not have placed the apples in the same bag with the chicken.

Slide 13
C-store
Matthew, the manager of the local convenience store, thought that food safety was the job of the health department. When the health inspector came in to review his establishment, he found five critical violations and shut them down until they were corrected. How could Matthew have avoided this situation?

Matthew should have performed routine self-audits on the store to ensure the food he was serving was safe and to ensure he was in accordance with local, state, and federal food safety guidelines. Matthew should implement a food safety management program to assure active managerial control of all food safety risks.

Slide 14
Cross contamination
Ron was preparing the rotisserie chickens to be cooked for the day. He was handling them on a cutting board. When he finished prepping the chicken, he put them in the rotisserie oven to cook. Ron wiped down his cutting board with a clean cloth. The cutting board looked clean. He then prepared sliced cucumbers for the salad bar. What was his mistake?

Ron did not realize microorganisms could not be seen by the naked eye. He thought since the cutting board looked clean, it was clean. He should have properly cleaned and sanitized the cutting board.

Making it Real

Foodborne Illness Deaths – Slide 6

Make a visual impact by showing them what 5,000 deaths caused by foodborne illness actually is.

- Supplies
  - 5,000 M&Ms (9.5 lbs. of plain M&Ms)
  - A large bowl
- Procedure
  - Give the learners the ability to see what 5,000 deaths look like in M&Ms. Pour them into a large bowl with dramatic flare!
- Teaching Tip – Use the M&Ms as snacks.

Posters – Slide 7

Posters on

1. Time and Temperatures
2. Proper Hand Washing
3. Proper Glove Usage
4. Avoiding Cross Contamination
5. Cleaning and Sanitizing

Physical Hazards – Slide 10 and 11

Show and touch various hazards and explain how they affect food safety and inadvertently enter the food supply.

- Supplies
  - Soil
  - Small stone
  - Staples
  - Any additional contaminants you may choose
- Procedure
  - Allow the learners to touch them and explain how and when the hazards could contaminate food.

Session 1:

Concepts to Keep

- Foodborne illness can have a major impact on a retail food establishment, its customers and its employees.
- Retail food managers must know how food becomes contaminated and how to prevent contamination.

Adapted from *Retail Best Practices and Guide to Food Safety and Sanitation* © 2003 RBP. Pg. 1 – 22 RSG. Pg. 2 - 19 18

Slide 18

Concepts to Keep

- Foodborne illness can have a major impact on a retail food establishment, its customers and its employees.
- Retail food managers must know how food becomes contaminated and how to prevent contamination.

Session 1:
More Concepts to Keep

- **Very important steps in avoiding foodborne illness:**
  - Controlling time and temperature
  - Practicing proper personal hygiene
  - Preventing cross contamination
  - Proper cleaning and sanitizing procedures
  - Ensuring effective food safety management systems.

Adapted from *Retail Best Practices and Guide to Food Safety and Sanitation* © 2003

RBP. Pg. 1 – 22
RSG. Pg. 2 - 19

19

Slide 19

More Concepts to Keep

- Very important steps in avoiding foodborne illness:
  - Controlling time and temperature
  - Practicing proper personal hygiene
  - Preventing cross contamination
  - Proper cleaning and sanitizing procedures
  - Ensuring effective food safety management systems.

"Take This Back To Your Team!"

- Food contaminants are everywhere!
- Foodborne illness can be prevented.
- Foodborne illness controls include:
  - Proper time and temperature controls
  - Good personal hygiene practices
  - Avoiding cross contamination
  - Proper cleaning and sanitizing procedures.

Adapted from *Retail Best Practices and Guide to Food Safety and Sanitation* © 2003

RBP. Pg. 1 – 22
RGG. Pg. 2 - 19

20

Slide 20

Take This Back to Your Team!

- Food contaminants are everywhere!
- Foodborne illness can be prevented.
- Foodborne illness controls include:
  - Proper time and temperature controls
  - Good personal hygiene practices
  - Avoiding cross contamination
  - Proper cleaning and sanitizing procedures.

Session 2

Hazards to Food Safety – Session 2 Key Points

Slide 1

Training Tip:
The content of each slide provides the most essential information in the session. The written text below each slide is provided to help you understand and better teach the essential information. The written text may also be used to answer questions that may be posed during the session. Depending on the length of your training session, you may not be able to address all the information contained beneath each slide.

Overview:
Session Two will overview all the foodborne hazards that could be encountered in a retail food establishment including physical hazards, chemical hazards, and an in-depth review of biological hazards. It will discuss how these hazards can arise as well as preventive measures that can be taken to reduce the possibility of their occurrence.

Suggested Timeline:
8-hour course: 1 hour

- 42 ***Slides.***

16-hour course: 1 hour 45 minutes using the 59 PowerPoint© slides and 6 ***Safety Stop!*** slides containing:

- 28 ***Sample Test Questions***
- 10 ***Real-life*** scenarios
- 8 ***Making it Real*** interactive exercises.

deo suggestions and supplemental resources:

Videos:
Invisible Challenge, Food Marketing Institute, 1-301-843-3084
Grime Fighters, Food Marketing Institute, 1-301-843-3084

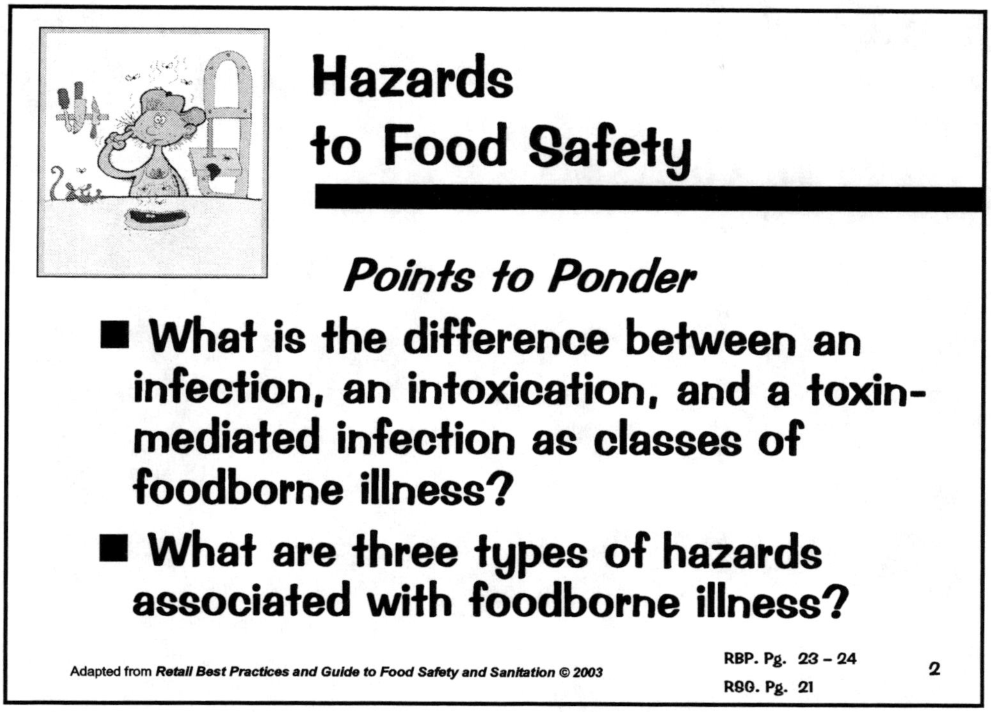

Slide 2

Points to Ponder

- What is the difference between an infection, intoxication, and a toxin-mediated infection as classes of foodborne illness?

- What are three types of hazards associated with foodborne illness?

Hazards to Food Safety

More Points to Ponder

- What are the major types of potentially hazardous foods?
- What conditions do bacteria need to grow?
- What is the food temperature danger zone?
- How can foodborne hazards be controlled?

Adapted from ***Retail Best Practices and Guide to Food Safety and Sanitation*** © 2003

RBP. Pg. 23 – 24
RSG. Pg. 21

3

Slide 3

More Points to Ponder

- What are the major types of potentially hazardous foods?
- What conditions do bacteria need to grow?
- What is the food temperature danger zone?
- How can foodborne hazards be controlled?

Foodborne Illness

Foodborne illnesses are classified as:

- Infections
- Intoxications
- Toxin-mediated Infections.

Foodborne illness
Adapted from ***Retail Best Practices and Guide to Food Safety and Sanitation*** *© 2003*

RBP. Pg. 24 – 27
RSG. Pg. 22 - 23

4

Slide 4

General symptoms of foodborne illness can include:

- Headache
- Nausea
- Vomiting
- Dehydration
- Abdominal pain
- Diarrhea
- Fatigue
- Fever.

Classifications of foodborne illness are:

1. Infection – Caused by eating food that contains a living disease-causing microorganism
- Microbe continues to live and reproduce in the victim's digestive tract
- The microbe causes the illness.

2. Intoxication – Caused by eating food that contains a harmful chemical or toxin produced by a bacteria or other source
- The toxin causes the illness.

3. Toxin-mediated infection – Caused by eating a food that contains harmful microorganisms that will produce a toxin once inside the human body
- The living microbe ingested produces harmful toxins while inside the victim
- The toxin causes the illness.

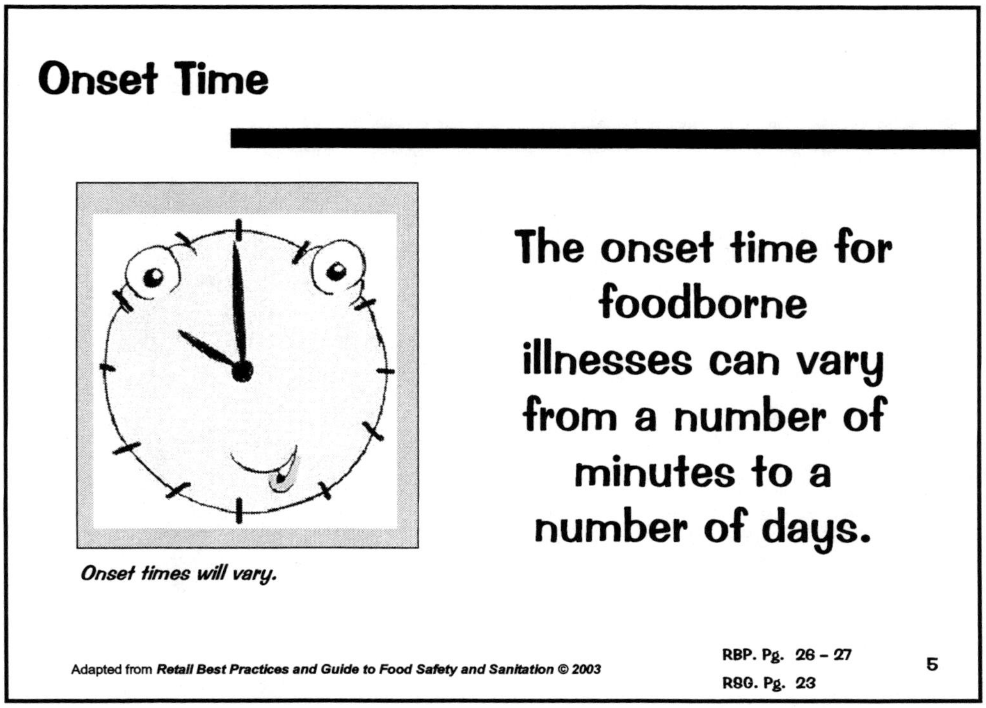

Slide 5

Onset time is the elapsed time from when a person first eats a contaminated food to when the symptoms first occur.

The onset time of foodborne illness is expressed in a range of hours.

Onset time will vary based on a number of factors including:

- The victim's age
- The victim's health status
- The victim's body weight
- The amount of contaminant ingested.

Slide 6

Some people are highly susceptible to foodborne illness including:

- Pregnant women
- Infants and young people
- Those with suppressed immune systems due to acquired immune deficiency syndrome (AIDS), cancer, diabetes
- People taking certain medications
- The elderly.

The symptoms and duration of the illness can be much more severe, even life threatening for these people.

In most cases, healthy adults will have flu-like symptoms and recover within a few days.

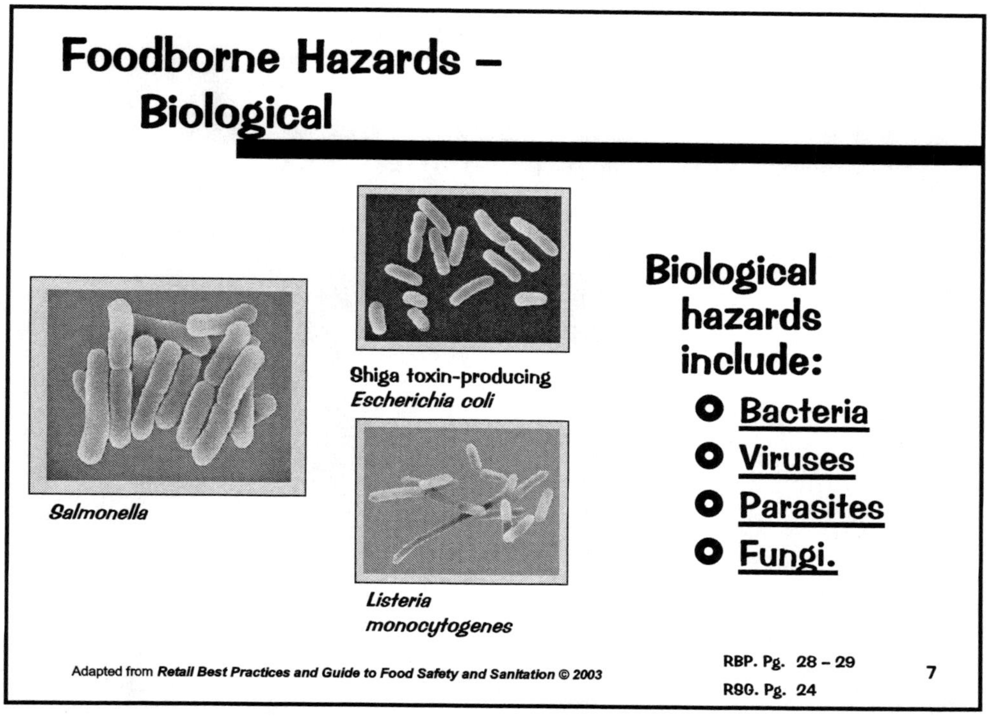

Slide 7

Foodborne hazards can cause illness or injury when consumed along with the food. There are three types of foodborne hazards:
1) Biological.
2) Chemical.
3) Physical.

Biological hazards are microscopic organisms that can cause foodborne illness. Biological hazards:
- Can only be seen with the aid of a microscope
- Are the primary target of food safety programs
- Are the most important foodborne hazards in any retail food establishment.

There are three types of biological hazards:
- **Bacteria**
  - Single-celled microorganisms that require food, moisture, and warmth to multiply which can cause foodborne infections, intoxications and toxin-mediated infections
- **Viruses**
  - A group of infectious microorganisms that require a living host in order to reproduce
- **Parasites**
  - An animal or plant that lives in or on another and from whose body it obtains nourishment.

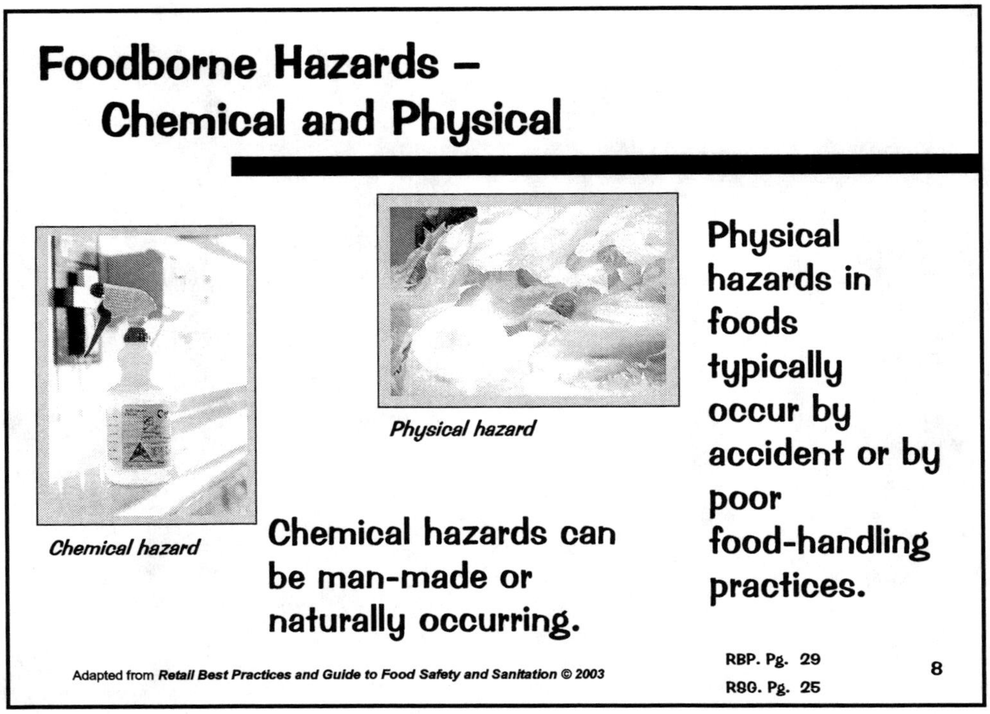

Slide 8

<u>Chemical hazards</u> are toxic substances that may occur naturally or may be added during the processing of food.

Naturally occurring chemical hazards (discussed in greater detail in slides 47 – 52) include:

- Food allergens
- Ciguatoxin
- Mycotoxin
- Scombrotoxin
- Shellfish toxins.

Man-made chemical hazards include:

- Cleaning solutions
- Food additives
- Pesticides
- Heavy metals.

<u>Physical hazards</u> are hard or soft foreign objects in food that can cause illness or injury such as:

- Toothpicks
- Metal shavings
- Glass fragments
- Jewelry
- Adhesive bandages.

Slide 9

Safety Stop! 2.1

Sample Test Questions

1. Which of the following is not a foodborne illness classification?
   A. Infection.
   B. Intoxication.
   C. Intoxicated infection.
   D. Toxin-mediated infection.

2. Which of the following factors will affect the onset time of a foodborne illness?
   A. The victim's health status.
   B. The victim's body weight.
   C. The amount of contaminant ingested.
   D. All of the above.

3. Toxic substances that may occur naturally or may be added during the processing of food are called
   A. Biological hazards.
   B. Chemical hazards.
   C. Foodborne illness.
   D. Viruses.

eal-life

ide 8
hysical hazards:
ckson was the shift manager in the bakery. He noticed that Lisa, who was mixing cake batter, was wearing everal earrings. Lisa was working hard, so Jackson wanted to wait until her break to ask her to remove them. A hort time later, Jackson saw Lisa, and a couple other employees, looking on the floor for something. He asked hat they were looking for. Lisa told him she lost one of her earrings. What should Jackson do now, and what hould he have done to prevent this from happening in the first place?

 they do not find the missing earring, Jackson will need to handle the food product Lisa has been working with s being potentially contaminated and follow the proper procedures with the product. He should document the ituation as policy calls for. Jackson should have performed a pre-shift uniform inspection and asked Lisa to emove any extra jewelry during his inspection, or at the very least, asked Lisa to remove her earrings as soon s he noticed them. He should not have waited until her break to have her remove the earrings.

Making it Real

Physical hazards – Slide 8
Bring in various physical hazards and commonly associated foods and explain how they affect food safety and inadvertently enter the food supply.

- Supplies
  - Stones or rocks which can be found in beans, rice, or flour
  - Hair which can be found in deli or bakery items (anything prepared on-site)
  - Staples which can be found in meats or produce
  - Any additional contaminants you may choose
- Allow the learners to touch them and explain how and when the hazards could contaminate food.

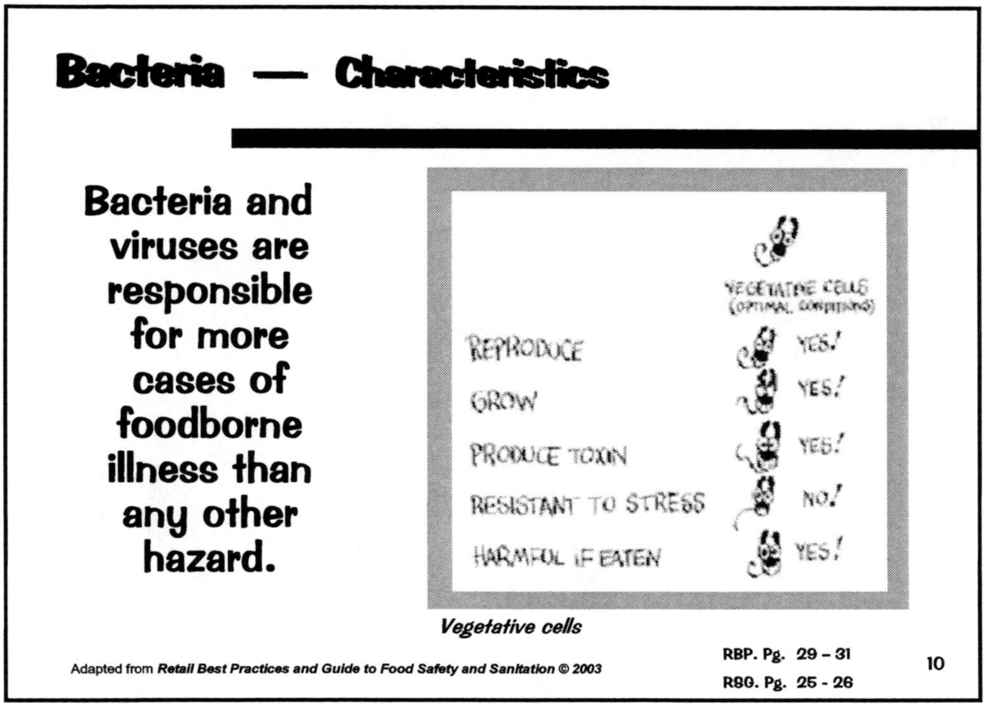

Slide 10

Vegetative cells — Cells that take in nourishment, grow, and produce wastes

- All bacteria exist in a vegetative state.

All bacteria are classified as either:

Spoilage bacteria:

- Degrade foods so they look, taste, and smell bad
- Deteriorate quality to unacceptable levels
- Cause foods to have to be thrown away creating a loss in profits.

Pathogenic bacteria:

- Disease-causing microorganisms
- Make people ill if they or their toxins are consumed with food
- May not alter appearance, taste, or smell.

A **toxin** is a poisonous substance produced by microorganisms, plants, and animals that causes diseases

- Cooking may not destroy all toxins.

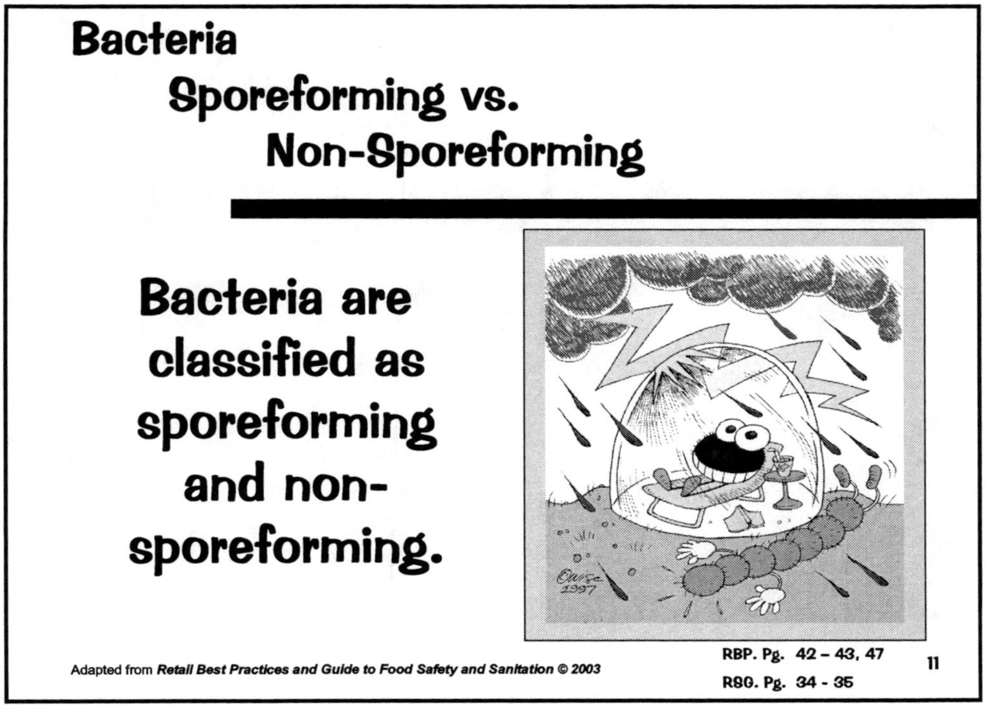

Slide 11

<u>Spores</u> — The inactive or dormant state of some bacteria:

- Bacteria can survive for months as a spore
- Some bacteria can form spore structures to help them survive during non-growth periods
- Bacterial spores can germinate, or return to their vegetative state and grow again
- Spores are not able to grow or reproduce
- It is harder to destroy bacteria when they are in spore form.

Spores become vegetative again when:

- They are heat shocked
- Optimum conditions exist for growth
- Food is held between 41°F (5°C) and 135°F (57°C) for more than four hours.

Bacterial Growth

Bacteria reproduce through <u>binary fission</u> – each cell divides to form two new cells.

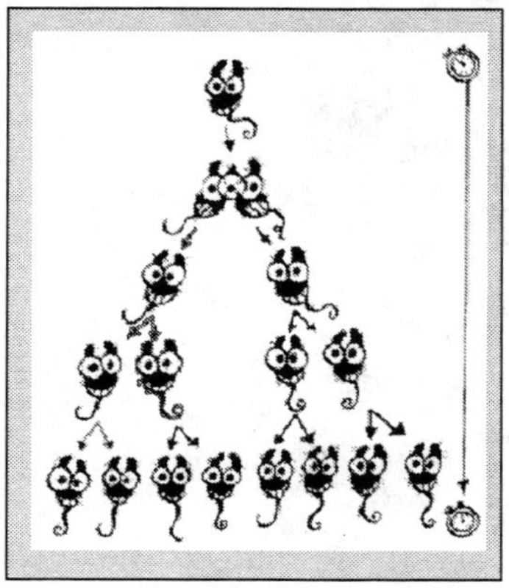

Adapted from ***Retail Best Practices and Guide to Food Safety and Sanitation* © 2003**

RBP. Pg. 31 - 32
RSG. Pg. 26 - 27

12

Slide 12

<u>Binary fission</u> – When bacteria reproduce, one bacterial cell divides to form two new cells

- Under ideal conditions, bacteria can double every 15 – 30 minutes.

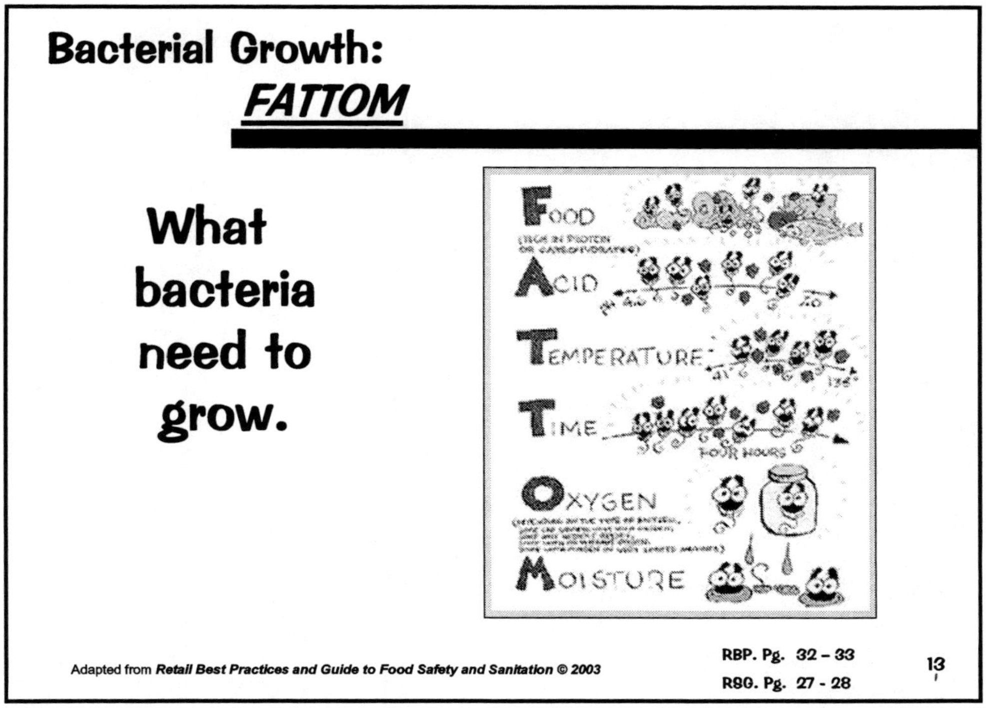

Slide 13

Six conditions for bacteria to multiply:
FATTOM is an acronym used to help remember:

1. **F**ood – High in protein or carbohydrates
2. **A**cid – Mildly acidic environment (pH of 4.6 – 7.0)
3. **T**emperature – Between 41°F (5°C) and 135°F (57°C)
4. **T**ime – Bacteria need about 4 hours to grow to high enough numbers to cause illness
5. **O**xygen – Different bacteria grow at different oxygen levels
6. **M**oisture – Disease-causing bacteria only grow in foods with a **water activity** **(Aw)** higher than .85.

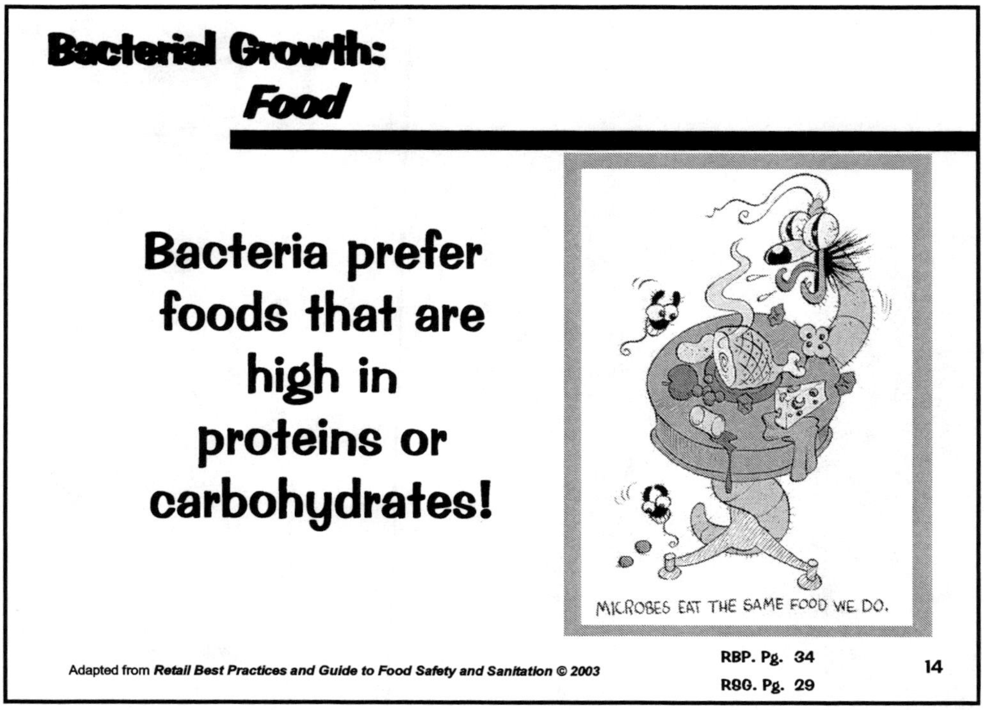

Slide 14

Food

Microbes eat the same foods we do!

A suitable food supply is the most important condition for bacterial growth.

Bacteria prefer foods that are high in protein or carbohydrates such as:

- Meats
- Poultry
- Seafood
- Dairy products
- Cooked rice, beans, and potatoes.

Bacterial Growth:
Acid

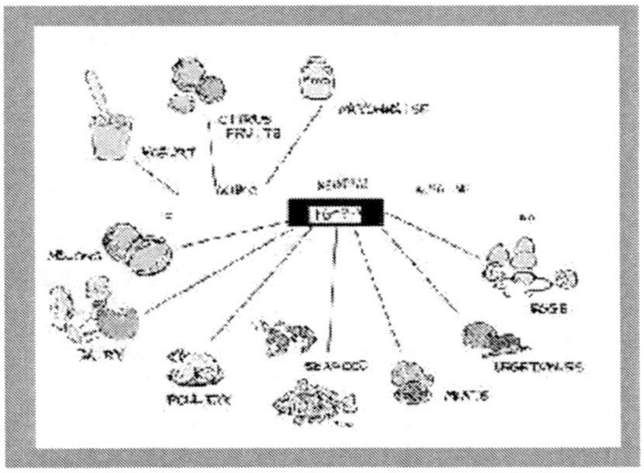

Disease-causing bacteria grow best in a mildly acidic environment (pH level of 4.6 to 7.0).

pH levels of food vary.

Adapted from ***Retail Best Practices and Guide to Food Safety and Sanitation* © 2003**

RBP. Pg. 34 – 35
RSG. Pg. 29

15

Slide 15

Acid

pH is a symbol used to designate the acidity or alkalinity of a food:
- pH is measured on a scale that runs from 0 to 14
- Most foods are acidic and have a pH less than 7.0.

Acidic foods have a pH of less than 7.0:
- Bacterial growth is best supported in a mildly acidic environment (pH of 4.6 – 7.0)
- Highly acidic foods (4.5 or below) like lemons, limes, and tomatoes typically do not support bacterial growth.

Alkaline foods have a pH above 7.0:
- Very few foods are alkaline foods
- Alkaline foods are capable of growing bacteria and include such foods as olives, egg whites, and soda crackers.

Disease-causing bacteria grow best in foods with a pH range of 4.6 to 7.0 such as meat, poultry, fish, and dairy products.

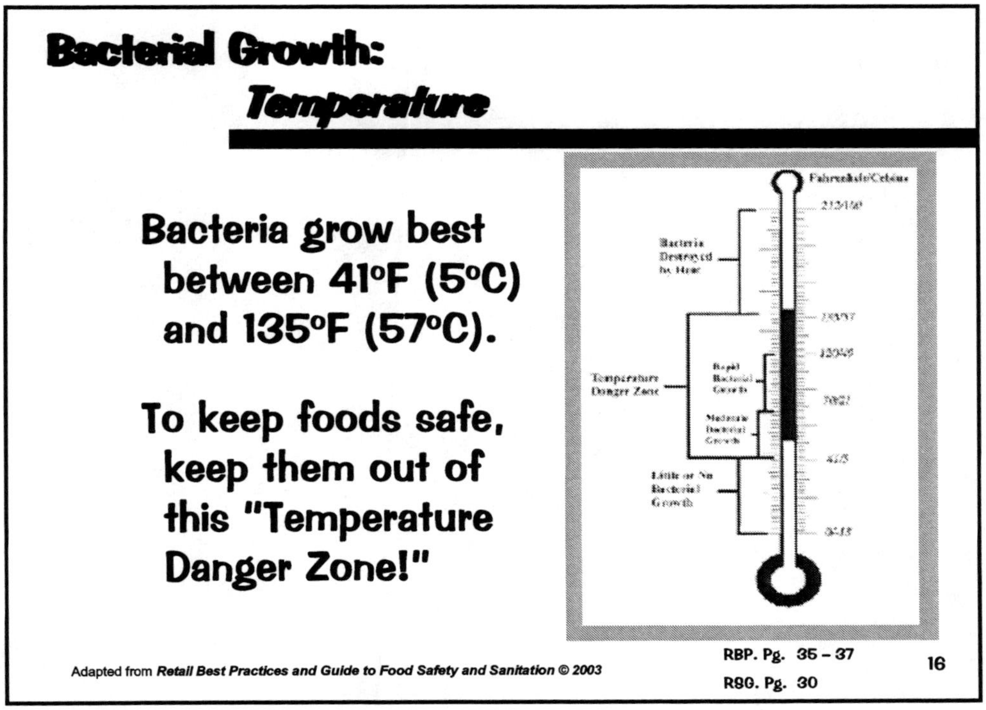

Slide 16

Temperature

Most bacteria grow best between 41°F (5°C) and 135°F (57°C). This temperature range is known as the "Temperature Danger Zone."

Different bacteria grow at different temperatures:

- Psychrophilic bacteria grow from 32°F (0°C) to 70°F (21°C)
    - Most dangerous because they can grow in refrigerated and room temperatures
    - Mostly spoilage organisms, but some can cause disease.
- Mesophilic bacteria grow from 70°F (21°C) to 110°F (43°C)
    - Most rapid growth occurs at human body temperature 98.6°F (37°C).
- Thermophilic bacteria grow at temperatures above 110°F (43°C)
    - Thermophilic bacteria are spoilage organisms.

<u>Temperature abused</u> foods are highly susceptible to foodborne illness. These are foods:

- That have not been cooked to a safe temperature
- That have been held at improper temperatures.

Some bacteria, such as *Listeria monocytogenes*, can grow at temperatures below 41°F (5°C), however their growth rate is very slow.

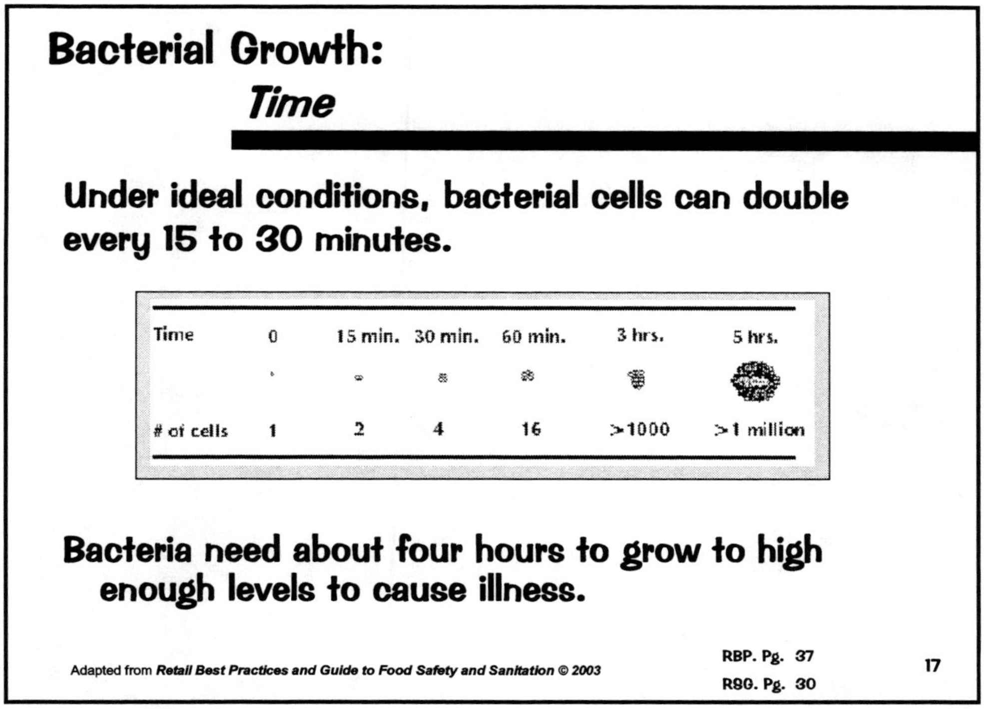

Slide 17

Time

Under ideal conditions:

- Most bacterial cells can double every 15 to 30 minutes
- A single cell could generate over one million cells in five hours
- Bacteria need about four hours to grow to high enough numbers to cause illness.

The industry rule of thumb is:

- "Keep it hot, Keep it cold, or Don't keep it!"

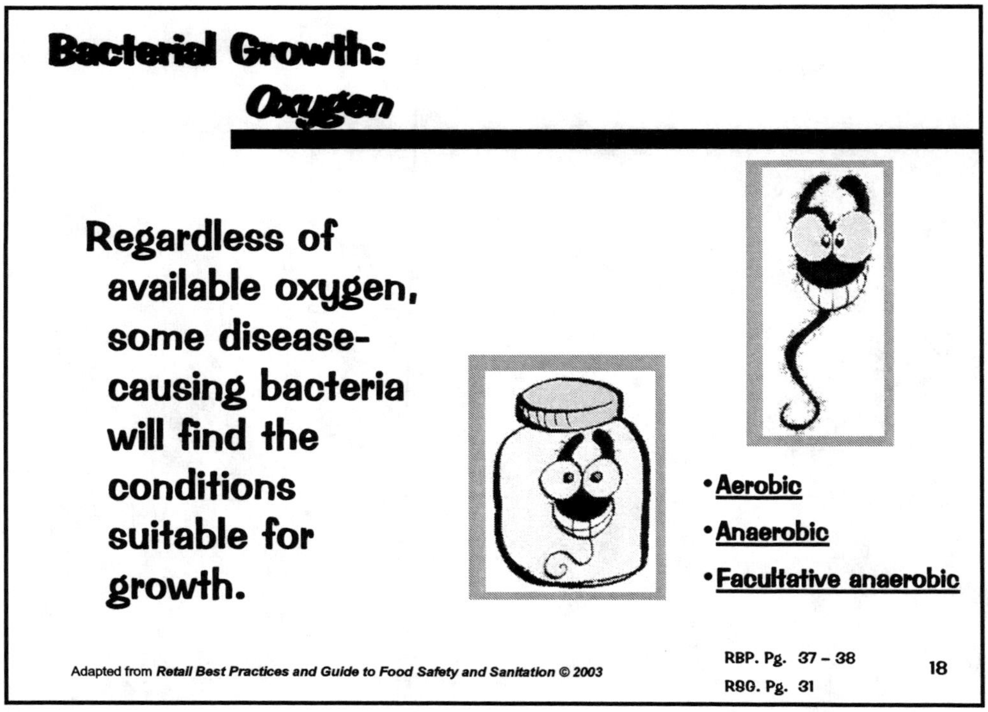

Slide 18

Oxygen

Different bacteria need different levels of oxygen to survive

- **<u>Aerobic</u>** must have oxygen to grow
  - Normally grow on the outside of food where oxygen is available.
- **<u>Anaerobic</u>** can survive only without oxygen
  - Typically grow in vacuum-packed foods and in the centers of cooked masses of foods (large stockpot).
- **<u>Facultative anaerobic</u>** can survive with or without oxygen
  - Most disease-causing microbes are facultative anaerobes.

Controlling oxygen conditions is not an effective way to prevent foodborne illness.

Regardless of available oxygen, some disease-causing bacteria will find the conditions suitable for growth.

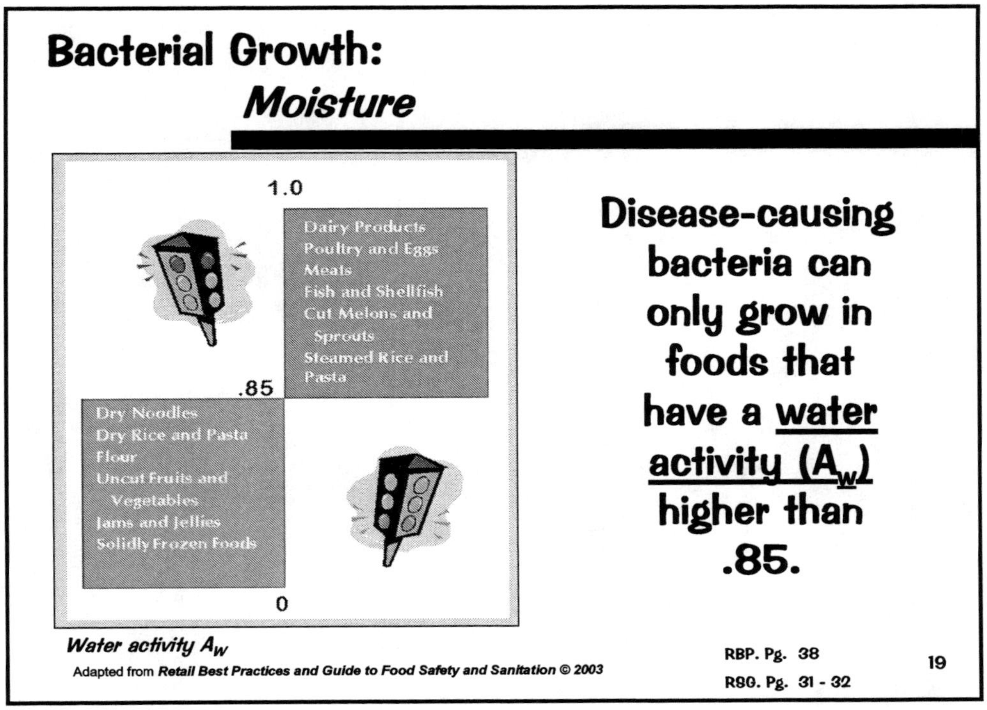

Slide 19

Moisture

Water Activity (Aw) is a measure of the amount of water that is not bound to the food and is available for bacterial growth.

Disease-causing bacteria can only grow in foods that have a water activity (Aw) higher than .85.

Freezing or drying foods are effective ways of reducing the amount of water activity (Aw) and preserving foods.

Example:

- Fresh chicken has 60% water by volume
- Its water activity is approximately .98%
- The same chicken, when frozen, still has 60% water by volume, but its water activity is nearly zero.

Bacterial Growth Curve

Bacteria life cycle

1. Lag phase.
2. Log phase.
3. Stationary phase.
4. Death phase.

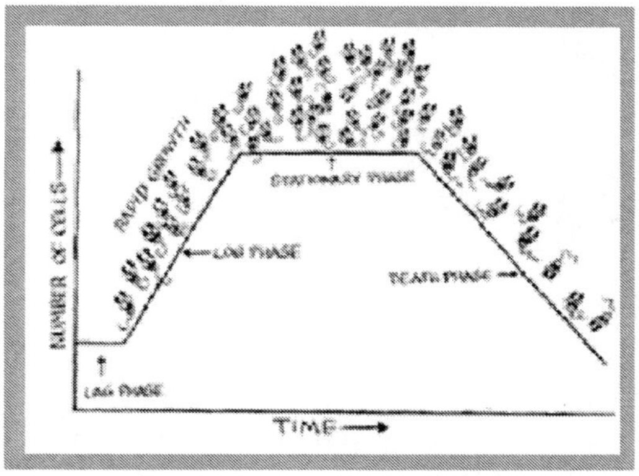

Adapted from ***Retail Best Practices and Guide to Food Safety and Sanitation* © 2003**

RBP. Pg. 31 – 32
RSG. Pg. 27

20

Slide 20

Bacteria grow in four phases:

- Phase 1: Lag phase
  - Bacteria exhibit little or no growth
  - Bacteria adjust to surroundings
  - Last a few hours at room temperature, but can be increased by keeping foods at or below 41°F (5°C).
- Phase 2: Log phase
  - Bacterial growth is rapid, doubling every few minutes
  - Keeping food out of this phase is critical to food safety.
- Phase 3: Stationary phase
  - The number of new bacteria being produced is equal to the number of bacteria dying off
  - Bacteria has used up most of the space, nutrients, and moisture needed to grow.
- Phase 4: Death phase
  - Final phase where bacteria die off rapidly
  - Die off due to:
    - A lack of nutrients
    - Poisoned by their own toxic wastes.

Potentially Hazardous Foods

Potentially hazardous foods are:

- High in protein or carbohydrates
- High in moisture
- Low in acidity.

Potentially hazardous foods

Adapted from ***Retail Best Practices and Guide to Food Safety and Sanitation*** © 2003

RBP. Pg. 38 – 40
RGG. Pg. 32 - 33

21

Slide 21

Potentially hazardous foods (PHFs) are foods that support the rapid and progressive growth of disease-causing bacteria (pathogenic), specifically:

- Growth of infectious or toxin-producing microbes
- Growth and toxin production of *Clostridium botulinum* bacteria
- Growth of *Salmonella enteritidis* bacteria in raw shell eggs.

Potentially hazardous foods:

- Can only be in the temperature danger zone for a total of four hours
- Have been associated with most foodborne disease outbreaks
- Include shell eggs that have not been treated to destroy all viable *Salmonella* bacteria, red meats, poultry, fish and shellfish, dairy products, cooked rice and potatoes, and cut fruits.

The ***FDA Food Code*** classifies the following as potentially hazardous foods:

- Foods of animal origin that are raw or heat-treated
- Foods of plant origin that are heat-treated or consist of raw seed sprouts
- Cut melons
- Garlic in oil mixtures that are not modified in a way to inhibit the growth of disease-causing microorganisms.

Ready-To-Eat Foods

Ready-to-eat foods can become contaminated if they are not handled properly.

Adapted from ***Retail Best Practices and Guide to Food Safety and Sanitation* © 2003**

RBP. Pg. 40 – 41
RSG. Pg. 33

22

Slide 22

Ready-to-eat foods are food items that are edible without washing, cooking or further preparation by the customer or the retail food establishment.

The *FDA Food Code* identifies the following types of food as ready-to-eat:

- Raw animal foods that are cooked (i.e., rotisserie chicken) or frozen (i.e., sushi) as prescribed in the *FDA Food Code*
- Raw fruits and vegetables that are washed
- Fruits and vegetables that are cooked for hot-holding
- All potentially hazardous foods that are cooked and then cooled as prescribed in the *FDA Food Code*
- Plant foods for which further washing, cooking, or other processing is not required for food safety, and from which rinds, peels, husks, or shells, if naturally present, are removed
- Substances derived from plants such as spices, seasonings, and sugar
- Bakery items such as bread, cakes, pies, fillings, or icing for which further cooking is not required for food safety
- Dry, fermented sausages (i.e., dry salami or pepperoni), salt-cured meat and poultry products (i.e., prosciutto ham, country cured ham, and Parma ham), and dried meat and poultry products (i.e., jerky or beef sticks) that are produced in accordance with USDA guidelines and have been treated to destroy pathogens
- Thermally processed low acid foods (i.e., smoked fish or meat) packaged in hermetically sealed containers.

Slide 23

Safety Stop! 2.2

Sample Test Questions

1. The inactive or dormant state of some bacteria is a:
   A. Toxin.
   B. Pathogen.
   C. Spore.
   D. Vegetative.

2. Bacteria prefer foods that are high in:
   A. Fats and oils.
   B. Sugar and salt.
   C. Proteins and carbohydrates.
   D. Fats and salt.

3. Disease-causing bacteria grow best in temperatures:
   A. Below 41°F (5°C).
   B. Above 140°F (60°C).
   C. Between 41°F (5°C) and 135°F (57°C).
   D. Between 0°F (-18°C) and 41°F (5°C).

4. How much time does it take for bacteria to multiply to high enough numbers to cause illness?
   A. 2 hours.
   B. 4 hours.
   C. 6 hours.
   D. 12 hours.

Disease-causing bacteria can only grow in foods that have a water activity higher than:

A. 65.

B. .85.

C. 3.5.

D. .085.

Which of the following is the correct order of phases of the bacterial growth curve?

A. Lag, log, stationary, death.

B. Log, lag, stationary, death.

C. Lag, log, death, stationary.

D. Stationary, lag, log, stationary.

. Food items that are edible without washing, cooking or further preparation by the customer or by the retail food establishment are called:

A. Potentially hazardous foods.

B. Canned foods.

C. Fruits and vegetables.

D. Ready-to-eat foods.

Real-Life

Slide 15
Temperature Control:
Jamie was cleaning out the seafood display case. She removed the fresh seafood and set it on a table by the case. She cleaned out the ice, cleaned the case, put in new ice and replaced the seafood in the display case. During the cleaning process, she took her lunch break and handled customer requests. The entire process took five hours. What was Jaime's mistake and how can it be corrected in the future?

Jaime set the raw seafood product out at room temperature and kept it there for more than four hours. The seafood should have been held at 41°F (5°C) or below during the cleaning process. Jaime should have held the product in the cooler during the cleaning process and should have cleaned the case at a time that would create fewer interruptions.

Making it Real

Acidity – Slide 14
Bacteria grow best in mildly acidic foods with a pH of 4.6 to 7.0.

- Supplies
  - pH meter and buffer solution for calibration
  - Lemon
  - Olives
  - Any additional foods of which you would like to test the pH level.
- Using the operating instructions from the pH meter, test the different foods pH level
- Have the learners explain the significance of the pH level of each food tested.

ater Activity – Slide 19

ater activity can be a difficult concept to grasp. To better explain the concept of A_w, bring ice to class.

- Supplies
    - Ice
    - Bowl
- When you begin Slide 19, set the ice in the bowl. Review the slide in its entirety. Once you are finished reviewing the information on water activity, the ice will have begun to melt.
- Explain to the learners how the water that is still in the frozen state is bound, however, the water that is now in the liquid state is no longer bound and thus available for bacterial growth.

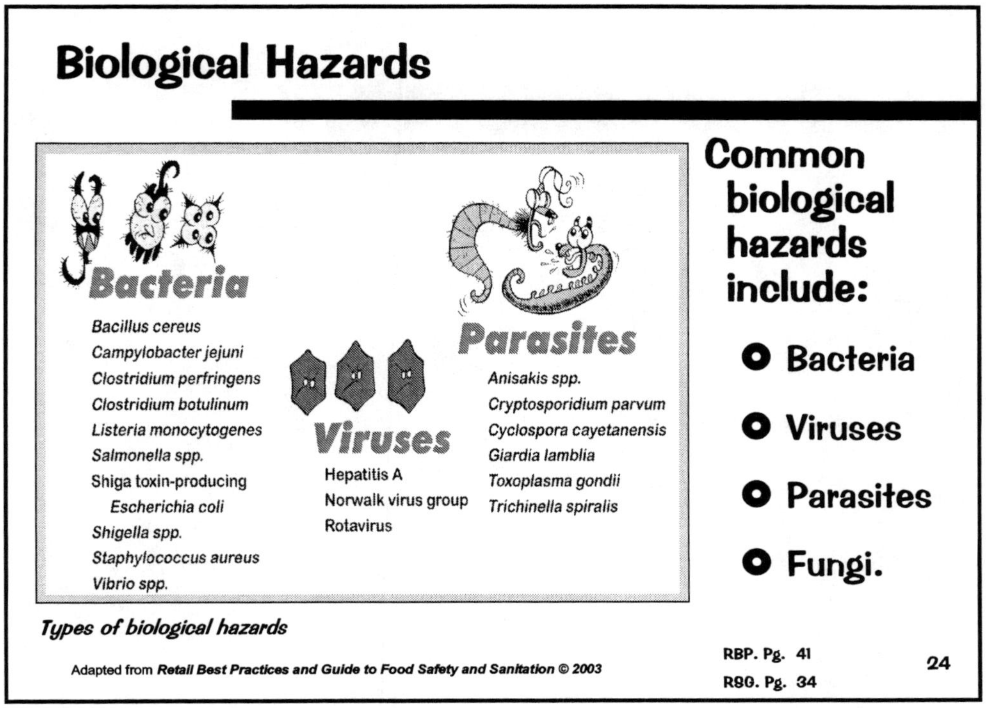

Slide 24

Common biological hazards include:

- Bacteria
- Viruses
- Parasites
- Fungi.

Biological hazards lead to the majority of foodborne illnesses.

The next several slides will review the most common bacteria found in retail food establishments.

They will discuss each biological hazard and its:

- Characteristics
- Symptoms and onset time
- Common foods
- Causes
- Controls.

Bacillus cereus

Bacillus cereus **is commonly found in:**

- Rice
- Pasta
- Potatoes
- Meats and fish
- Milk
- Vegetables.

Cooked rice

Adapted from ***Retail Best Practices and Guide to Food Safety and Sanitation* © 2003**

RBP. Pg. 44
RSG. Pg. 36

25

Slide 25

Bacillus Cereus – Bacteria

Characteristics:

- Sporeforming
- Facultative anaerobic
- Intoxication or toxin-mediated infection
- Associated with two distinct illnesses: vomiting and diarrhea.

Onset time:

- Vomiting: 30 minutes to 6 hours onset, lasting a day or so
- Diarrhea: 8 to 16 hours onset, lasting 12 to 14 hours.

Found in the following foods:

<u>Vomiting Type:</u>

Grain products such as:

- Rice
- Pasta
- Potatoes.

<u>Diarrhea Type:</u>

- Meats
- Milk
- Vegetables.

Causes:

- Improperly cooled or hot-held foods allow spores to change into vegetative cells that can produce a toxin in food.

Controls:

- Cook food to required temperature
- Cool foods rapidly to 41°F (5°C) before storage
- Hold hot foods at 135°F (57°C).

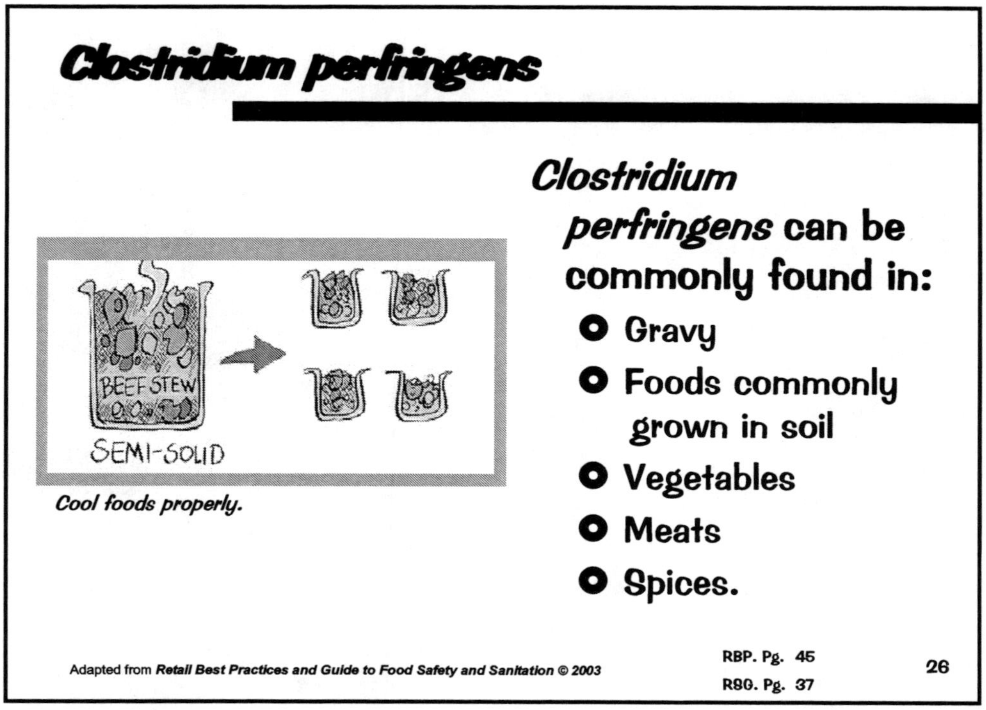

Slide 26

Clostridium perfringens – Bacteria

Characteristics:

- Sporeforming
- Anaerobic
- Toxin-mediated infection
- Symptoms include abdominal pain and diarrhea.

Onset time:

- 8 to 22 hours onset, lasting a day or less.

Found in the following foods:

- May be found in most foodstuffs
- Improperly cooled foods
- Vegetables, meats, and spices.

Causes:

- Improperly cooled or reheated foods.

Controls:

- Cook foods to 145°F (63°C) or above
- Cooked foods must be cooled from 135°F (57°C) to 70°F (21°C) within two hours and from 135°F (57°C) to 41°F (5°C) or below within six hours
- Reheat foods to 165°F (74°C) within two hours
- Hold hot foods at 135°F (57°C).

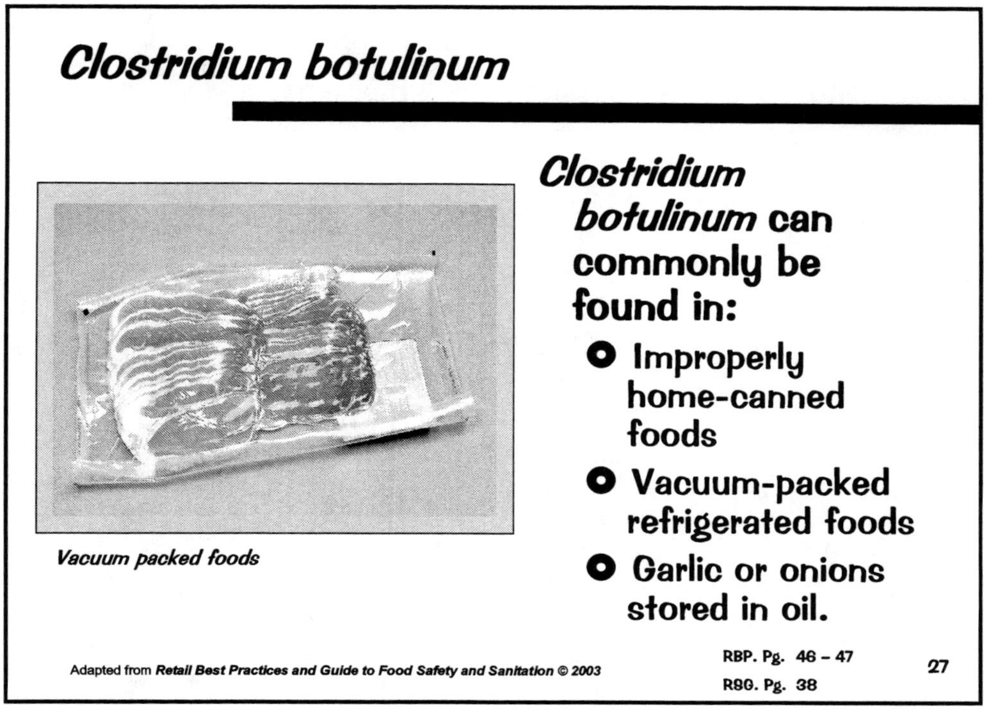

Slide 27

Clostridium botulinum – Bacteria

Characteristics:

- Sporeforming
- Anaerobic
- An intoxication
- Produces a neurotoxin that affects the central nervous system — one of the deadliest toxins known to man
- Symptoms include:
    - Headache
    - Visual disturbances
    - Inability to swallow
    - Respiratory paralysis.

Onset time:

- 12 to 36 hours, lasting several days up to a year.

Found in the following foods:

- Low acidic foods
- Foods packed in metal cans
- Vacuum-packed foods
- Garlic or onions stored in oil.

Causes:

- Improperly heat-processed, low acid foods (pH of 4.6 or above) that have been packaged anaerobically.

Controls:

- Purchase foods from approved sources
- Discard or refuse damaged cans
- Do not rely on normal cooking processes to destroy botulism toxins.

Campylobacter jejuni

It is estimated that 100% of raw poultry is infected with *Campylobacter jejuni.*

Raw poultry

Adapted from ***Retail Best Practices and Guide to Food Safety and Sanitation* © 2003**

RBP. Pg. 47
RGG. Pg. 39

28

Slide 28

Campylobacter jejuni — Bacteria

Characteristics:

- Non-sporeforming
- Microaerophilic (has very strict air requirement for growth – 3 to 6%)
- An infection
- Symptoms include abdominal pain, slight to severe bloody diarrhea.

Onset time:

- 2 to 5 days, lasting 2 to 7 days.

Found in the following foods:

- Raw poultry (it is estimated that 100% of all raw poultry is infected with this microorganism)
- Raw meats
- Raw milk.

Causes:

- Cross contamination – typically by a contaminated food-contact surface or a food employee's hands.

Controls:

- Cook raw meats and poultry properly
- Properly clean and sanitize food-contact surfaces between raw and cooked or ready-to-eat foods
- Wash hands after touching raw animal foods.

Shiga toxin-producing *Escherichia coli*

Shiga toxin-producing *Escherichia coli* is commonly found in:

- Unpasteurized fruit juices
- Improperly pasteurized milk
- Raw and undercooked red meats.

Unpasteurized apple cider

Adapted from ***Retail Best Practices and Guide to Food Safety and Sanitation* © 2003**

RBP. Pg. 48
RSG. Pg. 40

29

Slide 29

Shiga toxin-producing *Escherichia coli* – Bacteria

Characteristics:

- Non-sporeforming
- Facultative anaerobic
- An infection or toxin-mediated infection
- Symptoms include:
  - Abdominal pain
  - Nausea
  - Vomiting
  - Bloody diarrhea
  - Kidney failure
  - Death.

Onset time:

- 12 to 72 hours, lasting 1 to 3 days.

Found in the following foods:

- Raw and undercooked beef and other red meats
- Unpasteurized fruit juices
- Raw milk and improperly pasteurized milk.

Causes:

- Transferred to foods through contact with the intestines of animals during slaughter
- Transmission can occur if infected employees do not wash their hands after using the restroom.

Controls:

- Cook ground meats to at least 155°F (68°C)
- Store foods at proper temperatures
- Practice proper hygiene
- Prevent cross contamination
- Consumer education such as "Safe Handling" labels.

Listeria monocytogenes

Hot dogs

Listeria monocytogenes **is commonly found in:**

- Ready-to-eat meats (luncheon meats and hot dogs)
- Raw meats
- Raw poultry
- Dairy products
- Raw vegetables
- Raw seafood
- Seafood salads.

Adapted from ***Retail Best Practices and Guide to Food Safety and Sanitation* © 2003**

RBP. Pg. 49
RSG. Pg. 41

30

Slide 30

Listeria monocytogenes — Bacteria

Characteristics:

- Non-sporeforming
- Facultative anaerobic
- An infection
- Can survive and grow slowly at temperatures below 41°F (5°C)
- Symptoms include:
  - Nausea
  - Vomiting
  - Fever
  - Chills
  - Life threatening for highly susceptible populations.

Onset time:

- 1 day to 3 weeks, lasting indefinitely depending on treatment.

Found in the following foods: The organism is everywhere, but common in:

- Ready-to-eat luncheon meats and hot dogs
- Raw poultry
- Dairy products
- Raw meats
- Raw vegetables
- Seafood salads.

Causes:

- Transmission to foods can occur by cross contamination
- Improperly cooked foods.

Controls:

- Cook foods to proper temperature
- Practice safe food-handling techniques
- Prevent cross contamination
- Rotate processed refrigerated foods using FIFO (First-in—First-out) to ensure timely use.

Salmonella spp.

Salmonella spp. **is commonly found in:**

- Intestinal tracts of humans and animals
- Raw meat and raw poultry
- Pork
- Dairy products
- Chocolate
- Cream-filled desserts.

Eggs

Adapted from ***Retail Best Practices and Guide to Food Safety and Sanitation © 2003***

RBP. Pg. 50
RSG. Pg. 42

31

Slide 31

Salmonella spp. – Bacteria

Characteristics:

- Non-sporeforming
- Facultative anaerobic
- An infection
- Symptoms include:
    - Abdominal pain
    - Headache
    - Nausea
    - Vomiting
    - Fever
    - Diarrhea.

Onset time:

- 6 to 48 hours, lasting 2 to 3 days.

Found in the following foods:

- Raw and undercooked meat and poultry
- Raw and undercooked eggs
- Raw and undercooked pork
- Dairy products
- Chocolate
- Cream-filled desserts.

Causes:

- Transmission to foods is through cross contamination where fecal material is transferred to food by contact with:
    - Raw foods
    - Food-contact surfaces
    - Food employees.

ontrols:

- Cook foods to proper temperature
- Properly clean and sanitize food-contact surfaces
- Wash hands properly before working with food and after using the restroom
- Prevent cross contamination.

Shigella spp.

Ready-to-eat salads

Shigella **accounts for about 10% of foodborne illnesses in the United States and are commonly found in:**

- Intestines of humans
- Ready-to-eat salads
- Poultry
- Milk and dairy products
- Raw vegetables.

Adapted from *Retail Best Practices and Guide to Food Safety and Sanitation* © 2003

RBP. Pg. 51
RSG. Pg. 43

32

Slide 32

Shigella spp. – Bacteria

Characteristics:

- Non-sporeforming
- Facultative anaerobic
- An infection
- Symptoms include:
  - Abdominal pain
  - Chills
  - Dehydration
  - Fever
  - Diarrhea.

Onset time:

- 1 to 7 days, duration depends on treatment.

Found in the following foods:

- Poultry
- Milk and dairy products
- Ready-to-eat salads
- Raw vegetables.

Causes:

- Water contaminated with fecal material
- Food and utensils contaminated by employees who carry the bacteria.

Controls:

- Do not allow individuals who have been diagnosed with shigellosis to handle food
- Cook foods to proper temperatures
- Prevent cross contamination
- Wash produce and other foods with **potable water** (water suitable for drinking).

Staphylococcus aureus

Pre-cooked, ready-to-eat foods

Staphylococcus aureus **can commonly be found in:**

- Pre-cooked, ready-to-eat foods that have been re-contaminated by food employees
- Vegetable and egg salads
- Milk and dairy products
- Foods that require considerable food preparation and handling.

Adapted from ***Retail Best Practices and Guide to Food Safety and Sanitation*** © 2003

RBP. Pg. 52 – 53
RSG. Pg. 44

33

Slide 33

Staphylococcus aureus — Bacteria

Characteristics:

- Non-sporeforming
- Facultative anaerobic
- An intoxication
- Can be found living in and on 30 to 50 % of people including those who are otherwise healthy
- Can grow in foods that contain high salt or high sugar and a low water activity, such as hams and luncheon meats
- Symptoms include:
  - Vomiting
  - Acute abdominal pain
  - Diarrhea.

Onset time:

- 2 to 6 hours, lasting 1 to 2 days.

Found in the following foods:

- Pre-cooked, ready-to-eat foods
- Vegetable and egg salads
- Custards
- Milk and dairy products
- Foods that require considerable food preparation and handling.

Causes:

- Cross contamination from food employees
- Saliva from talking, coughing, or sneezing near food
- Food employees who improperly taste food
- Commonly found on cooked foods that are recontaminated by poor handling practices.

ontrols:

- Avoid touching ready-to-eat food with bare hands
- Ensure infected wounds are properly covered
- Proper handwashing
- Proper tasting techniques
- Proper temperature control.

Vibrio spp.

Vibrio spp. **is commonly found in seafood such as:**

- Fish
- Oysters
- Crabs
- Shrimp
- Clams
- Lobster.

Handle seafood carefully.

Adapted from ***Retail Best Practices and Guide to Food Safety and Sanitation* © 2003**

RBP. Pg. 53
RSG. Pg. 45

34

Slide 34

Vibrio spp. – Bacteria

Characteristics:

- Non-sporeforming
- Facultative anaerobic
- An infection
- Symptoms include:
    - Vomiting
    - Abdominal pain
    - Nausea
    - Headache
    - Fever/ Chills
    - Diarrhea.

Onset time:

- 2 to 48 hours, lasting 2 to 3 days.

Found in the following foods:

- Fish
- Oysters
- Crabs
- Shrimp
- Clams
- Lobster.

Causes:

- Cross contamination
- Eating of raw and undercooked seafood.

ontrols:

- Buy seafood from approved sources
- Cook seafood to the proper temperatures
- Avoid consuming raw, or lightly cooked seafood
- Practice proper personal hygiene and handwashing procedures between raw and ready-to-eat foods.

Slide 35

Safety Stop! 2.3

Sample Test Questions

1. *Bacillus cereus* is not found in which of the following foods?
   A. Rice.
   B. Meats.
   C. Milk.
   D. Poultry.

2. *Listeria monocytogenes* can survive and grow slowly at temperatures:
   A. Above 140°F (60°C).
   B. Above 165°F (74°C).
   C. Below 41°F (5°C).
   D. Below 32°F (0°C).

3. *Clostridium botulinum* is commonly found in which of the following foods?
   A. Brown Rice.
   B. Garlic stored in oil.
   C. Juice.
   D. Poultry.

4. It is estimated that what percentage of poultry is infected with *Campylobacter jejuni*?
   A. 10%.
   B. 50%.
   C. 95%.
   D. 100%.

Staphylococcus aureus can be found in what percentage of otherwise healthy humans?

A. None.
B. 30 to 50.
C. 50 to 80.
D. 100.

Shigella spp. is **not** commonly found in which of these foods?

A. Dairy products.
B. Red meat.
C. Raw vegetables.
D. Ready-to-eat salads.

Real-Life

Slide 27
Home-canning

Mark managed a small grocery. He had a friend that was trying to start her own business selling fresh fruits and vegetables and canned vegetables. She ran her business out of her home. He decided to stock some of her home-canned vegetables in his market. What potential foodborne illness could result from this decision?

Clostridium botulinum is common in improperly heat-processed, low acidic foods that have been packaged anaerobically which includes canning. Mark should only stock products from reputable sources that are inspected regularly and deemed safe.

Slide 30
Cross contamination

Maria was cutting chicken. She took a short break to use the bathroom. When she went to dry her hands, she realized there were no towels. She dried her hands on her soiled apron, restocked the towels and returned to cutting the chicken. What did she do wrong, what are some possible foodborne illnesses that could result from her mistake, and what could she have done to keep the food safe?

Maria used a soiled apron to dry her hands. This could create cross contamination from the apron to the chicken. This could cause several foodborne illnesses including *Campylobacter jejuni, Salmonella,* and *Listeria monocytogenes*. Maria should not have used her soiled apron to dry her hands. She should have restocked the paper towels and then rewashed her hands and dried them properly.

Slide 33
Sneezing

Hector was asked to stock the potato salad for the day. As he was performing his task, he sneezed near the potato salad. He changed his gloves, washed his hands, and continued to stock the product. The next day the manager received three calls from people who had been diagnosed with foodborne illness. They had all purchased and eaten the potato salad from his store. What could have happened, what illness may the customers have been diagnosed with, and what could have been done to prevent it?

Staphylococcus aureus is present in 30 to 50% of people who may otherwise seem healthy. Chances are Hector is infected with *Staphylococcus aureus*. The bacteria could have contaminated the potato salad from his sneeze. Although Hector changed his gloves and washed his hands properly, he did not discard the potato salad. By discarding the potato salad, he could have prevented the outbreak.

Making it Real

***Clostridium botulinum* – Slide 27**

- Supplies
    - Bulging end can
    - Springer can
    - Flipper can
- When discussing *Clostridium botulinum*, show the cans to the learners. Pass them around and allow them to tell you what is wrong with the can, how it can present a problem, and what they should do with the can.
- Any swellers, bulging ends, springers, or flippers should be treated as if they were caused by *Clostridium botulinum*.

***Staphylococcus aureus* – Slide 33**

- Supplies
    - Class
- Get the class to stand up and separate into two equal groups. Explain to them that statistically one of the two groups is infected with *Staphylococcus aureus* bacteria. Have them tell you when and where the *Staphylococcus aureus* bacteria might be found and what the infected employees would need to do in order to keep food safe while they are working.

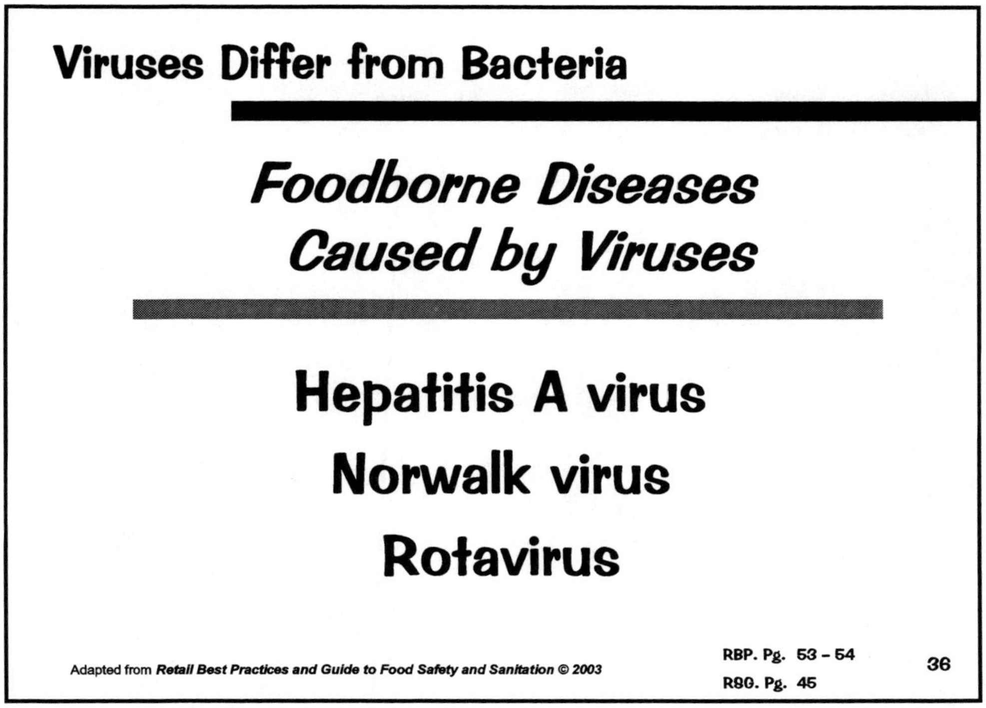

Slide 36

Viruses

Viruses differ from bacteria. Viruses:

- Are smaller in size
- Require a living host to grow
- Do not need a potentially hazardous food
- Do not multiply in foods
- Need only a few viral particles to cause an infection in a susceptible person.

Foodborne diseases caused by viruses include:

- Hepatitis A
- Norwalk virus
- Rotavirus.

Viruses are transmitted to food by:

- Infected workers spreading fecal material from unwashed or poorly washed hands
- Shellfish harvested from polluted waters.

Viruses can be controlled by:

- Handling foods properly
- Cooking foods to recommended temperatures
- Avoiding consumptions of raw and lightly cooked seafood
- Keeping raw foods separate from ready-to-eat foods
- Washing hands thoroughly before working with food and after using the restroom
- Avoiding direct bare-hand contact with ready-to-eat foods.

Slide 37

Hepatitis A virus – Virus

Characteristics:
Hepatitis A virus causes a viral infection of the liver. Symptoms of infectious hepatitis include:

- Fever
- Nausea
- Vomiting
- Abdominal pain
- Fatigue
- Jaundice.

Onset time:

- 10 to 50 days, lasting several weeks to several months.

Found in the following foods:

- Raw and lightly cooked oysters and clams harvested from polluted waters
- Raw vegetables irrigated or washed with polluted water
- Foods contaminated by a food employee who is infected with the virus
- Contaminated water.

Controls:

- Buy shellfish and seafood from approved sources
- Cook seafood to the proper temperatures
- Proper handwashing
- Avoid direct bare-hand contact with ready-to-eat foods
- Exclusion of ill employees
- Keep raw and ready-to-eat foods separate during preparation, storage, and display.

Norwalk Virus & Rotavirus

Clams

Norwalk virus and Rotavirus are commonly found in raw and undercooked seafood.

Adapted from *Retail Best Practices and Guide to Food Safety and Sanitation* © 2003 RBP. Pg. 55 - 56 RSG. Pg. 47 38

Slide 38

Norwalk virus and **Rotavirus** — Viruses

Characteristics:

- An infection
- Infectious dose is presumed to be low
- Norwalk and Rotavirus are transmitted by the fecal-oral route via contaminated water and foods. Foods other than shellfish are contaminated by ill food handlers.
- Symptoms
  - Nausea
  - Vomiting
  - Diarrhea
  - Abdominal pain.

Onset time:

- Norwalk virus – 24 to 48 hours, lasting 2 to 3 days
- Rotavirus – 1 to 3 days, lasting 4 to 8 days

Found in the following foods:

- Contaminated salad ingredients
- Raw seafood harvested from polluted waters.

Controls:

- Use potable water (water that is safe to drink)
- Practice good personal hygiene and wash hands thoroughly before working with food
- Cook food to the proper temperatures
- Keep raw and ready-to-eat foods separate during preparation, storage, and display
- Purchase molluscan shellfish from approved sources.

Slide 39

Safety Stop! 2.4

Sample Test Questions

1. Which of the following diseases is not commonly caused by a virus?
   A. Infectious hepatitis.
   B. Viral gastroenteritis.
   C. Salmonella spp.
   D. Rotavirus.

2. Symptoms of Hepatitis A may not be present for up to:
   A. 8 weeks.
   B. 10 weeks.
   C. 7 weeks.
   D. 12 weeks.

3. Norwalk and Rotavirus are most commonly caused by:
   A. Transmission of feces from contaminated water or food handlers.
   B. Not holding foods at the proper temperature.
   C. Improper cooling procedures.
   D. Inadequate cooking temperatures.

eal-life

ide 34. 37, 38
pproved Suppliers

at operated a small meat and seafood grocery. His usual supplier of oysters had run out and shorted him everal bushels. He needed the oysters to fill an order that had been placed a week prior for a customer who was lanning on serving oysters on the half shell at a party. Pat called a friend who harvested oysters for a hobby nd purchased several bushels from him. Several weeks later several customers who purchased the oysters from at were diagnosed with a foodborne illness. What did Pat do wrong, what illness could the customers have een diagnosed with, and how could the problem have been avoided?

he customers were diagnosed with Hepatitis A virus that can be found in shellfish harvested from polluted vaters. Pat should have found another reputable supplier that was regulated by government agencies to urchase the oysters from rather than buying them from a friend.

Making it Real

Shellfish Tag – Slide 34, 37, 38

- Supplies
  - Shellfish tag(s)
- Bring in a shellfish tag (or a copy of a shellfish tag) and describe the various parts of the tag to the learners. Allow them to inspect the tag.

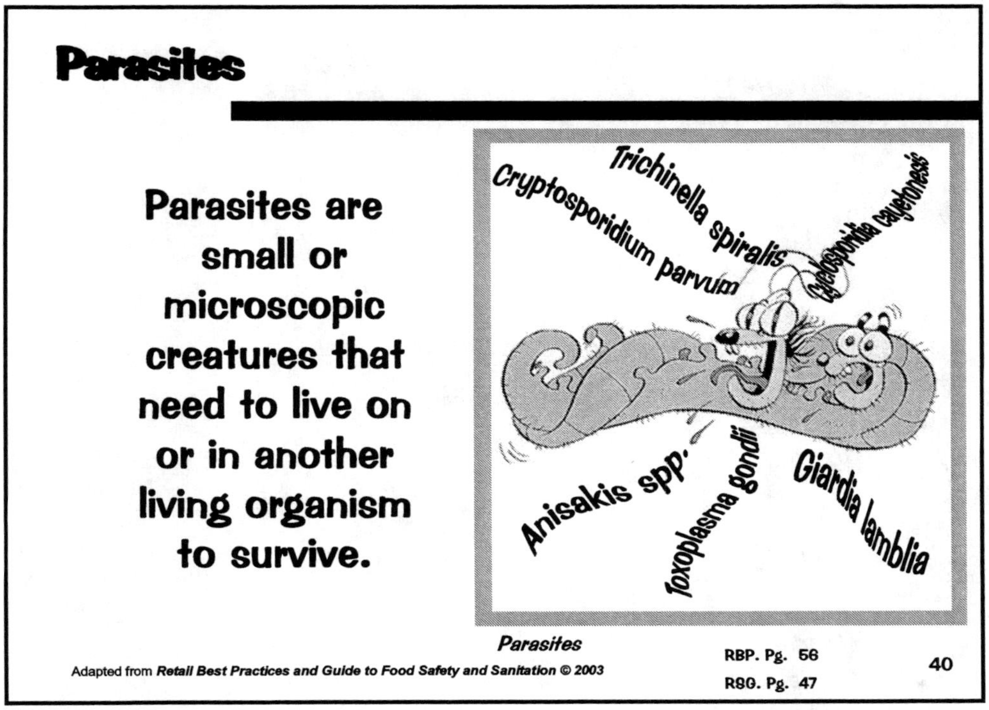

Slide 40

Parasites

Parasites are small or microscopic creatures that need to live on or inside a living host to survive.

Parasitic infection is far less common than either bacterial or viral foodborne illness.

Foodborne diseases caused by parasites:

- *Anisakis spp.*
- *Cyclospora cayetanensis*
- *Cryptosporidium parvum*
- *Giardia lamblia*
- *Toxoplasma gondii*
- *Trichinella spiralis.*

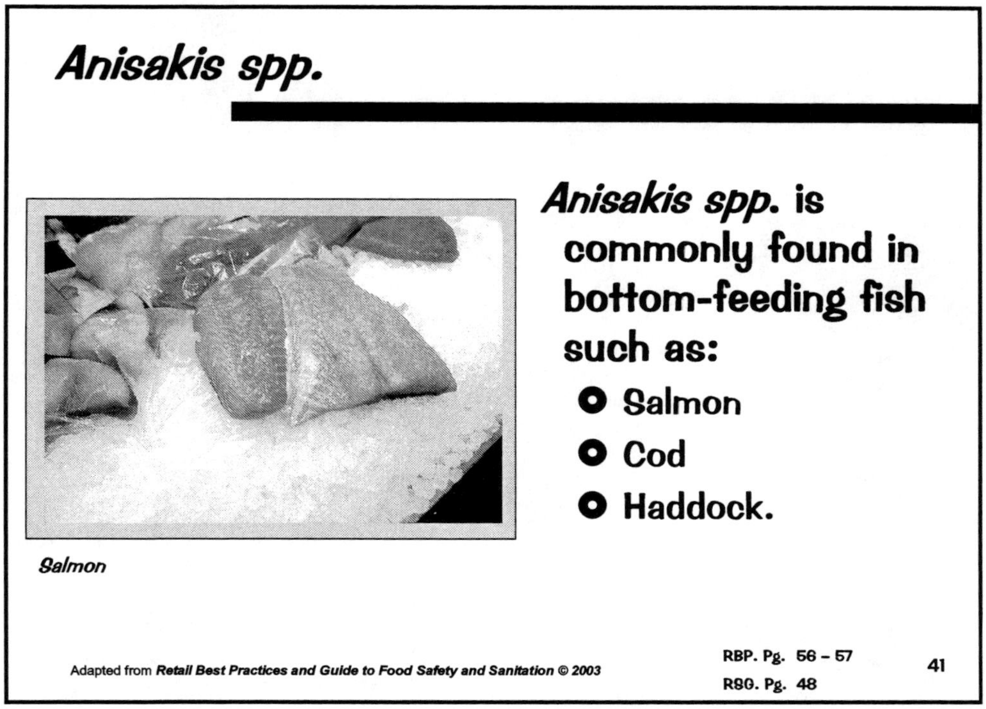

Slide 41

Anisakis spp. – Parasite

Characteristics:

- *Anisakis spp.* are roundworms (also known as cod worm and herring worm)
- Symptoms
  - Coughing
  - Abdominal pain
  - Vomiting.

Onset time:

- 1 hour to 2 weeks.

Found in the following foods:

- Commonly found in bottom-feeding fish such as:
  - Cod
  - Haddock
  - Flounder
  - Salmon
  - Tuna.

Controls:

- Buy seafood from approved sources
- Freeze seafood to proper *FDA Food Code* specifications
- Cook seafood to the proper safe temperature
- Inspect seafood to detect the presence of worms.

Cyclospora cayetanensis

Strawberries and raspberries

Cyclospora cayetanensis is commonly found in contaminated:

- Water
- Raspberries
- Strawberries
- Fresh produce.

Adapted from ***Retail Best Practices and Guide to Food Safety and Sanitation*** © 2003

RBP. Pg. 57
RSG. Pg. 49

42

Slide 42

Cyclospora cayetanensis — Parasite

Characteristics:

- An infection
- A pathogen that has been linked to outbreaks of foodborne illness in school children
- Humans are the only known carriers of this parasite
- Symptoms
  - Watery and explosive diarrhea
  - Loss of appetite
  - Bloating.

Onset time:

- Within one week, lasting weeks to months.

Found in the following foods:

- Strawberries
- Raspberries
- Fresh produce.

Causes:

- Fresh fruits and vegetables that were contaminated at the farm
- Transmitted by infected employees or contaminated water touching foods.

Controls:

- Wash berries thoroughly with a potable water supply
- Follow proper personal hygiene techniques.

Cryptosporidium parvum & *Giardia lamblia*

Cryptosporidium parvum **and *Giardia lamblia* are transferred by:**

- **Contaminated water**
- **Infected employees.**

Employees need to be clean.

Adapted from ***Retail Best Practices and Guide to Food Safety and Sanitation* © 2003**

RBP. Pg. 58
RSG. Pg. 51

43

Slide 43

Cryptosporidium parvum and ***Giardia lamblia*** – Parasites

Characteristics:

- An infection
- Symptoms
    - Severe watery diarrhea.

Onset time:

- Within 1 week of ingestion or contact, lasting up to a month.

Found in the following foods:

- Contaminated water supplies
- Ready-to-eat foods touched by an infected worker
- Salad vegetables fertilized with manure.

Causes:

- Transmitted by water contaminated by feces
- Fecal contamination by an infected employee

Controls:

- Provide a potable water supply
- Prevent contamination and cross contamination
- Proper personal hygiene, especially handwashing.

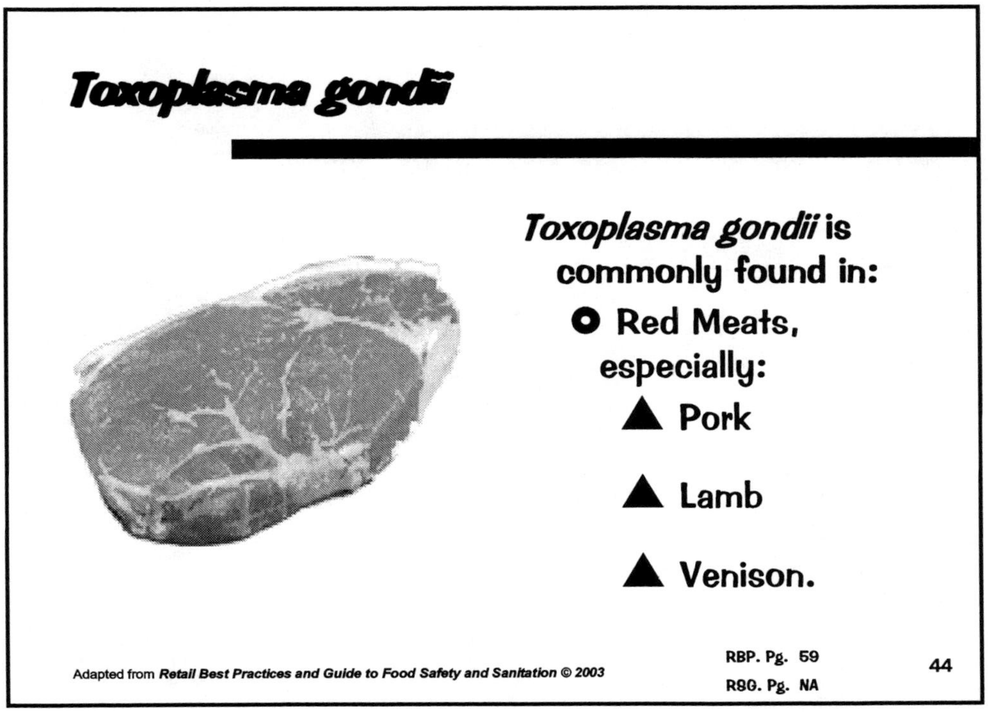

Slide 44

Toxoplasma gondii – Parasite

Characteristics:
- Causes an infection called toxoplasma
- Found in warm-blooded animals
- Does not affect healthy children and adults
- Symptoms of those with impaired immune systems and newborns have experienced:
    - Eye and brain damage
    - Mental retardation.

Onset time:
- 10 to 13 days.

Found in the following foods:
- Red meat
- Pork
- Lamb
- Venison
- Fruits and vegetables contaminated by feces.

Causes:
- Domestic cat is the primary carrier
- Food contacted by fecal material.

Controls:
- Cook meats to proper temperatures
- Proper personal hygiene and handwashing
- Prevent cross contamination.

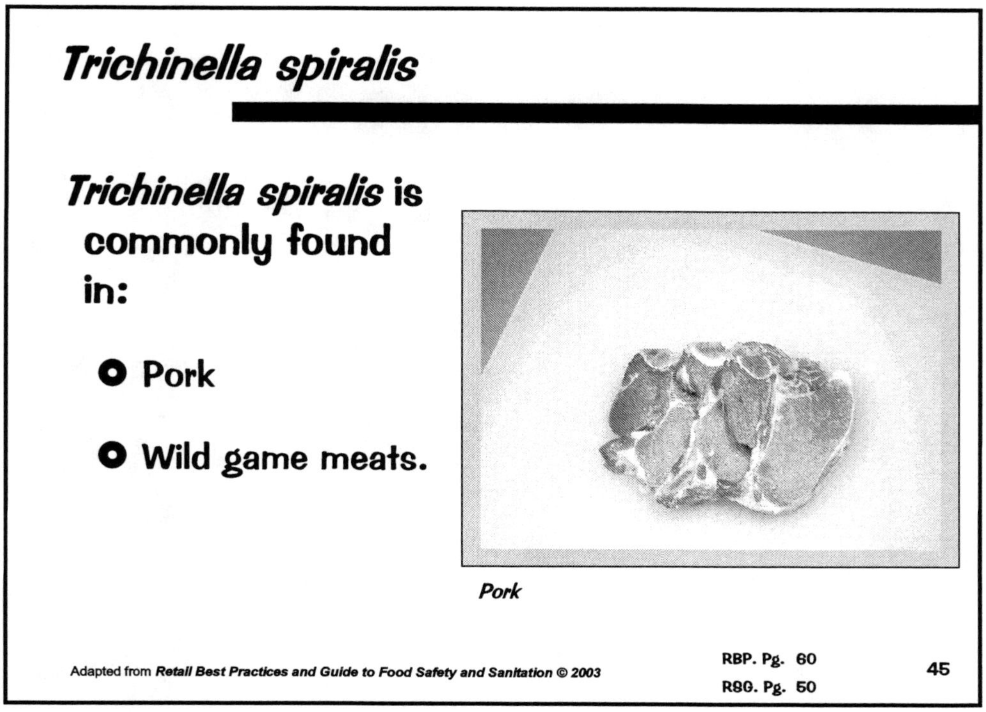

Slide 45

Trichinella spiralis — Parasite

Characteristics:

- A roundworm that causes a parasitic infection
- Symptoms
  - Abdominal pain
  - Nausea
  - Vomiting
  - Diarrhea
- Later stages are characterized by
  - Fever
  - Swelling of tissues around the eyes
  - Muscle stiffness
  - Death.

Onset time:

- 2 to 28 days.

Found in the following foods:

- Pork
- Wild animals (bear, wild boar).

Causes:

- Undercooked pork products and wild game meats.

Controls:

- Cook foods to proper temperature.

Slide 45

Safety Stop! 2.5

Sample Test Questions

1. Which of the following diseases is commonly caused by parasites?
   A. *Anisakis spp.*
   B. *Giardia lamblia.*
   C. *Trichinella spiralis.*
   D. All of the above.

2. *Cyclospora cayetanensis* is commonly found in:
   A. Red meat.
   B. Garlic stored in oil.
   C. Strawberries.
   D. Seafood.

3. *Trichinella spiralis* is commonly found in what food?
   A. Shellfish.
   B. Red meat.
   C. Pork.
   D. Poultry.

eal-life

ide 41
nisakis spp.
au was receiving the fresh fish order for the day. He was running behind in his duties and did not inspect all e fish for wholesomeness. Later that day, the manager received a call from a customer who had purchased ome haddock from the seafood department claiming the fish had worms in it. What parasite could have been fecting the haddock and what could have been done to prevent this from happening?

he haddock was infected with *Anisakis spp.*, a roundworm that is found in bottom feeding fish. Lau should ave taken the time to properly inspect the fish for wholesomeness prior to accepting it.

lide 42
yclospora cayetanensis
Melissa, a local day care operator, had purchased some fresh strawberries and raspberries from her local upermarket. She used them in a fruit salad for the children. A few days later, several of the children were ill vith diarrhea and loss of appetite. What parasitic infection could the children have gotten, and how could it ave been avoided at any stage in the flow of food?

The children were infected with *Cyclospora cayetanensis* that is commonly found on fresh fruits and produce hat were contaminated on the farm. Ready-to-eat foods such as fruits and vegetables need to be thoroughly vashed with potable water to rid them of these possible parasites. The berries could have been washed at the farm, in the supermarket or at the day care to avoid this problem.

Making it Real
None

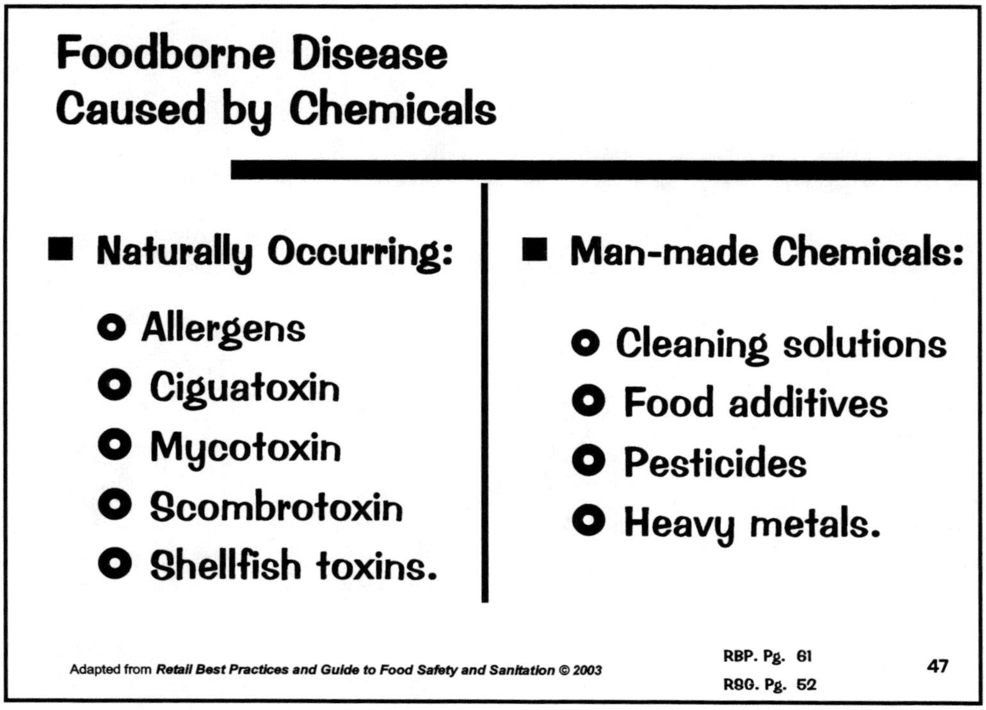

Slide 47

<u>Chemical hazards</u> are classified in two ways –

Naturally occurring chemicals – produced by a biological organism

- Allergens
- Ciguatoxin
- Mycotoxin
- Scombrotoxin
- Shellfish toxins.

Man-made chemicals – may be present in a food after processing

- Pesticides
- Cleaners
- Food additives
- Heavy metals.

Methods to control chemical hazards:

- Purchase foods from approved sources
- Use separate utensils for foods that are potential allergens
- Properly label foods and chemicals
- Properly store chemicals away from foods and utensils
- Use all chemicals according to manufacturer's directions.

Food Allergens

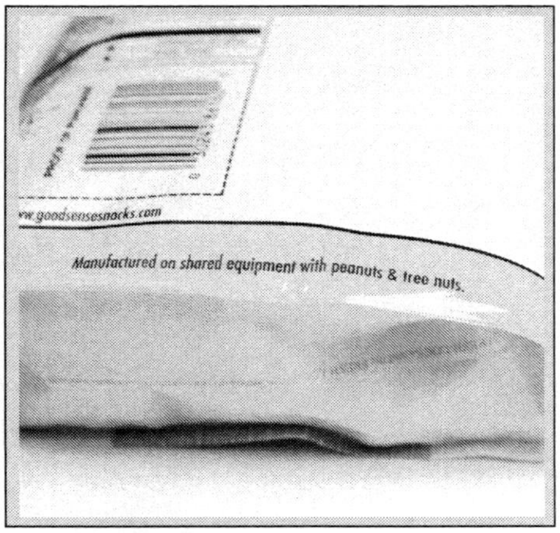

90% of all allergies are caused by:

- Milk products
- Egg products
- Wheat proteins
- Peanuts
- Soy products
- Tree nuts
- Fish
- Shellfish.

Label products correctly.

Adapted from ***Retail Best Practices and Guide to Food Safety and Sanitation* © 2003**

RBP. Pg. 62 – 63
RSG. Pg. 53

48

Slide 48

<u>Food Allergens</u> – Naturally occurring chemical

Characteristics:

- Food allergens are chemicals that cause a person's immune system to overreact
- 5 - 8% of children and 1 - 2% of adults have food allergies
- Symptoms
  - Hives
  - Difficulty breathing
  - Diarrhea
  - Swelling of the lips, tongue and mouth
  - Vomiting
  - Cramps
- Symptoms can onset in as few as five minutes
- In severe cases, a life-threatening allergic reaction called anaphylaxis can occur
- Symptoms of anaphylaxis include:
  - Itching and hives
  - Lowered blood pressure
  - Swelling of the throat
  - Unconsciousness
- 90% of all allergies are caused by:
  - Milk products
  - Egg products
  - Wheat proteins
  - Peanuts
  - Soy products
  - Tree nuts
  - Fish
  - Shellfish.

Control:

- The only way for a person who is allergic to a food to avoid a reaction is to avoid the food product
- Proper cleaning and sanitizing of food-contact surfaces
- Proper usage of utensils
- Proper labeling of food products.

Ciguatoxin

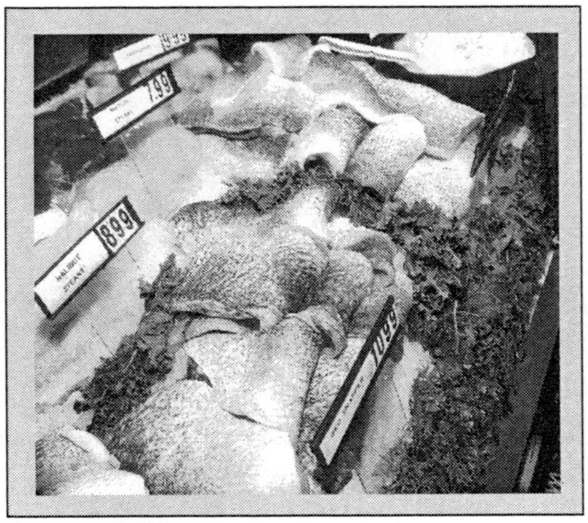

Red snapper

Ciguatoxins are commonly found in marine finfish such as:

- Snapper
- Grouper
- Barracuda
- Jack
- Mackerel
- Reef fish.

Adapted from ***Retail Best Practices and Guide to Food Safety and Sanitation*** © 2003

RBP. Pg. 64
RSG. Pg. 54

49

Slide 49

Ciguatoxin – Naturally occurring chemical

Characteristics:

- An intoxication caused by consumption of contaminated tropical fish that eat algae containing ciguatoxin
- There is no reliable method to determine presence of ciguatoxin
- Symptoms:
  - Nausea
  - Vomiting
  - Diarrhea
  - Dizziness
  - Shortness of breath
  - Reversal of hot and cold sensations.

Onset time:

- 15 minutes to 24 hours, GI symptoms usually within 6 hours and neurological systems follow 1 to 2 days, lasting a few days up to a month.

Found in the following foods (these are considered predatory fish):

- Grouper
- Barracuda
- Snapper
- Reef fish
- Jack
- Mackerel
- Triggerfish.

Causes:

- Eating fish that contain ciguatoxin.
- Ciguatoxin is heat stable and cannot be destroyed by cooking

ontrols:

- Purchasing seafood from a reputable supplier
- Avoid eating large predatory fish.

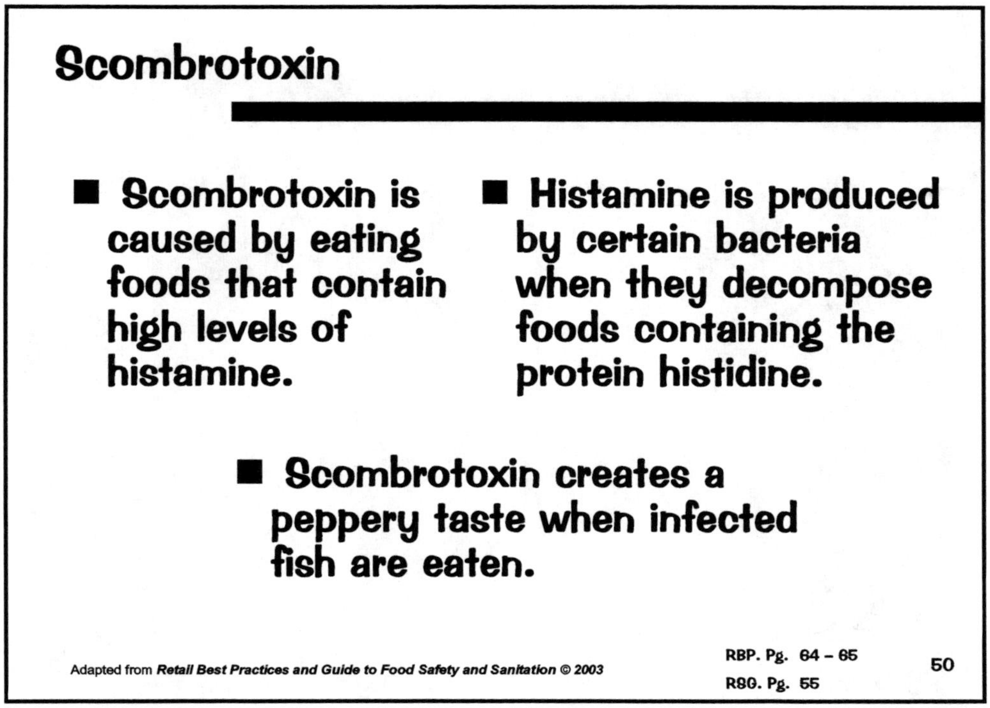

Slide 50

Scombrotoxin – Naturally occurring chemical

Characteristics:

- Also called histamine poisoning
    - Histamine:
        - Is produced by certain bacteria when they decompose foods containing the protein histidine
        - Creates a peppery taste in mouth when fish is eaten
- Symptoms
    - Dizziness
    - Burning sensation in the mouth
    - Hives
    - Shortness of breath
- Severe cases can cause loss of muscle control, the inability to speak, swallow, or breathe, and can lead to death.

Onset time:

- Within minutes, lasting 8 to 12 hours.

Found in the following foods:

- Tuna
- Anchovies
- Amberjack
- Mahi-mahi
- Other dark meat fishes
- Swiss cheese.

Causes:

- Eating foods that contain high levels of histamine.

Controls:

- Purchase seafood from a reputable supplier
- Reject seafood that has been thawed and refrozen or temperature abused
- Store "fresh" seafood between 32°F and 39°F (0°C and 4°C).

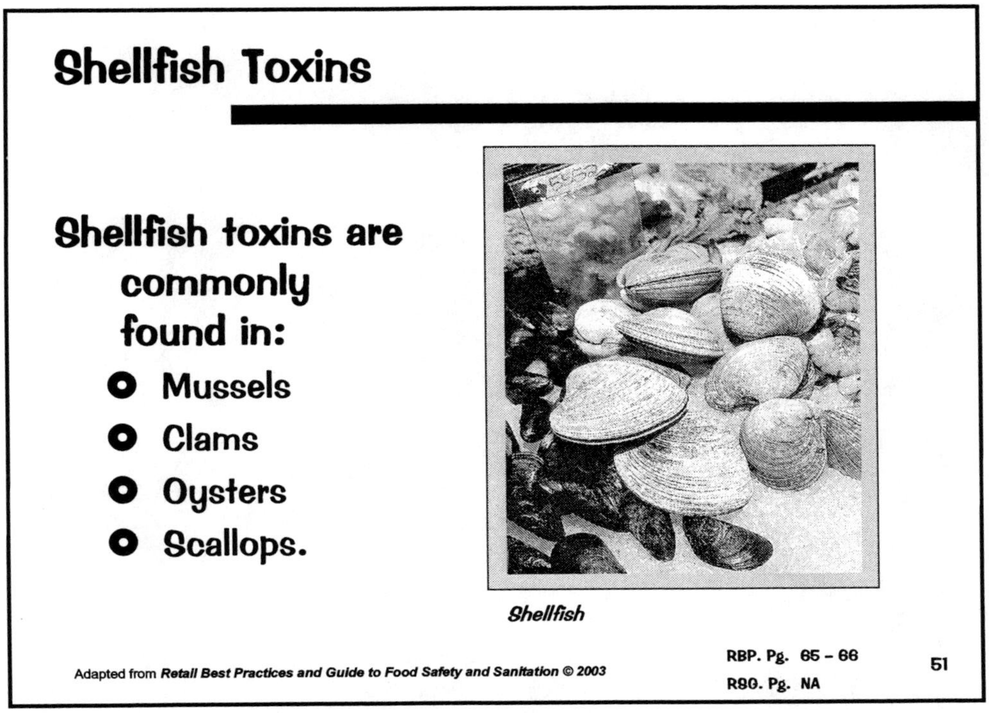

Slide 51

Shellfish toxins – Naturally occurring chemical

Characteristics:
- Intoxication
- Most common shellfish illnesses:
  - Paralytic Shellfish Poisoning (PSP)
  - Diarrhetic Shellfish Poisoning (DSP)
  - Domoic Acid Poisoning (DAP)
  - Neurotoxic Shellfish Poisoning (NSP)
- Symptoms:
  - Numbness of lips, tongue, arms, legs, and neck
  - Lack of muscle coordination.

Onset time:
- Within minutes, lasting only a short period of time.

Found in the following foods:
- Mussels
- Clams
- Oysters
- Scallops.

Causes:
- Filter-feeding shellfish feed on toxic algae (dinoflagellates).

Control:
- Purchase seafood from a reputable supplier.

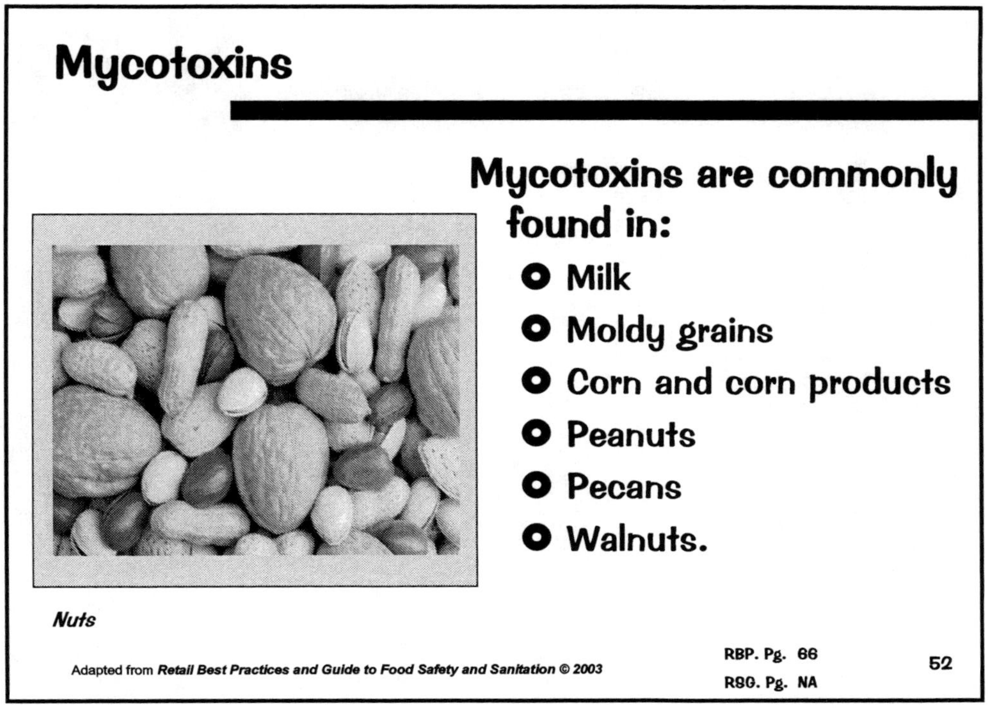

Slide 52

Mycotoxins – Naturally occurring chemical

Characteristics:

- Mycotic organisms or fungi are molds, yeasts and mushrooms, some of which are capable of causing foodborne illnesses by producing toxic chemicals called mycotoxins
- Fungi are larger than bacteria
- Fungi prefer foods that are high in sugar or starches
- Cooking does not destroy many mycotoxins
- Symptoms:
    - Acute onset – hemorrhage, fluid buildup
    - Chronic onset – cancer from small doses over time.

Found in the following foods:

- Moldy grains
- Corn and corn products
- Peanuts
- Pecans
- Walnuts
- Milk.

Controls:

- Purchase from a reputable supplier
- Keep grains and nuts dry
- Visually inspect foods for signs of mold growth.

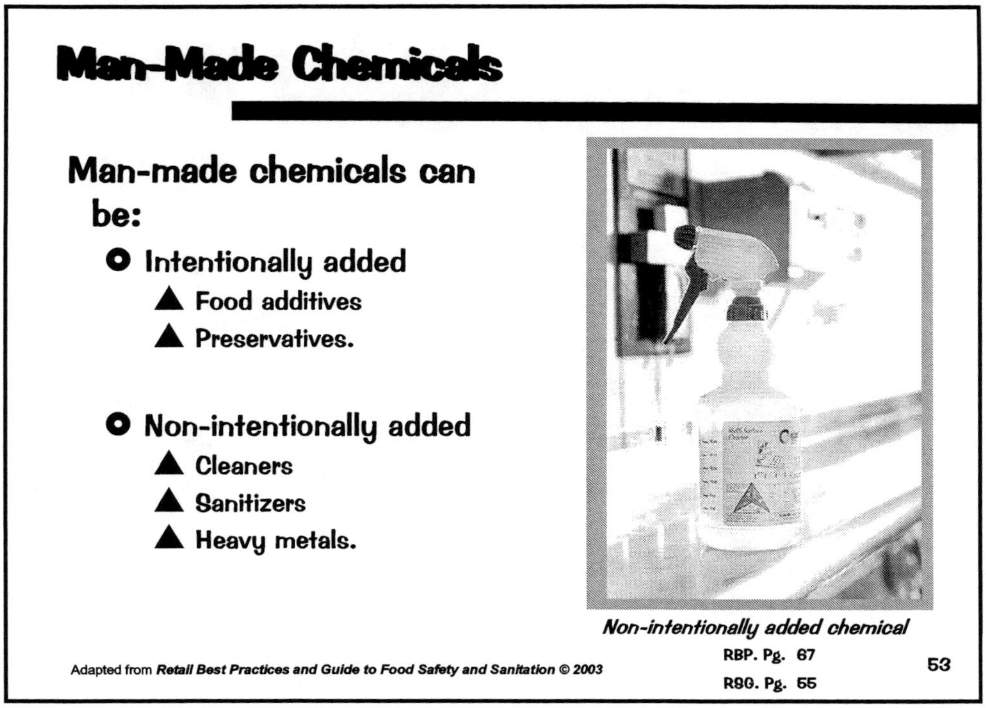

Slide 53

Man-made chemicals

- There are many chemicals added to foods that may pose a potential health risk. Intentionally added chemicals include:
    - o Food additives (some food colorings)
    - o Food preservatives (nitrates, nitrites, sulfites)
    - o Pesticides (can be removed with thorough cleaning)
- Non-intentionally added chemical examples:
    - o Cleaners
    - o Sanitizers
    - o Food containers or food-contact surfaces made from inferior metals that are misused may lead to heavy metal or inferior-metal poisoning.

Physical Hazards

Physical hazards can create injury.

Physical hazards are objects in food that can cause injury or illness such as:

- Toothpicks
- Stones
- Human hair
- Jewelry.

Adapted from ***Retail Best Practices and Guide to Food Safety and Sanitation* © 2003**

RBP. Pg. 67 – 68
RSG. Pg. 56

54

Slide 54

Physical hazards are foreign objects in food that can cause illness or injury such as:

- Fragments of broken glass
- Metal shavings
- Toothpicks
- Human hair
- Jewelry
- Bandages
- Stones
- Wood particles.

Physical hazards commonly result from accidental contamination and poor food-handling practices.

Physical hazards controls:

- Wash raw fruits and vegetables thoroughly
- Visually inspect foods
- Handle food and ice safely to avoid physical hazards
- Food employees should not wear jewelry when involved in food production with the exception of a plain wedding band.

Slide 55

Safety Stop! 2.6

Sample Test Questions

1. Which of the following is a man-made chemical?
   - **A. Pesticides.**
   - B. Allergens.
   - C. Shellfish Toxins.
   - D. Ciguatoxin.

2. The most effective way to control the growth of bacteria in a retail food establishment is by controlling:
   - **A. Time and temperature.**
   - B. pH and oxygen conditions.
   - C. Temperature and water activity.
   - D. Time and food availability.

3. Which of the following naturally occurring chemicals is associated with filter-feeding shellfish that feed on toxic algae?
   - A. Mycotoxins.
   - **B. Shellfish toxins.**
   - C. Scombrotoxin.
   - D. Allergens.

4. Which of the following is a histamine poisoning?
   - A. Ciguatoxin.
   - **B. Scombrotoxin.**
   - C. Mycotoxin.
   - D. Paralytic Shellfish Poisoning (PSP).

Food allergens are chemicals that cause a person's immune system to:

A. Shut down.
B. Under react.
C. Overreact.
D. Slow down.

Which of the following is not an example of an intentionally added chemical?

A. Food additives.
B. Food allergens.
C. Food preservatives.
D. Pesticides.

Real-Life

Slide 47 – 48
Allergen
Jessica was refilling the bulk bins of nuts and seeds. She noticed there were two scoops in the peanuts bin and a missing scoop from the sunflower seed bin. She took one of the scoops from the peanut bin and placed it in the sunflower seed bin. What did Jessica do wrong, what could result from her mistake, and what should Jessica have done?

Jessica could have contaminated the sunflower seeds with peanuts or oils from the peanut. Someone who is allergic to peanuts could use the bulk bin for sunflower seeds and have an allergic reaction to the peanut oils. Jessica should have removed the second scoop from the peanut bin and taken it to be properly cleaned and sanitized. She should have put a properly cleaned and sanitized scoop into the sunflower seed bin.

Slide 52
Storage and Rotation
A small market that was known for its ready-to-eat hot lunches had little room for storage in the stock room. They stored their nuts used in their fresh salads above the range oven where they boiled their pastas. The market used the Last In, First Out rotation method for their nuts. What problems could occur and what could be done better.

There are several issues with the handling of the nuts. They should be rotated on the First In, First Out inventory method to be sure the old product is used first. They should not be stored above the range oven where there is very moist air. This could allow mycotic organisms to grow on the nuts. Since the nuts are not cooked and used raw in salads, these organisms cause foodborne illness by producing toxic chemicals called mycotoxins. The nuts should be stored in a cool, dry place to prevent this from occurring.

Making it Real

Allergen - Slide 47 – 48

- Supplies
  - Properly labeled product that does not list a specific ingredient (i.e., peanuts), however lists that ingredient as a potential allergen that may be in the product.
- Pass the product or product label around the classroom and ask them to identify all the potential allergens in the product. See if anyone notices the disclosure of the potential allergen in the product that is not a part of the ingredient list.

ıysical Hazards - Slide 54

- Supplies
  - Learners
- Have the class remove any physical hazards they may be wearing and place them on their desk. Pick and choose individuals to explain why the items are hazards and tell you whether or not they wear them to work on a regular basis.
- Pay specific attention to any hazards that may not have been removed (i.e., medical alert bracelets, earrings, watches, piercings, etc.) and ask those learners to tell you why they believe the item is not a hazard.

Session Two:
Concepts to Keep

- Foodborne illnesses are classified as infections, intoxications, or toxin-mediated infections.
- The three types of food hazards are biological, chemical, and physical.
- The six conditions required for bacterial growth are:
  - Food
  - Acidity
  - Temperature
  - Time
  - Oxygen
  - Moisture.

Adapted from *Retail Best Practices and Guide to Food Safety and Sanitation* © 2003

RBP. Pg. 23 – 73
RSG. Pg. 20 - 60

56

Slide 56

Concepts to Keep

- Foodborne illnesses are classified as infections, intoxications, or toxin-mediated infections.
- The three types of food hazards are biological, chemical, and physical.
- Biological hazards are responsible for more cases of foodborne illness than any other hazard.
- The six conditions required for bacterial growth are:
  - **F**ood
  - **A**cidity
  - **T**emperature
  - **T**ime
  - **O**xygen
  - **M**oisture.

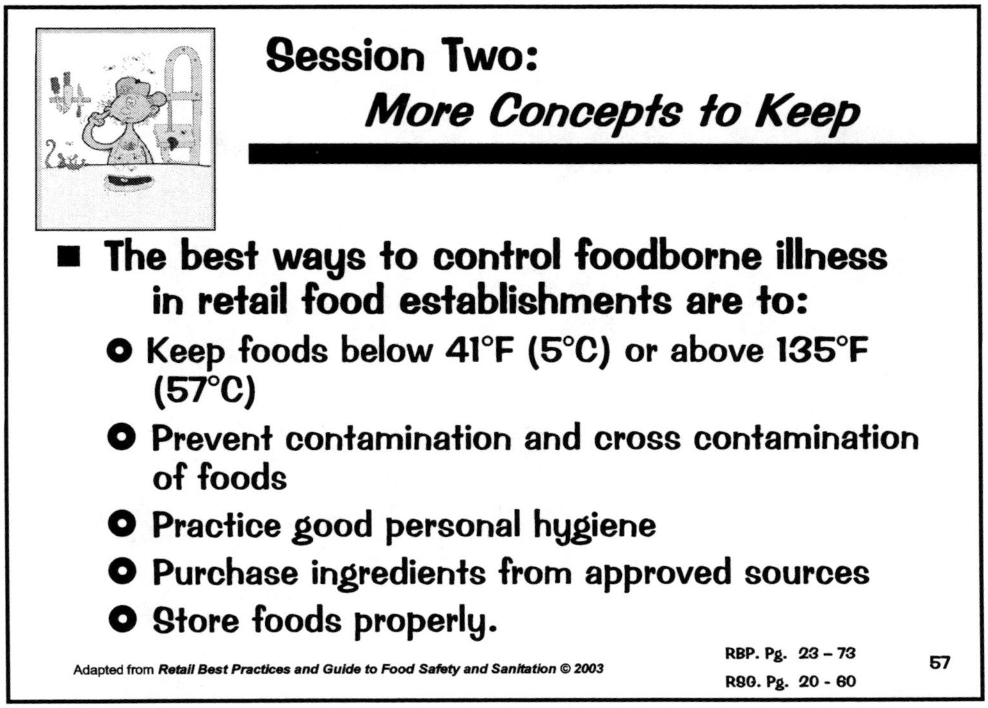

Slide 57

<u>More Concepts to Keep</u>

- The best ways to control foodborne illness in retail food establishments are to:
    - Keep foods below 41°F (5°C) or above 135°F (57°C)
    - Prevent contamination and cross contamination of foods
    - Practice good personal hygiene
    - Purchase ingredients from approved sources
    - Store foods properly.

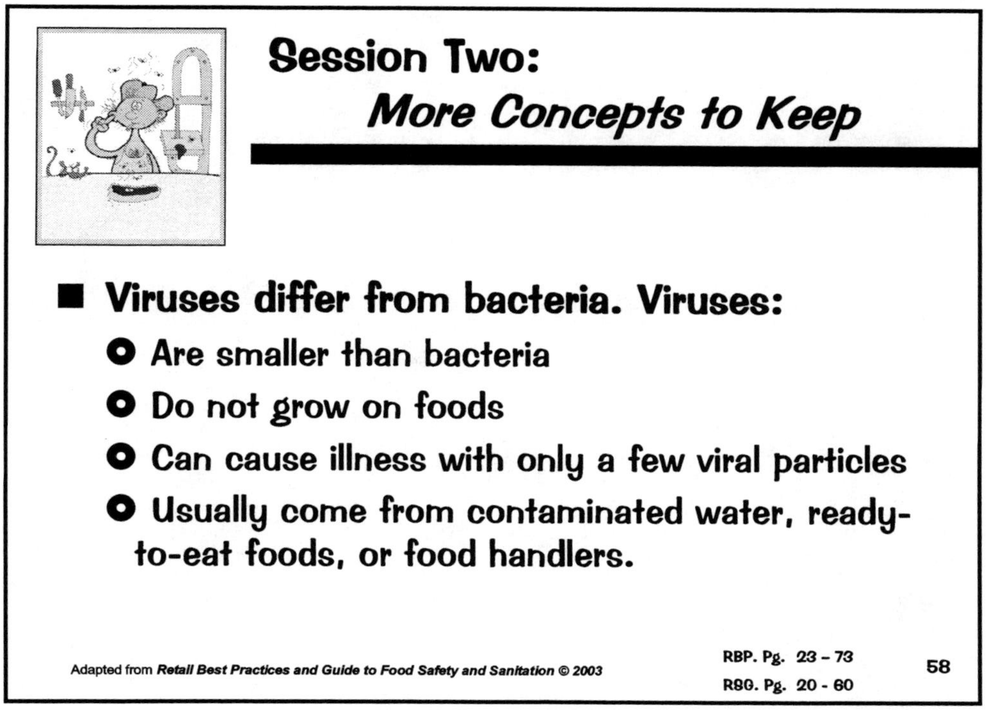

Slide 58

<u>More Concepts to Keep</u>

- Viruses differ from bacteria. Viruses:
  - Are smaller than bacteria
  - Do not grow on foods
  - Can cause illness with only a few viral particles
  - Usually come from contaminated water, ready-to-eat foods, or food handlers.

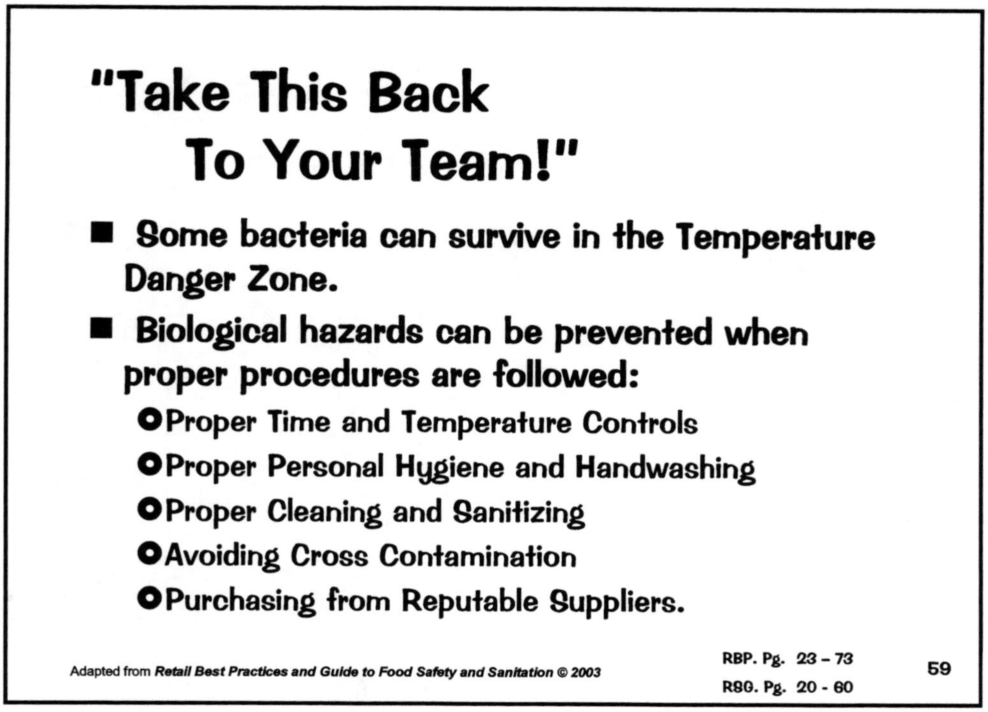

Slide 59

<u>**Take This Back to Your Team!**</u>

- Some bacteria can survive in the Temperature Danger Zone.
- Biological hazards can be prevented when proper procedures are followed:
  - Proper Time and Temperature Controls
  - Proper Personal Hygiene and Handwashing
  - Proper Cleaning and Sanitizing
  - Avoiding Cross Contamination
  - Purchasing from Reputable Suppliers.

Session 3

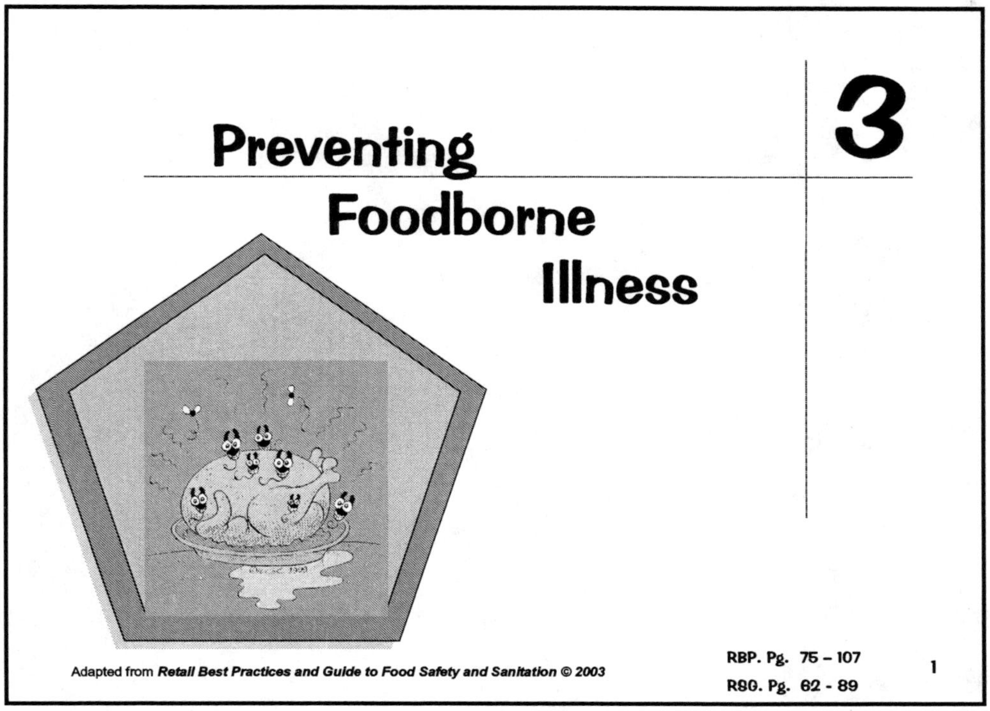

Preventing Foodborne Illness – Session 3 Key Points

Slide 1

<u>Training Tip:</u>

The content of each slide provides the most essential information in the session. The written text below each slide is provided to help you understand and better teach the essential information. The written text may also be used to answer questions that may be posed during the session. Depending on the length of your training session, you may not be able to address all the information contained beneath each slide.

Overview:

Session 3 is filled with extremely important, and highly testable food safety and sanitation information. This chapter discusses:

- The importance of measuring temperatures of food product throughout the flow of food
- Approved temperature-measuring devices
- How and when to measure food temperatures
- Personal hygiene including hand washing and disposable glove usage
- Contamination and cross contamination including how to avoid them.

Session 3 will also briefly overview cleaning and sanitation, however, little time will be spent on this topic as it is discussed in detail in Session 6.

Suggested Timeline:
8-hour course: 1 hour using 37 PowerPoint© slides and the following:

- ***Safety Stop!*** 3.2 ***Making it Real*** Calibration of Temperature-measuring Devices exercise
- ***Safety Stop!*** 3.4 ***Making it Real*** Proper Hand Washing Procedures exercise

16-hour course: 2 hours using the 49 PowerPoint© slides and the 5 ***Safety Stop!*** slides containing:

- 28 ***Sample Test Questions***
- 22 ***Real-life*** scenarios
- 13 ***Making it Real*** interactive exercises.

Video suggestions and supplemental resources:

Videos:
Hand Washing for Life Video Series, Infocus Learning Systems, 1-800-446-3628.

Factors That Affect Foodborne Illness

Points to Ponder

- What is the temperature danger zone, and how does it affect food safety?
- How can time and temperature abuse be prevented?
- How do you properly measure the temperature of foods?
- What are some good personal hygiene practices for food employees?
- What practices can be used to prevent cross contamination of foods?

Adapted from ***Retail Best Practices and Guide to Food Safety and Sanitation*** © 2003

RBP. Pg. 76
RSG. Pg. 63

2

Slide 2

Points to Ponder

- What is the temperature danger zone, and how does it affect food safety?
- How can time and temperature abuse be prevented?
- How do you properly measure the temperature of foods?
- What are some good personal hygiene practices for food employees?
- What practices can be used to prevent cross contamination of foods?

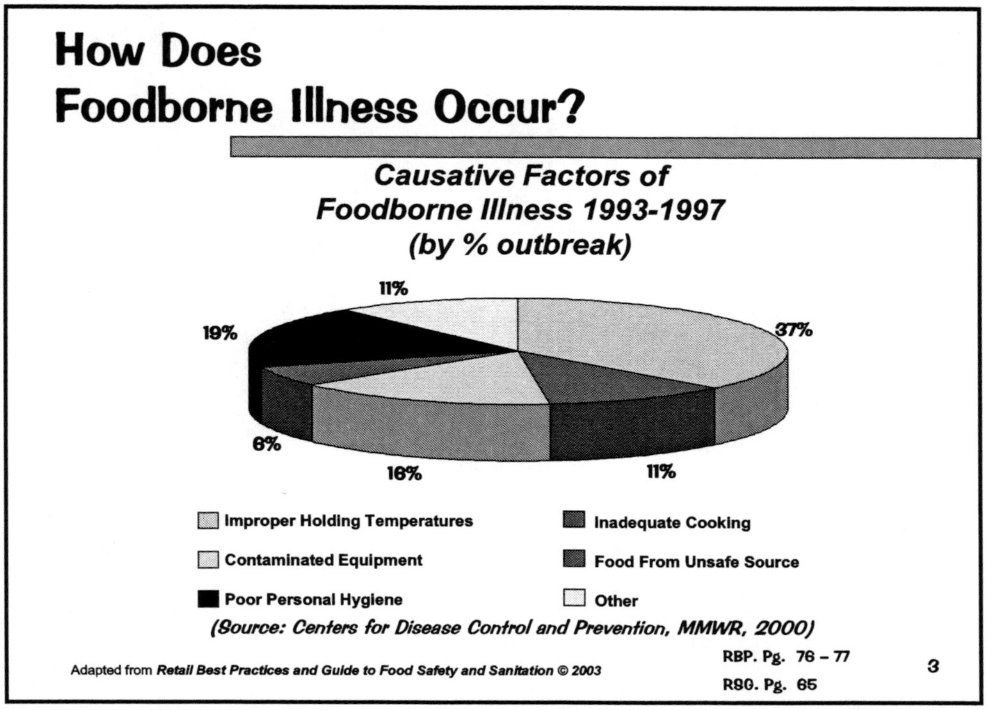

Slide 3

The leading factors that contribute to foodborne illness are:

- Improper cooling of foods
- Large amount of time between preparing and serving food
- Infected persons handling food
- Inadequate cooking
- Improper hot storage
- Inadequate reheating
- Contaminated raw food
- Cross contamination
- Improper cleaning and sanitizing.

Controlling time and temperature are the most critical ways to ensure food safety.

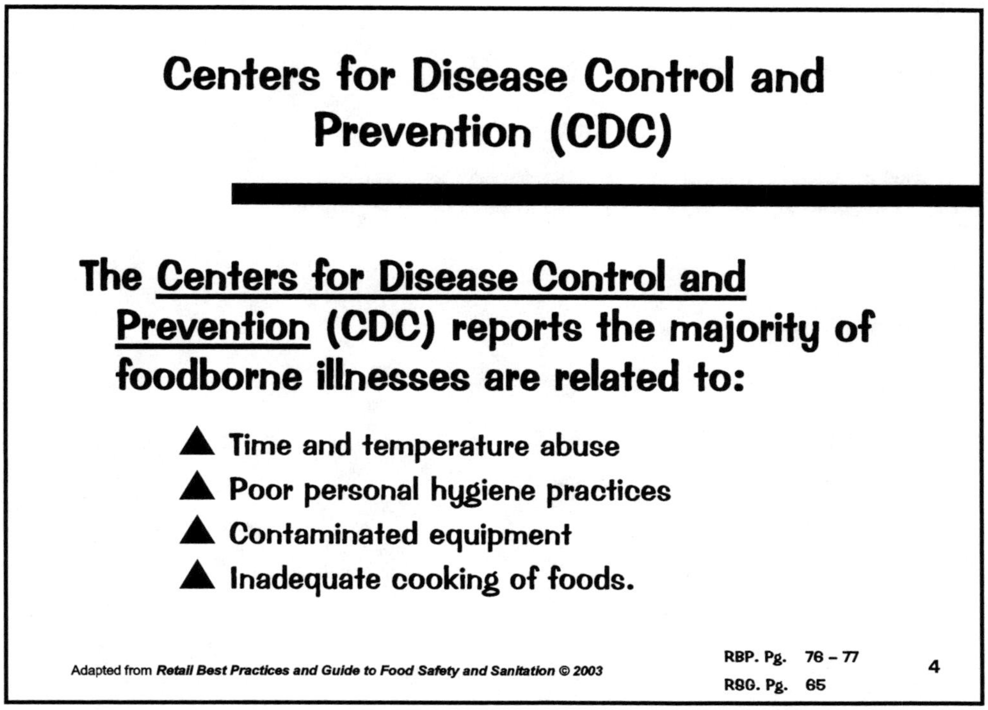

Slide 4

<u>The Centers for Disease Control and Prevention</u> – (CDC) is a federal government agency that collects statistics on diseases that affect people in America including foodborne illness.

The Centers for Disease Control and Prevention (CDC)

- Reports the majority of foodborne illnesses are related to:
  - Time and temperature abuse
  - Poor personal hygiene including hand washing
  - Contaminated equipment
  - Inadequate cooking.

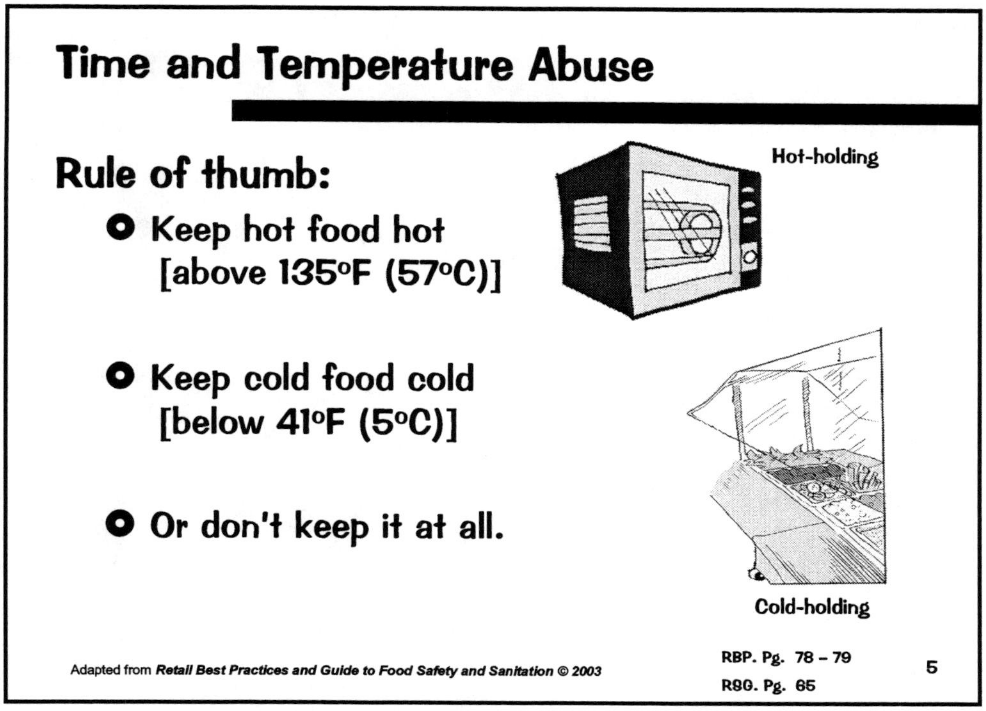

Slide 5

Controlling time and temperature are critical ways to ensure food safety.

Time and **temperature abuse** is allowing foods to remain in the temperature danger zone for an unacceptable period of time.

Temperature abuse occurs when foods are:

- Exposed to temperatures in the danger zone for enough time to allow growth of harmful microorganisms
- Not cooked or reheated sufficiently to destroy harmful microorganisms.

Rule of Thumb: Keep Hot Foods Hot, Keep Cold Foods Cold, or Don't Keep the Food at All!

Temperature Danger Zone
41°F (5°C) to 135°F (57°C)

Keeping foods above 135°F (57°C) will

- Prevent growth of disease-causing microbes
- Destroy microbes.

Keeping foods below 41°F (5°C) will

- Prevent or slow the growth of microbes.

Adapted from *Retail Best Practices and Guide to Food Safety and Sanitation* © 2003

RBP. Pg. 78 – 79
RSG. Pg. 65 - 66

6

Slide 6

The temperature danger zone is a range of temperatures between 41°F (5°C) and 135°F (57°C) at which disease-causing bacteria grow best.

Keep the inner core of food temperatures above 135°F (57°C) in order to:

- Prevent harmful microbes from growing
- Destroy most microbes, however, some toxins produced by microbes may not be affected by heat.

Keep the inner core of food temperatures below 41°F (5°C) in order to:

- Prevent most microbes from growing
- Slow the rate of bacterial growth.

Slide 7

There are unavoidable times during food production when foods must be in the temperature danger zone such as:

- Cooking
- Reheating
- Cooling
- Food handling (i.e., sandwich assembly).

Methods to prevent time and temperature abuse during these times include:

- Reheat foods to 165°F (74°C) within two hours
- Cool hot foods from 135°F (57°C) to 70°F (21°C) within two hours and from 135°F (57°C) to 41°F (5°C) within six hours
- Pass foods through the temperature danger zone as quickly as possible
- Pass foods through the temperature danger zone as few times as possible
- Hot foods should be cooled and reheated as few times as possible to assure safety and quality.

Slide 8

Safety Stop! 3.1

Sample Test Questions

1. The federal agency that collects statistics on disease that affect people in the United States is the:
   A. Centers for Disease Control and Prevention (CDC).
   B. Critical Disease Control Center.
   C. Center for Disease Causes.
   D. Coalition of Disease Controls.

2. According to the CDC, what is the leading cause of foodborne illness?
   A. Inadequate cooking.
   B. Infected employees touching foods.
   C. Improper holding temperatures.
   D. Use of leftovers.

3. Allowing foods to remain in the temperature danger zone for an unacceptable amount of time is considered:
   A. Time and temperature abuse.
   B. Cross contamination.
   C. Poor cooking procedures.
   D. Poor personal hygiene.

4. The temperature danger zone is:
   A. 40°F (4°C) to 120°F (49°C).
   B. 45°F (7°C) to 145°F (63°C).
   C. 41°F (5°C) to 135°F (57°C).
   D. 45°F (7°C) to 165°F (74°C).

5. Foods should be passed through the temperature danger zone:
   A. Only before lunch.
   B. As few times as possible.
   C. Slowly.
   D. Only during the cooking process.

Real-life

Slide 4
CDC
In a large case of foodborne illness, the CDC had identified 37 people who got sick in eight states. All had fallen ill between August 2 and November 30. All were infected by *Listeria monocytogenes*. The CDC's research had found a common link: Many of the victims had eaten the same brand of hot dogs.

Slide 6
Receiving
Louis was receiving the meat order for the day. He propped the door of the cooler open so he could move the meat around more easily. What was Louis's mistake?

By propping the door of the cooler open, he subjected the meat in the front of the cooler to temperatures above 41°F (5°C), and allowed them to be in the temperature danger zone. He let cold air escape which could cause all foods to be in the temperature danger zone.

Slide 6
Receiving
Mark was receiving Saturday's dairy order in the walk-in refrigerator at 6 a.m. He had to move some of the old product out of the cooler at 6:15 a.m. to get the new product behind it. He went on break at 10:45 and did not tell anyone he had pulled the old product out of the cooler. He returned at 11 a.m. and put the old product in front of the new product. What was Mark's mistake?

Mark left the old product out of the cooler at room temperature for more than four hours.

Food Temperature-measuring Devices:
Dial Face, Metal Stem Thermometer

Dial face metal stem thermometer

Insert the probe of the dial face thermometer at least 2 inches into the core of the food product.

Adapted from ***Retail Best Practices and Guide to Food Safety and Sanitation*** © 2003

RBP. Pg. 81 - 83
RSG. Pg. 67 - 69

9

Slide 9

<u>**Food temperature-measuring devices**</u> typically measure food temperatures in degrees Fahrenheit (°F) and Celsius (°C).

- It is important to know which temperature scale your device is measuring.

The next several slides will discuss the different types of food temperature-measuring devices.

<u>**Dial face, metal stem thermometer (bi-metallic)**</u>

- Display is analog
- Is the most commonly used thermometer
- Measures temperatures ranging from 0°F (-18°C) to 220°F (104°C)
- Sensing probe must be inserted at least 2 inches into food products
- Wait for the temperature to stabilize for an accurate reading.

Sensing device:

- Dial face thermometers have a dimple on the stem of the probe
- The sensing device extends from the tip of the probe to the dimple on the stem
- The temperature indicated is an average as measured over this distance.

Temperature-measuring Devices: *Digital Thermometer*

Insert the probe of the digital thermometer at least 1 inch into the core of the food product.

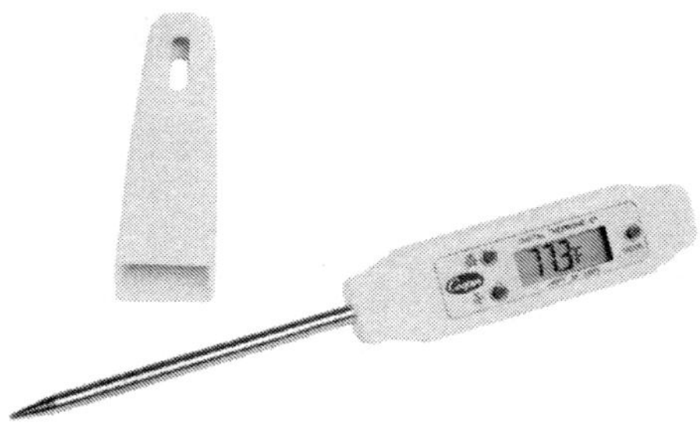

Digital thermometer

Adapted from *Retail Best Practices and Guide to Food Safety and Sanitation* © 2003

RBP. Pg. 81 - 83
RSG. Pg. 67 - 69

10

Slide 10

Digital thermometer

- Display is numerical (digital)
- Temperature ranges vary based on model but provide a wider range than dial face thermometer
- Insert probe at least 1 inch into food products to obtain an accurate reading
- Wait for the temperature to stabilize for an accurate reading.

Sensing device:

- The sensor on a digital thermometer is on the tip of the stem of the probe.

Temperature-measuring Devices:
Thermocouple

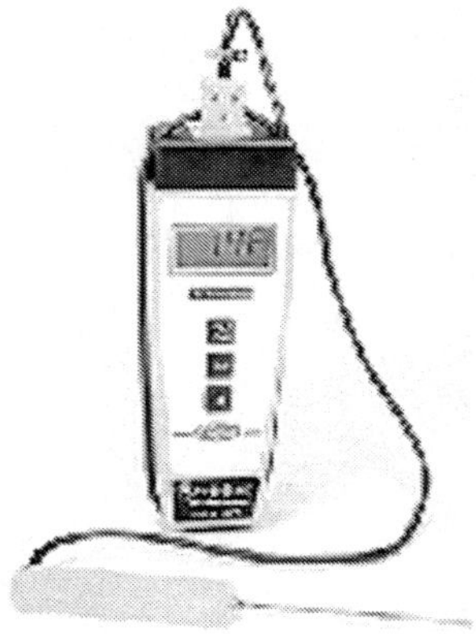

Thermocouple

The thermocouple has a variety of interchangeable probes for a variety of uses.

Adapted from ***Retail Best Practices and Guide to Food Safety and Sanitation © 2003***

RBP. Pg. 81 - 83
RSG. Pg. 67 - 69

11

Slide 11

<u>Thermocouple</u>

- Display is numerical (digital)
- Temperature ranges vary based on model but provide a wider range than a dial face thermometer
- Insert probe at least 1 inch into food products to obtain an accurate reading
- Wait for the temperature to stabilize for an accurate reading
- Has a variety of interchangeable probes for different applications.

Sensing device:

- The sensor on a digital thermometer is on the tip of the stem of the probe.

Temperature-measuring Devices: *Infrared*

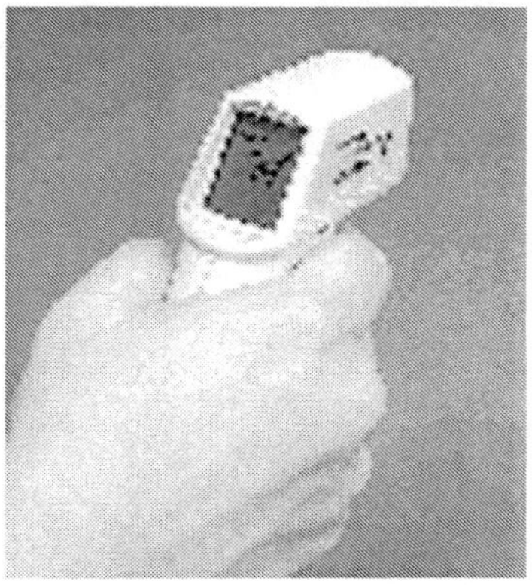

Infrared

The infrared thermometer does not measure the internal temperature of food products, only the surface temperature.

Adapted from ***Retail Best Practices and Guide to Food Safety and Sanitation* © 2003**

RBP. Pg. 81 - 83
RSG. Pg. 67 - 69

12

Slide 12

Infrared

- Display is numerical (digital)
- Measures the surface temperature of foods
- Does not measure the internal temperature of foods
- Measures food temperatures without touching food products
- Can measure many different products without causing cross contamination
- Accuracy must be checked frequently.

Potential for thermal shock:

- Cannot adjust to different ambient temperatures quickly (i.e., dry storage to the walk-in freezer)
- Allow 20 minutes before use when changing ambient temperatures.

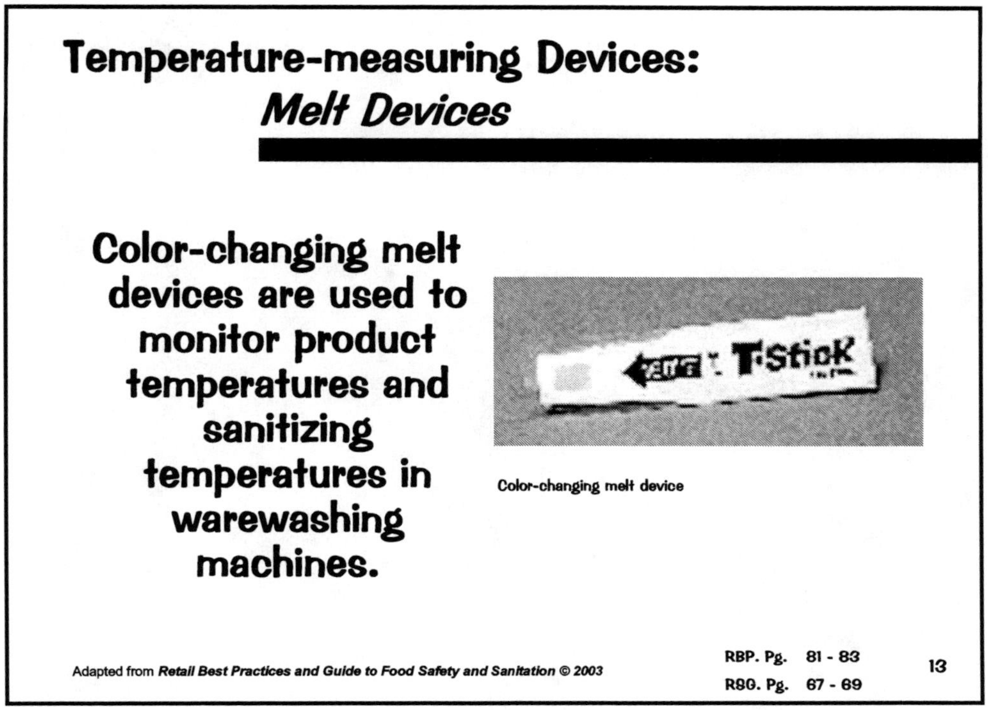

Slide 13

<u>**Color-changing melt devices (T-stick/Thermolabel)**</u>

- Single-use disposable temperature changing device
- No display of actual temperature
- Change color when the temperature reaches or exceeds the indicated temperature
- Used to monitor product temperatures and washing/sanitizing temperatures in dishwashing machines.

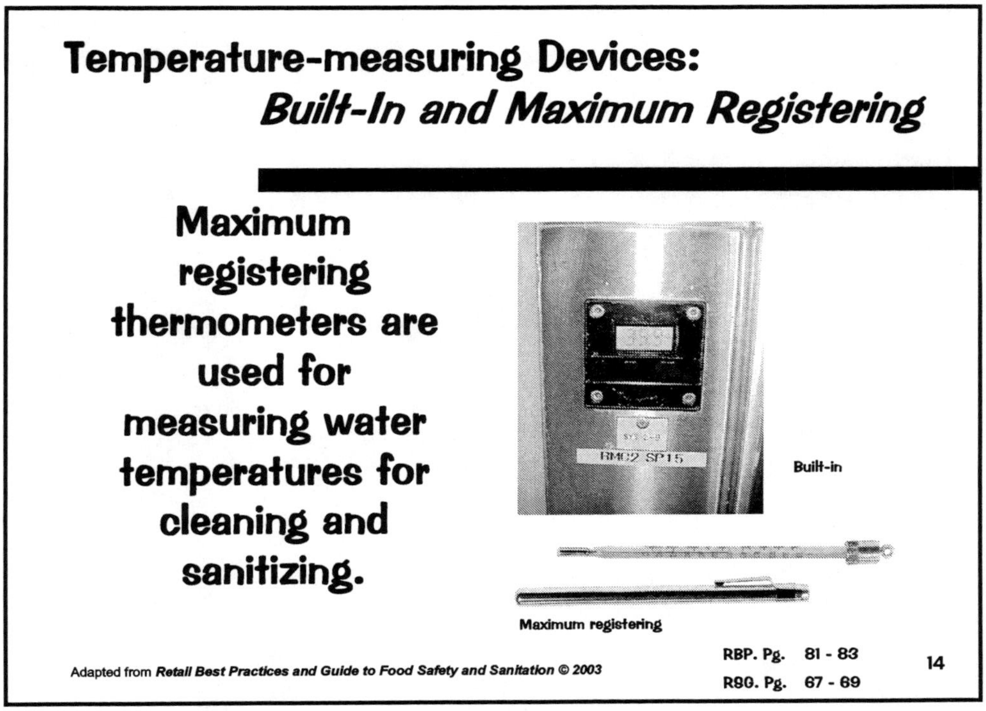

Slide 14

Built-in

- Refrigerated and frozen food cases contain built-in thermometers for constant measurement of temperatures of food storage.

Maximum registering

- Measures the temperature of hot water used to sanitize dishware and utensils in mechanical warewashing machines and three-bay sinks
- This device is becoming less popular in retail food establishments because it contains mercury, which can be a contaminant of food and the general environment.

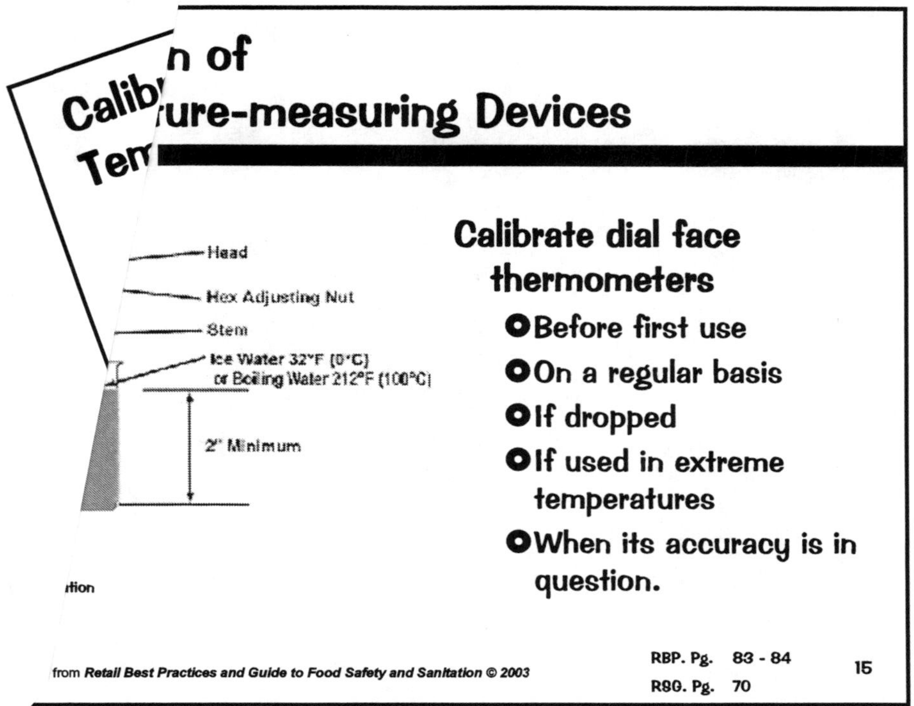

ate – to determine and verify the scale of a measuring instrument with a standard
- All thermometers must be checked for accuracy on a regular basis.

Dial face thermometers should be calibrated on a regular basis and:

- Before their first use
- If they are dropped
- If used to measure extreme temperatures
- Whenever accuracy is in question.

All other types of thermometers can only be calibrated by the manufacturer.

There are two ways to calibrate dial face thermometers:

1. Ice point method.
2. Boiling point method.

Steps to follow when calibrating a dial face thermometer:

1. Immerse the first 2 inches of the stem into crushed ice and water or boiling water.
2. Allow the indicator needle to stabilize.
3. Use pliers, an open-ended wrench or the holder on the case of the thermometer to grip the hex adjusting nut.
4. While holding the hex adjusting nut in place, rotate the head of the thermometer until the indicator needle points to 32°F (0°C) (ice point) or 212°F (100°C) (boiling point).
5. At higher altitudes, the temperature of the boiling point may vary. Consult your local health department for the boiling point temperature in your area.

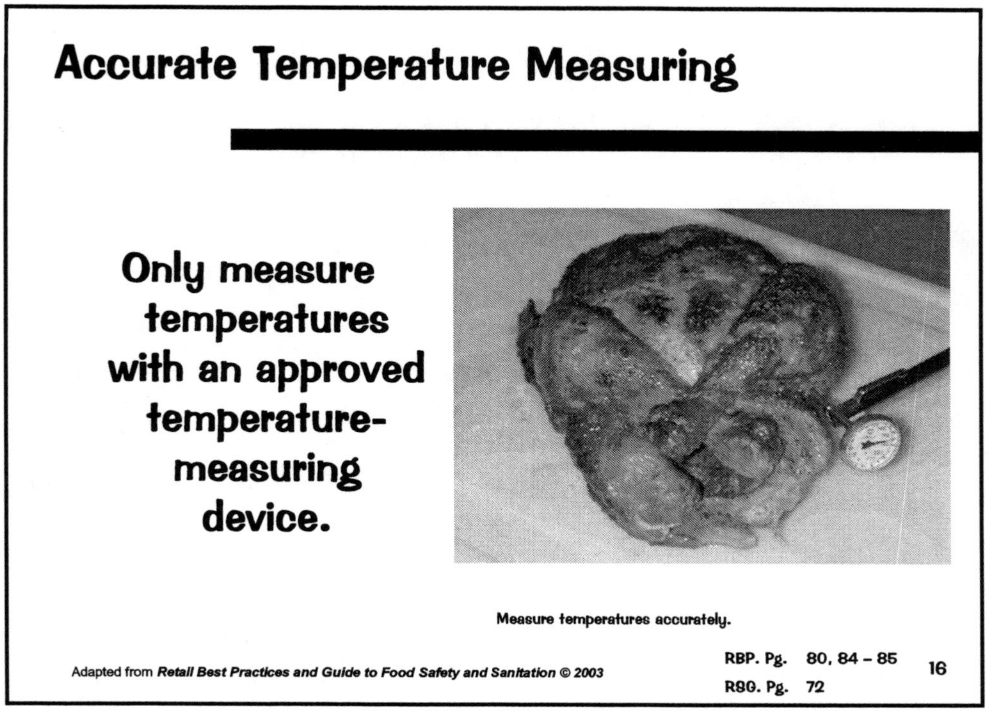

Slide 16

To accurately measure food temperatures, be sure to:

- Use an approved temperature-measuring device
- Locate the sensing portion of the device
- Test the measuring device using the ice or boiling point method; calibrate if necessary
- Clean and sanitize the probe of the temperature-measuring device according to procedure
- Measure the internal temperature of the food by inserting the probe into the center or thickest part of the item.

Slide 17

Measuring the temperature of unpackaged foods:

- Insert the probe of the thermometer into the core, or the thickest part of the food product
- Ensure the probe is inserted to the appropriate depth
- When possible stir the food prior to measuring the temperature
- Always wait for the temperature reading to stabilize.

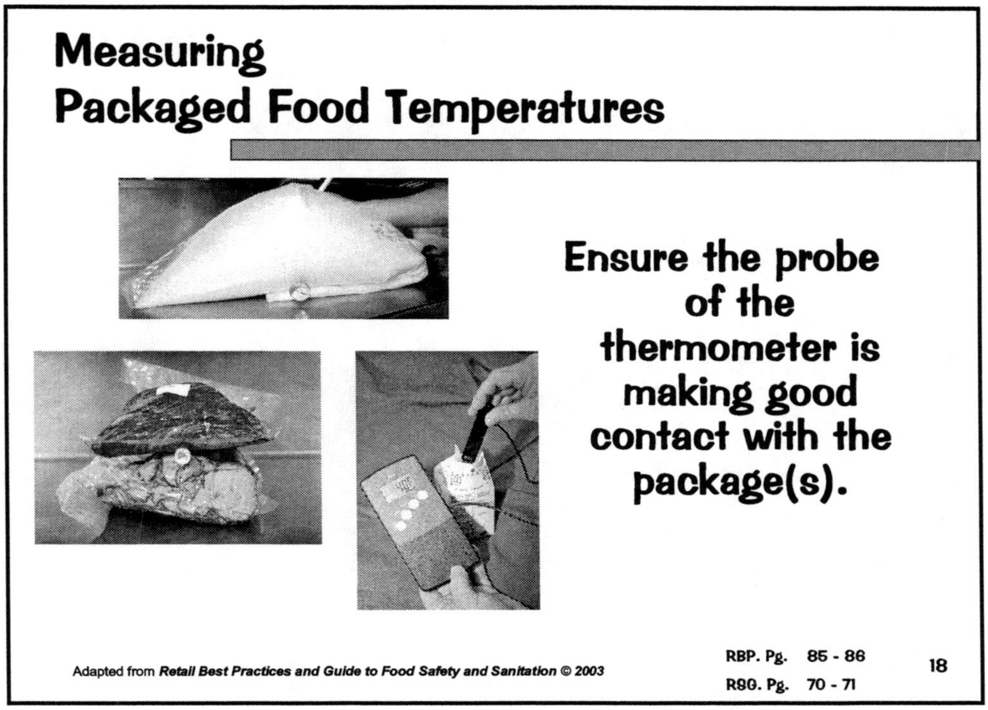

Slide 18

Measuring the temperature of packaged foods:

- Place the stem or probe of the thermometer between the two packages, or fold the package around the stem or probe of the thermometer
- Ensure the probe is making good contact with packages
- With some packaged foods it is best to open the package to obtain an accurate reading
- Always wait for the temperature reading to stabilize.

Slide 19

Safety Stop! 3.2

Sample Test Questions

1. The temperature-sensing device on a bi-metallic thermometer is:
    A. At the tip of the probe.
    B. At the top of the probe.
    C. At the dimple on the stem of the probe.
    D. From the dimple on the probe to the tip of the probe.

2. How far into the food product does the stem of the digital thermometer need to be inserted?
    A. Just the tip of the thermometer needs to be inserted.
    B. 1 inch.
    C. 2 inches.
    D. 3 inches.

3. Which food temperature-measuring device has interchangeable probes?
    A. Thermocouple.
    B. Digital thermometer.
    C. Dial face.
    D. Infrared.

4. Infrared thermometers measure:
    A. The surface temperature of the food.
    B. The ambient air surrounding the food.
    C. The internal temperature of the food.
    D. The temperature of the pan.

5. Which temperature-measuring device does not provide a specific temperature?
   A. Thermocouple.
   B. Digital thermometer.
   C. Color-changing melt device.
   D. Maximum registering thermometer.

6. Which of the following thermometers can be calibrated in house?
   A. Digital.
   B. Thermocouple.
   C. T-Stick.
   D. Dial face metal stem.

7. When measuring the temperature of unpackaged foods, it is important to:
   A. Insert the probe into the core of the product.
   B. Insert the probe into the bottom of the product.
   C. Insert the probe into the thinnest portion of the product.
   D. Insert the probe into the top of the product.

8. When measuring the temperature of packaged foods it is important to:
   A. Completely insert the probe into the package.
   B. Touch the tip of the probe to the package.
   C. Ensure the probe is making good contact with the packages.
   D. Use a new thermometer.

Real-life

Slide 9
Dial Face Metal Stem
Julie was filling out the temperature log for the salad bar. She checked the temperature of the potato salad by sticking the tip of her dial face thermometer into the top layer of the salad. What was Julie's mistake?

Julie should be sure to check the internal, or core temperature of the products by inserting the dial face thermometer at least 2 inches.

Slide 11
Thermocouple
Maurice was filling out the temperature log for the soup station. He inserted the probe of the thermocouple 1 inch into the soup. He quickly removed it and logged the temperature. What was his mistake?

Maurice should have waited for the temperature reading on the thermocouple to stabilize before he removed it.

Slide 14
Maximum Registering Thermometer
Jamie was using a maximum registering thermometer to check the temperature of the soup at the soup bar. What was her mistake?

Maximum registering thermometers are used to measure the water temperature of warewashing water, not for measuring the temperature of foods. The mercury in the thermometer is a potential contaminant and should never be placed in a food item. The thermometer has a glass case which could be a physical hazard too.

Slide 15
Calibration
Mike was going to fill out the Temperature Log for the salad bar. He could not find his thermometer. He asked the manager for a new dial face thermometer. Mike proceeded to take the thermometer out of its box and use it to check the temperatures of the salad bar. What was his mistake?

Mike should have cleaned, sanitized, and calibrated the new dial face thermometer before he used it.

Slide 15 – 16
Cleaning and Calibration
Michelle was checking the temperatures of the meat in the display case. She dropped her dial face metal stem thermometer on the ground, picked it up, and proceeded to check the temperatures of the meat. What was her mistake?

Michelle should have properly cleaned, sanitized, and calibrated the thermometer after she dropped it.

Slide 18
Measuring Packaged Food Temperatures
Jeff was checking in the packaged deli meats. He touched the tip of his thermocouple to a side of the packaged roast beef and logged its temperature. What was his mistake?

Jeff should have laid the probe of the thermocouple between two packaged meats and ensured the probe was making good contact with both packages.

Making it Real

Temperature-measuring Devices: Dial Face, Metal Stem Thermometer – Slide 9

- Supplies
  - Dial face, metal stem thermometer
- Allow the learners to handle, inspect, and ask questions regarding the thermometer.

Temperature-measuring Devices: Digital Thermometer – Slide 10

- Supplies
  - Digital thermometer
- Allow the learners to handle, inspect, and ask questions regarding the thermometer.

Temperature-measuring Devices: Thermocouple – Slide 11

- Supplies
  - Thermocouple
- Allow the learners to handle, inspect, and ask questions regarding the thermometer.

Temperature-measuring Devices: Infrared – Slide 12

- Supplies
  - Infrared thermometer
- Allow the learners to handle, inspect, and ask questions regarding the thermometer.

Temperature-measuring Devices: Color-changing Melt Type – Slide 13

- Supplies
  - Color-changing melt type thermometer
- Allow the learners to handle, inspect, and ask questions regarding the thermometer.

Temperature-measuring Devices: Maximum Registering – Slide 14

- Supplies
  - Maximum registering thermometer
- Allow the learners to handle, inspect, and ask questions regarding the thermometer
- Re-enforce the reduction of use of this thermometer in retail food establishments.

Calibration of Temperature-measuring Devices – Slide 15

- Calibrate a Dial face, metal stem thermometer
  - Supplies
    - Dial face, metal stem thermometer
    - Cup of crushed ice (at least 2 inches deep)
    - Water
    - Pliers
  - Procedure
    - Immerse the first 2 inches of the stem into shaved ice and slowly add cool water until the dimple is submerged in the water ice mixture
    - Allow the indicator needle to stabilize
    - Use pliers, an open-ended wrench, or the holder on the case to grip the hex adjusting nut
    - While holding the hex adjusting nut in place, rotate the head of the thermometer until the indicator needle points to 32°F (0°C).
- Teaching Tip – This is a good break exercise for students that are hands-on learners!

Measuring Unpackaged Food Temperatures – Slide 17

- Take the temperature of unpackaged foods (pick from list of supplies). Take the temperature properly and improperly to show the difference.
    - Supplies
        - Thermometer of choice
            - Dial face
            - Digital
            - Thermocouple
        - Product(s) of choice
            - Do this exercise before lunch and use your lunch
                - Solid – Chicken Salad
                - Liquid – Soup
    - Procedure
        - Insert the probe of the thermometer into the core, or the thickest part of the food product
        - Ensure the probe is inserted to the appropriate depth
        - When measuring fluid products stir the food prior to measuring the temperature
        - Always wait for the temperature reading to stabilize.
    - Teaching Tip – Temperature taking exercises is a great method to increase awareness of food temperatures!

Measuring Packaged Food Temperatures – Slide 18

- Take the temperature of packaged foods (pick from list of supplies). Take the temperature properly and improperly to show the difference.
    - Supplies
        - Thermometer of choice
            - Dial face
            - Digital
            - Thermocouple
        - Product(s) of choice
            - Do this exercise before lunch and use your lunch!!!
                - Sliced cheese in a zip-lock
                - Sliced meat in a zip-lock
                - Package of hot dogs
                - Package of cheese
                - Two yogurt cups
    - Procedure
        - Place the stem or probe of the thermometer between the two packages, or fold the package around the stem or probe of the thermometer
        - Ensure the probe is making good contact with packages
        - With some packaged foods it is best to open the package to obtain an accurate reading
        - Always wait for the temperature reading to stabilize
    - Teaching Tip – Temperature taking exercises is a great method to increase awareness of food temperatures!

Slide 20

Receiving and storing

- Frozen foods – keep frozen foods solidly frozen and maintain frozen until used
- Properly freezing foods will:
  - Help maintain food quality
  - Prevent the growth of spoilage and harmful microorganisms
  - Allow foods to be stored for long periods of time without losing wholesomeness and quality
  - Prevent the growth of disease-causing and spoilage microorganisms
- Refrigerated Foods – maintain at or below 41°F (5°C)
  - Raw shell eggs may be received and stored at or below 45°F (7°C) ambient temperature
- Properly refrigerating foods will:
  - Prevent or slow the growth of harmful microorganisms
  - Help retain quality and extend shelf life of the product.

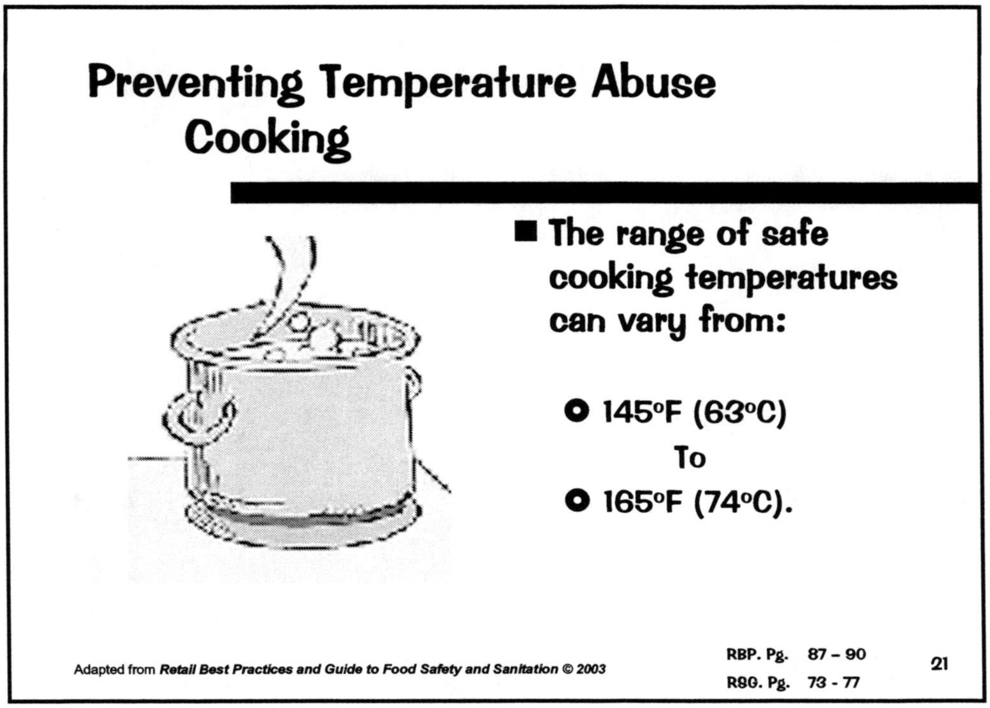

Slide 21

Cooking – Applying heat helps preserve food

- Time and temperature guidelines vary based on the food product being cooked
- Cooking temperatures can range from 145°F (63°C) to 165°F (74°C)
- Examples of proper cooking times and temperatures include:
  - Rare roast beef 130°F (54°C) for 112 minutes or 140°F (60°C) for 12 minutes
  - Beef, pork, and fish 145°F (63°C) for 15 seconds
  - Ground beef, ground pork and ground game animals 155°F (68°C) for 15 seconds
  - Beef roast (medium), pork roast, ham 145°F (63°C) for 4 minutes
  - Poultry, stuffed meats/food products 165°F (74°C) for 15 seconds

- Foods should reach the required final cooking temperature within two hours

- Proper cooking destroys harmful microorganisms that may be present in foods

- Different foods and the methods by which they are cooked require end point temperatures to be safe.

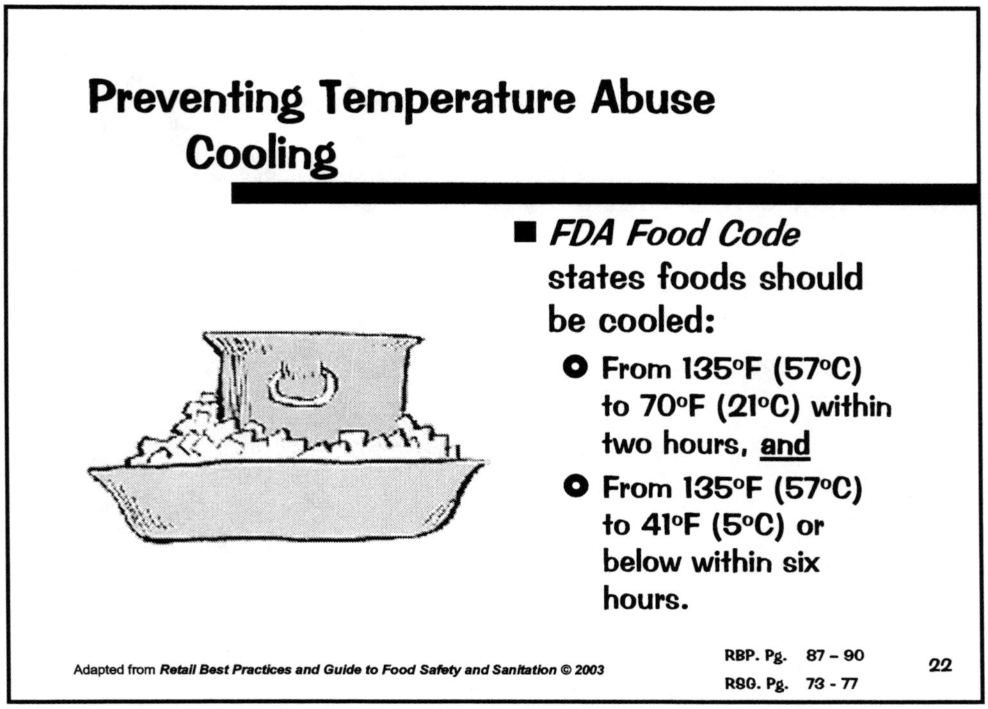

Slide 22

Cooling is a rapid reduction of temperature through and out of the temperature danger zone

- *FDA Food Code* states foods should be cooled:
  - From 135°F (57°C) to 70°F (21°C) within two hours <u>and</u> from 135°F (57°C) to 41°F (5°C) or below within six hours
- Proper cooling prevents:
  - The conversion of spore-forming bacterial cells to vegetative bacterial cells
  - The growth of vegetative bacterial cells.

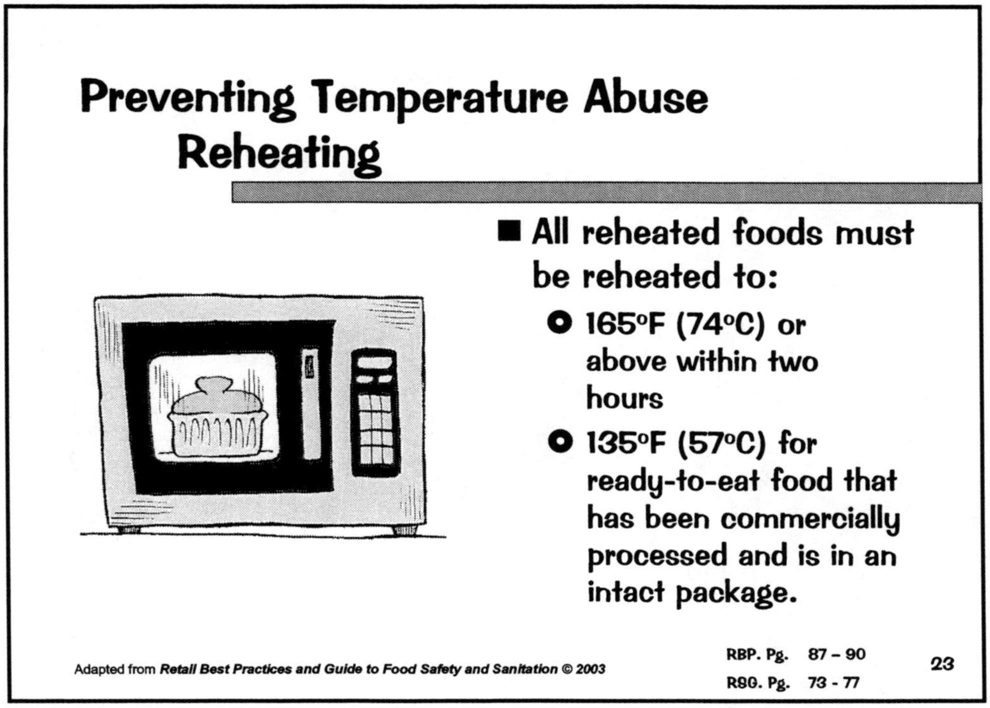

Slide 23

Reheating – bringing food back up to serving temperature

- All reheated foods must be reheated to 165°F (74°C) or above within 2 hours
- 135°F (57°C) for ready-to-eat food that has been commercially processed and is in an intact package.

Proper reheating destroys harmful bacteria that may be present.

Preventing Temperature Abuse
Hot-holding

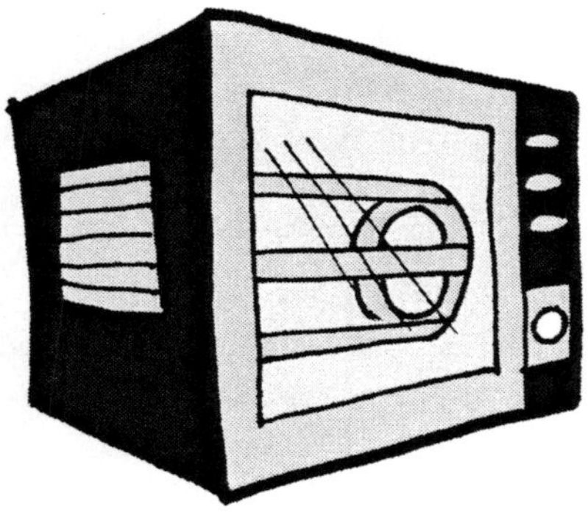

- Holding foods above 135°F (57°C) prevents the growth of harmful bacteria.

Adapted from ***Retail Best Practices and Guide to Food Safety and Sanitation* © 2003**

RBP. Pg. 87 – 90
RGG. Pg. 73 - 77

24

Slide 24

Hot-holding – maintaining hot foods for display and sale.

Improper holding of foods is the No. 1 contributing factor that leads to foodborne illness.

- Hold hot foods above 135°F (57°C)
- Holding foods above 135°F (57°C) prevents the growth of harmful bacteria
- Spores of some bacteria, specifically *Clostridium botulinum, Clostridium perfringens,* and *Bacillus cereus,* can survive cooking and become vegetative (grow) again when exposed to ideal conditions.

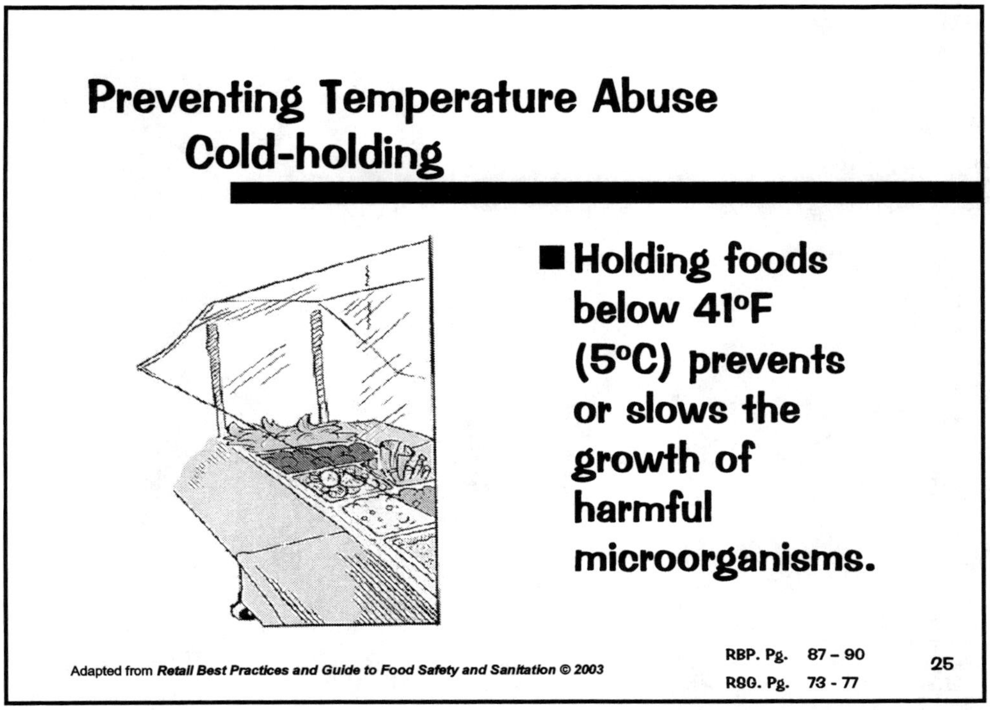

Slide 25

Cold-holding – maintaining cold foods for display and service.

Improper holding of foods is the No. 1 contributing factor that leads to foodborne illness.

- Hold cold foods below 41°F (5°C)
- Holding foods below 41°F (5°C) prevents or slows the growth of harmful microorganisms.

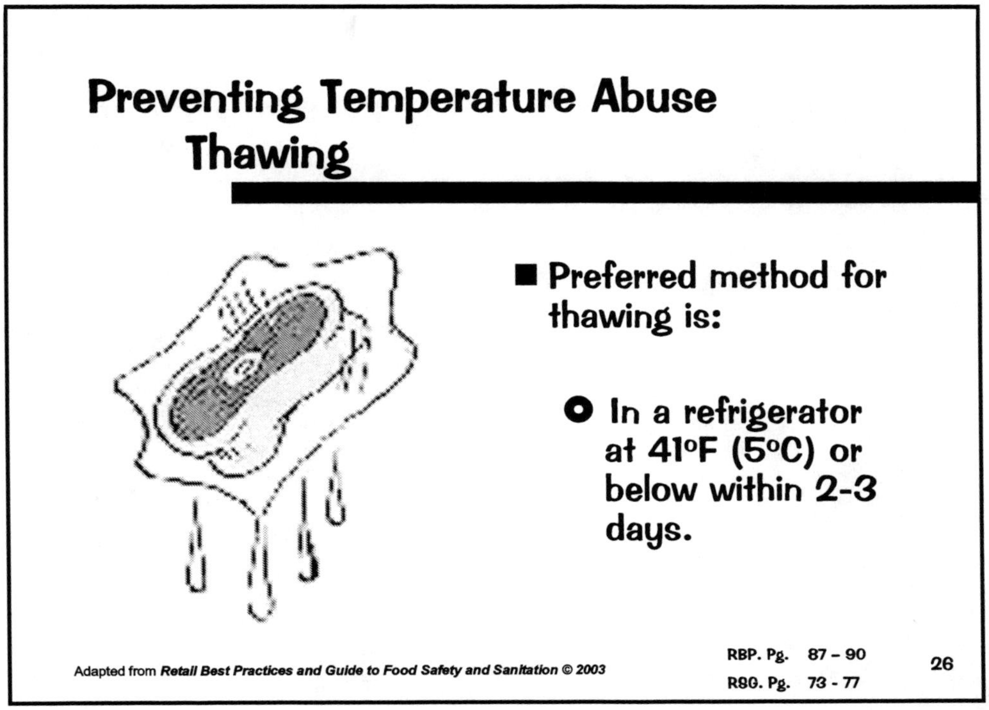

Slide 26

Thawing foods – taking food from a frozen state to a non-frozen state.

- The preferred thawing method is:
  - In a refrigerator at 41°F (5°C) or below within 2 to 3 days
  - This is the preferred method because the food product never enters the temperature danger zone
- Other acceptable methods for thawing are:
  - Completely submerged in cool [70°F (21°C)] running water
  - In the microwave oven after which the food goes directly into the cooking process
  - As part of the cooking process
- Thaw raw animal foods in less than 4 hours (this time includes cooking and preparation time or the time to lower the food temperature to 41°F (5°C).
- Ready-to-eat foods must never exceed 41°F (5°C).
- Proper thawing prevents or reduces the growth of harmful bacteria.

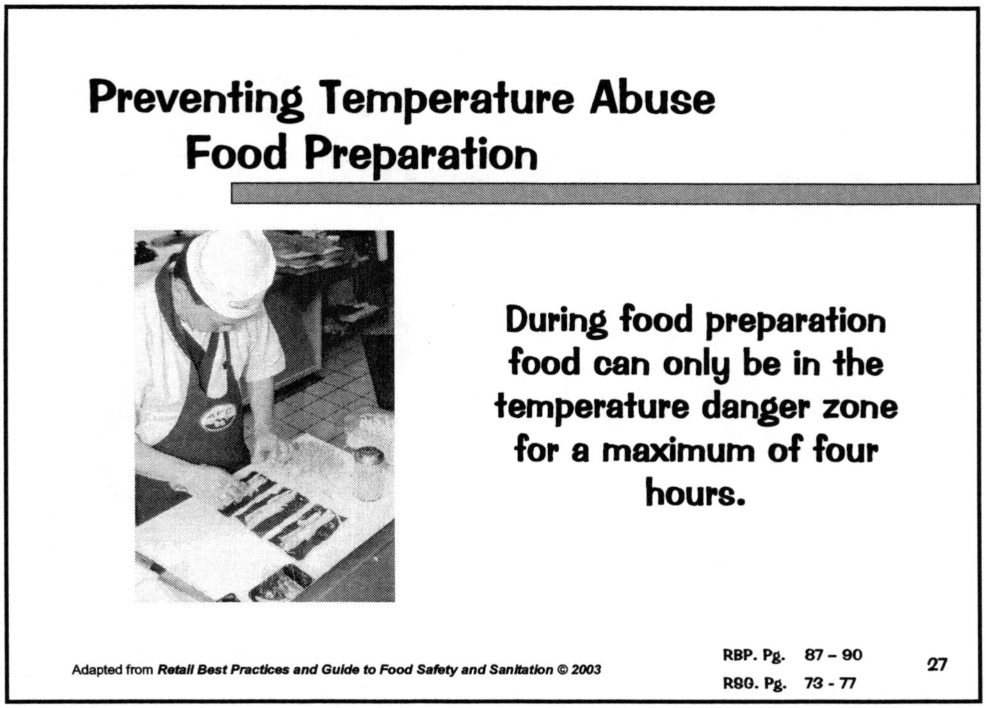

Slide 27

Food preparation

- There are times food must be in the temperature danger zone. This time should be kept to a minimum.

FDA Food Code allows for foods to be in the temperature danger zone for a total of four hours.

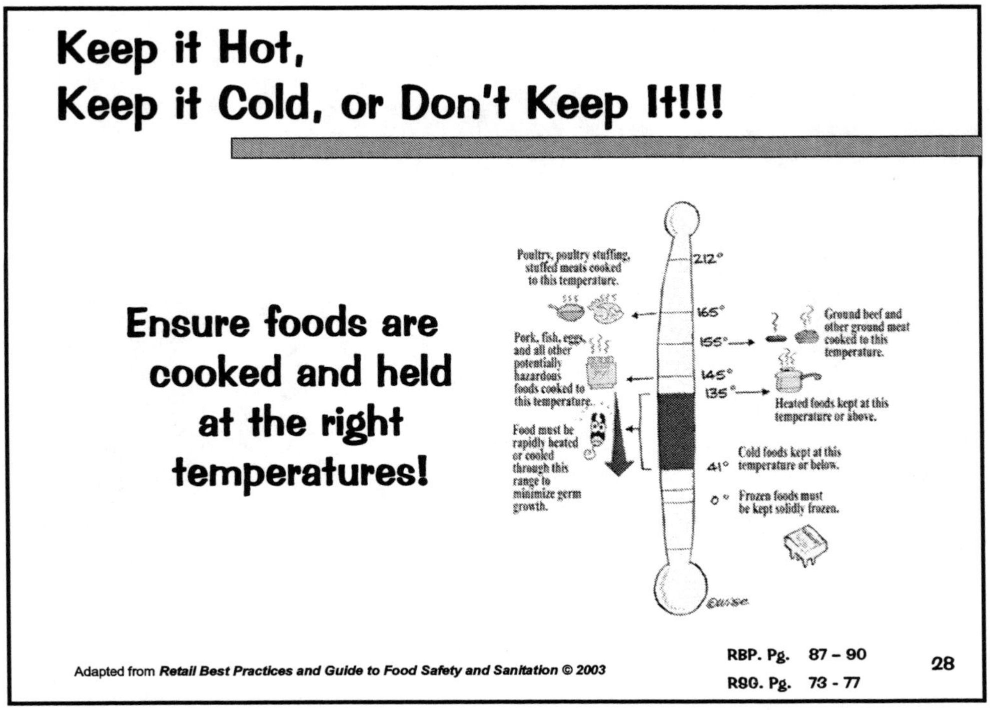

Slide 28

Hot Food Hot and Cold Food Cold!

In order to keep food safe:

- Move food through the temperature danger zone quickly
- Move food through the temperature danger zone as few times as possible
- Take the temperature of foods passing through the danger zone as often as possible.

Slide 29

Safety Stop! 3.3

Sample Test Questions

1. Raw shell eggs should be received at or below what ambient temperature?
   - A. 41°F (5°C).
   - **B. 45°F (7°C).**
   - C. 32°F (0°C).
   - D. 70°F (21°C).

2. Properly refrigerating foods will:
   - A. Kill all harmful microorganisms.
   - B. Prevent the growth of all harmful microorganisms.
   - **C. Prevent or slow the growth of all harmful microorganisms.**
   - D. All of the above.

3. *FDA Food Code* requires foods be cooled from 135°F (57°C) to 70°F (21°C) within two hours, and from 135°F (57°C) to 41°F (5°C) within:
   - A. 1 hour.
   - B. 2 hours.
   - C. 3 hours.
   - **D. 6 hours.**

4. When reheating foods the internal temperature of the food must reach 165°F (74°C) within:
   A. 1 hour.
   B. 2 hours.
   C. 3 hours.
   D. 4 hours.

5. Which of the following is the most preferred method to thaw food products?
   A. In the microwave oven.
   B. In the refrigerator.
   C. On the prep counter.
   D. As a part of the cooking process.

6. The temperature of foods passing through the temperature danger zone should be taken:
   A. Once.
   B. Twice.
   C. As often as possible.
   D. At the end.

Real-life

Slide 21
Cooking
John was cooking the ground beef for the taco bar. He cooked the meat to 140°F (60°C), added the seasonings, and put it on the hot line at the taco bar. What was his mistake?

John should have cooked the ground beef to 155°F (68°C) for a minimum of 15 seconds to ensure it was safe.

Slide 22
Cooling
Alicia was breaking down the self-service soup station. She took the leftover chicken noodle soup, covered it tightly, and put it in the cooler to be reheated in the morning. What was Alicia's mistake?

Alicia should have ensured the soup was properly cooled to the correct temperature in the proper amount of time before she covered it and put it in the cooler. [From 135°F (57°C) to 70°F (21°C) in two hours and from 135°F (57°C) to 41°F (5°C) in six hours.]

Slide 23
Reheating
Maria took out the chicken noodle soup that was saved from the prior evening. She put it in a large stockpot and put it on the stove to reheat. She set the burner to low so she could prep the salad for the day. She came back in three hours to check the soup and it was 160°F (73°C). What was Maria's mistake?

Maria should have heated the soup to 165°F (74°C) within two hours to keep it safe.

Slide 24
Hot-holding
Daniel was starting his 5 p.m. shift in the deli area. He began by filling out the temperature log for the steam table. Everything in the steam table had a temperature between 120°F (49°C) and 130°F (54°C). The temperature log had not been filled out during the rest of the day. What should Daniel do?

Since Daniel has no way of telling how long the food had been in the temperature danger zone, Daniel should discard the food.

Slide 26
Thawing
Marissa was making the seafood salad. She needed to thaw the precooked frozen shrimp that was going into the salad. She placed the bag of shrimp in the sink and ran cool water 70°F (21°C) over it for two hours until it was thawed. What was her mistake?

Marissa should have ensured the bag of shrimp was completely submerged under the cool running water.

Slide 30

Good **personal hygiene** is essential for those who handle foods.

If a food employee is not clean, the food can become contaminated.

Desirable hygiene behaviors of a food employee include:

- Knowing when and how to properly wash their hands
- Wearing clean clothing and aprons
- Maintaining good personal habits such as:
    - Bathing
    - Washing and restraining hair
    - Keeping fingernails short and clean
    - Washing hands after using the toilet
    - Not wearing jewelry except for a plain wedding band
    - Not smoking, eating or chewing gum in food preparation, service or warewashing areas
- Maintaining good health
- Reporting illnesses to managers to avoid spreading infections.

Slide 31

Food employees must know when and how to wash their hands.

Food employees should always wash their hands:

- Before preparing food
- After touching human body parts, other than clean hands and clean exposed forearms
- After using the toilet
- When returning to a work station
- After coughing, sneezing, or using a tissue/handkerchief
- After using tobacco
- After eating or drinking
- During food preparation as often as necessary to remove soil and contaminants
- During food preparation as often as necessary to avoid cross contamination
- Before putting on gloves for working with food
- When switching from preparing raw food to ready-to-eat food
- After engaging in any activities that may contaminate hands such as:
    - Taking out the garbage
    - Wiping counters or tables
    - Handling chemicals
    - Picking up dropped items.
- After caring for or touching service or aquatic animals.

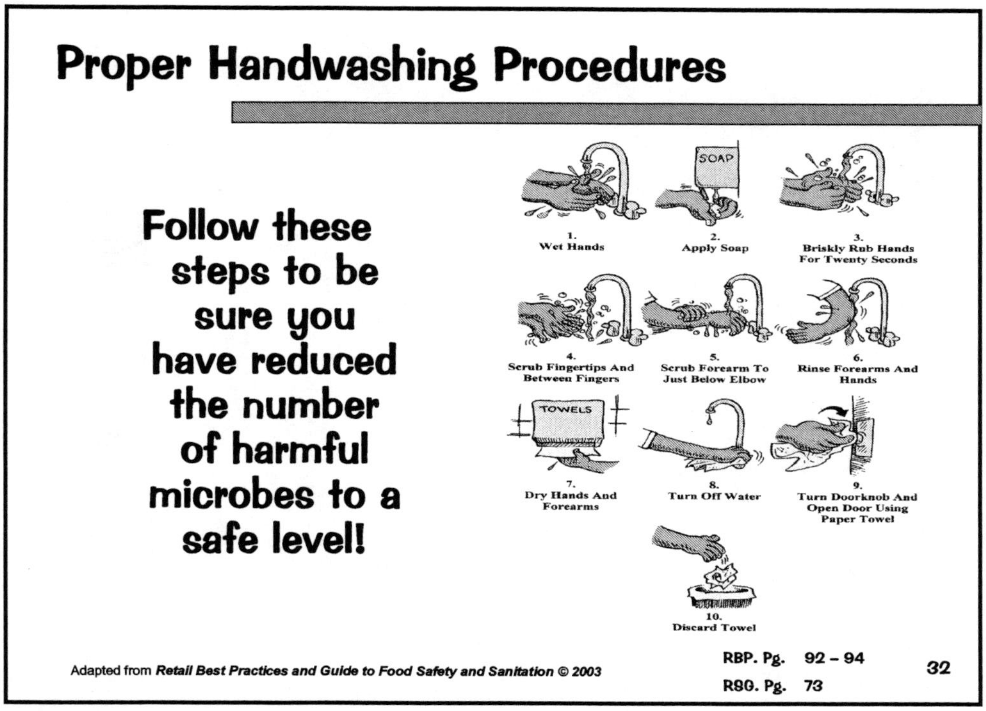

Slide 32

According to the *FDA Food Code*, hands shall be washed in a separate sink specified as a handwashing sink.

Employees must not clean their hands in a sink used for:

- Food preparation
- Warewashing
- Service sinks used for disposal of mop water or liquid waste.

Every handwashing sink must be provided with:

- Hand cleanser (soap or detergent) in a dispenser
- A suitable hand-drying device.

Steps in proper handwashing:

1. Wet hands.
2. Apply soap.
3. Briskly rub hands for 20 seconds.
4. Scrub fingertips and between fingers.
5. Scrub forearm to just below the elbow.
6. Rinse forearms and hands.
7. Dry hands and forearms.
8. Turn off water using towel.
9. Turn doorknob and open door using paper towel.
10. Discard towel.

Hand sanitizing lotions may be used, however they are not an acceptable replacement for hand washing.

Disposable Gloves

Gloves are like skin. . .

Change your gloves whenever you would wash your hands!!

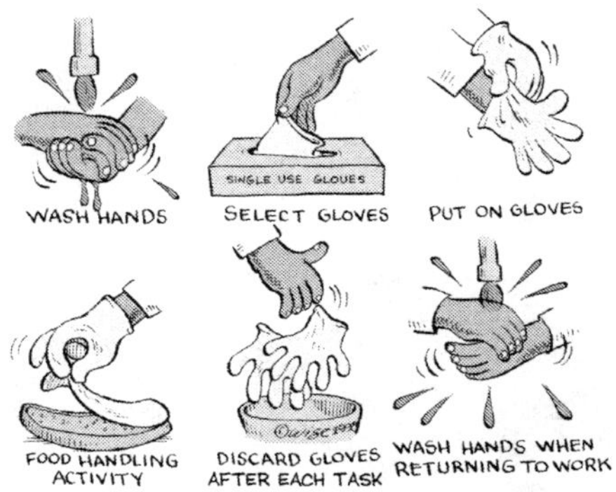

Adapted from ***Retail Best Practices and Guide to Food Safety and Sanitation* © 2003**

RBP. Pg. 95 - 96
RSG. Pg. 80 - 81

33

Slide 33

Except when washing fruits and vegetables, food employees should not contact exposed ready-to-eat foods with their hands. Instead, they should use suitable utensils such as:

- Deli tissue
- Spatulas
- Tongs
- Single-use disposable gloves.

Disposable glove usage:

- Gloves are like skin, change gloves whenever you would wash your hands
- Gloves must be impermeable (do not allow anything to penetrate the glove)
- Disposable glove usage is a good idea when handling foods extensively, such as when handling:
    - Deli sandwiches
    - Tacos
    - When conducting food demonstrations.

Proper disposable glove usage:

1. Wash hands.
2. Select gloves.
3. Put on gloves.
4. Engage in ONE (1) task.
5. Discard gloves.
6. Wash hands properly.

Never reuse a pair of disposable gloves.

Money handling contaminates gloves; treat money handling as a food-handling activity.

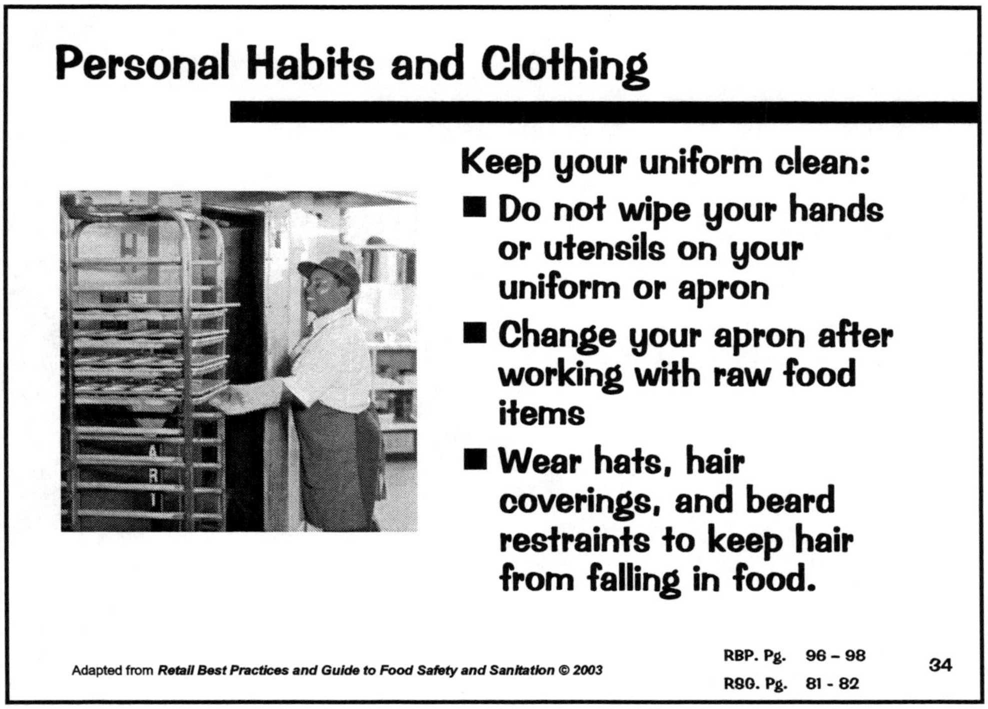

Slide 34

Keep your uniform clean:

- Do not wipe your hands or utensils on your uniform or apron
- Change your apron after working with raw food items
- Wear hats, hair coverings, and beard restraints to keep hair from falling in food.

Aprons:

- Do not dry or wipe hands on your apron
- Leave aprons in the department when you go on break or use the restroom
- Aprons, and all protective apparel, no longer protect food when they are contaminated.

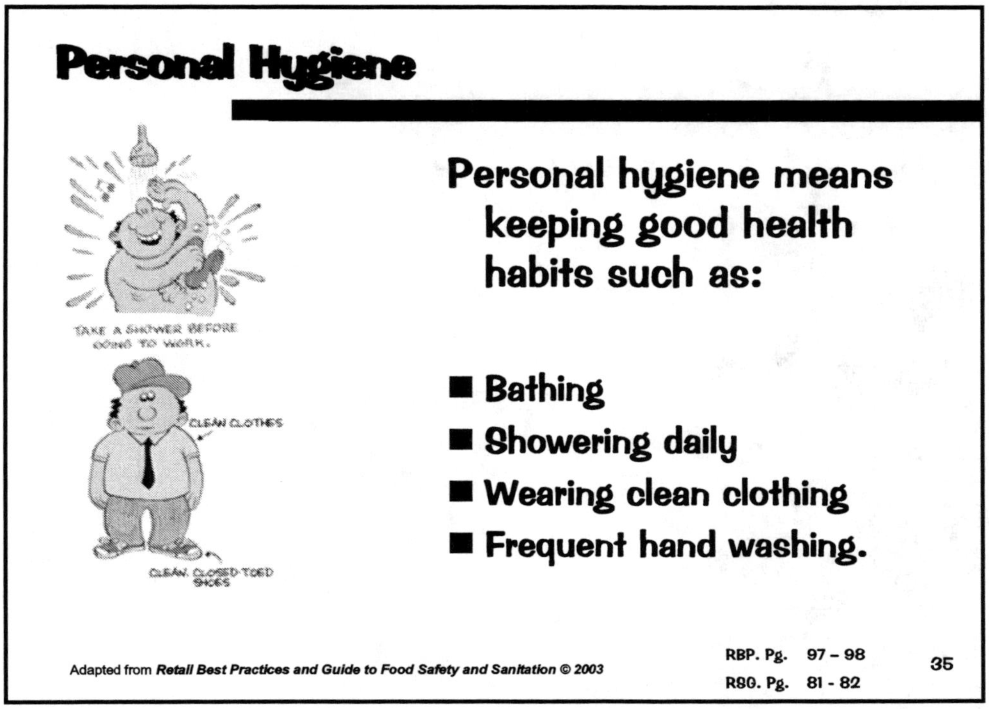

Slide 35

Personal hygiene means keeping good health habits such as:

- Bathing and showering daily
- Wearing clean clothing
- Frequent hand washing.

Fingers and hands:

- Fingernails should be trimmed and filed so hand washing can properly remove soil from under and around them
- Unless wearing intact gloves in good repair, a food employee must not wear:
  - Fingernail polish
  - Artificial fingernails.

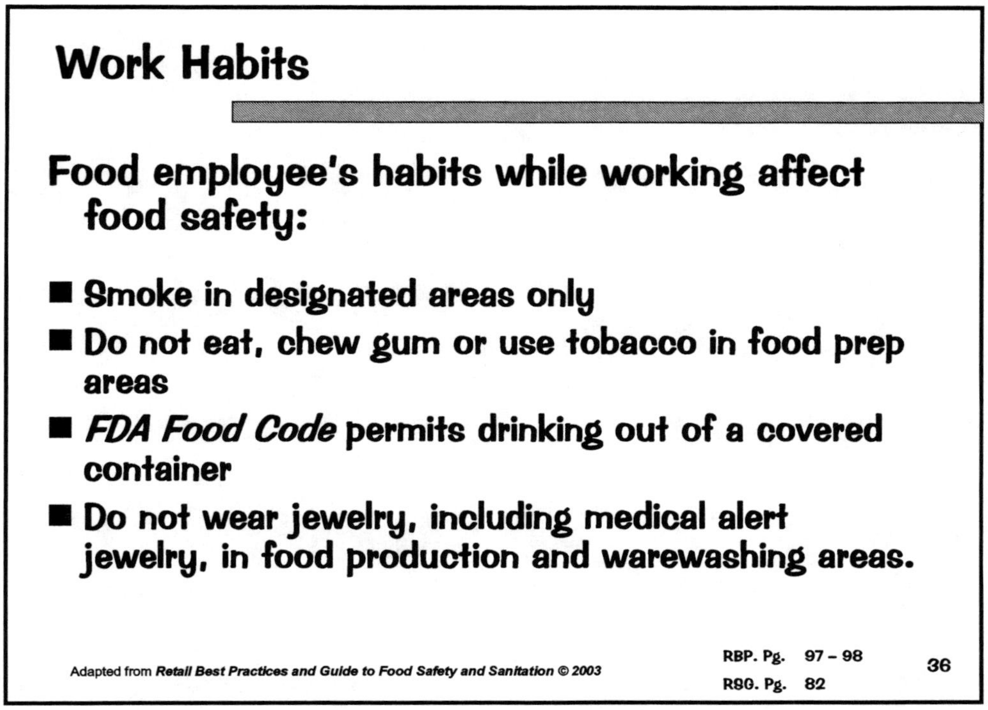

Slide 36

Food employee's habits while working affect food safety:

- Smoke in designated areas only
- Do not eat, chew gum or use tobacco in food prep areas
- ***FDA Food Code*** permits drinking out of a covered container
- Do not wear jewelry, including medical alert jewelry, in food production and warewashing areas.

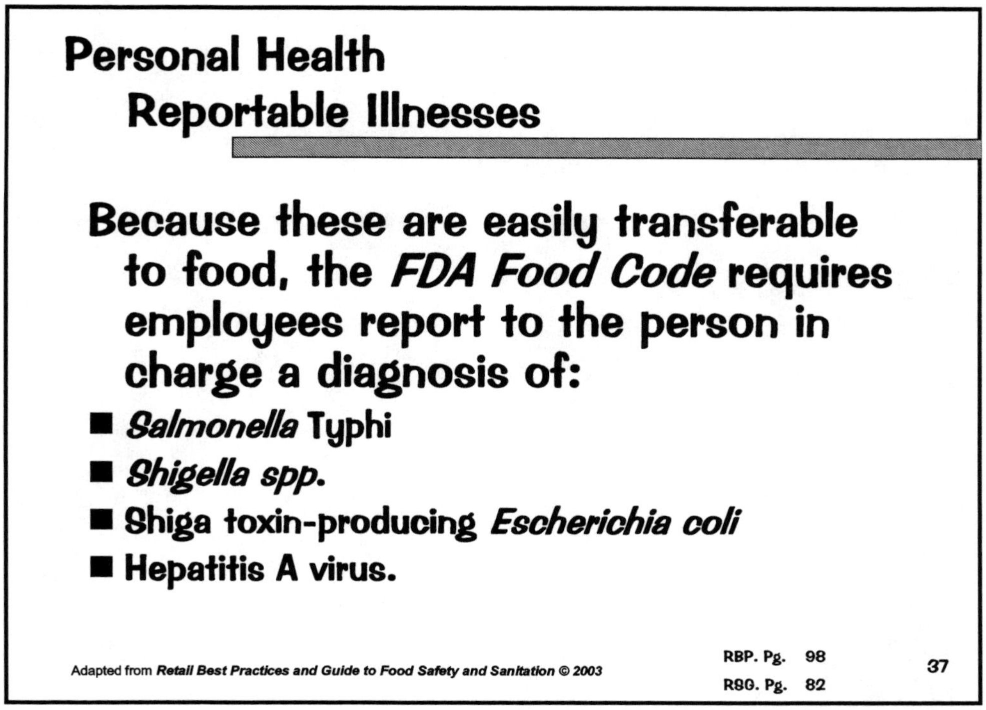

Slide 37

Employee health is extremely important in keeping food safe.

Because these are easily transferable to food, the ***FDA Food Code*** requires employees report to the person in charge a diagnosis of:

- *Salmonella* Typhi
- *Shigella spp.*
- Shiga toxin-producing *Escherichia Coli*
- Hepatitis A virus.

Slide 38

The person in charge must notify the regulatory authority if an employee is diagnosed with reportable illness.

If a food employee has been exposed to any of these, he or she:

- May be excluded from work
- May be assigned to restricted activities that do not involve contact with food
- Must not handle food or have contact with clean equipment, utensils, or single-service items.

Food employees must also report to the person in charge:

- Symptoms of intestinal illness or flu-like symptoms such as:
    - Vomiting
    - Diarrhea
    - Sore throat with fever
    - Coughing
    - Sneezing
    - Running nose
    - Jaundice
- A lesion containing pus such as a boil or infected wound that is open and draining.

Slide 39

Safety Stop! 3.4

Sample Test Questions

1. To help keep foods safe, it is important for food employees to:
   A. Show up to work everyday regardless of how they feel.
   B. Report certain illnesses and symptoms of illnesses to managers.
   C. Work on their day off.
   D. Go to the restroom if they are ill.

2. It is important to wash your hands:
   A. After using the toilet.
   B. Before preparing food.
   C. After eating or drinking.
   D. All of the above.

3. How long should workers spend briskly rubbing their hands during proper handwashing?
   A. 10 seconds.
   B. 15 seconds.
   C. 20 seconds.
   D. 30 seconds.

4. Food employees should change their apron:
   A. When it gets soiled.
   B. When switching from preparing raw foods to ready-to-eat foods.
   C. After taking trash outside to the dumpster.
   D. All of the above.

5. The *FDA Food Code* requires food employees report a diagnosis of each of the following except:
   A. Hepatitis A virus.
   B. Shigella spp.
   C. Rotavirus.
   D. Shiga toxin-producing *Escherichia Coli.*

Real-life

Slide 31
When to Wash Your Hands
Marsha was setting up a produce display. She saw some trash on the ground. Wanting to keep her area clean, she picked it up and threw it away. She returned to setting up her display. What was her mistake?

Marsha should have washed her hands after she picked up the garbage.

Slide 33
Glove Usage
Ichiro was preparing the day's sushi. He took off his gloves to use the bathroom. After using the bathroom, Ichiro washed his hands and returned to his station. He put his gloves back on and continued to work. What was his mistake?

Ichiro should have discarded his gloves before he went to the bathroom and washed his hands at the station, then put on a new pair when he returned.

Slide 34
Apron Usage
After Jose finished his smoke break, he washed his hands before he went back to his work station. Unfortunately, there were no towels left in the paper towel dispenser. Jose used his apron to dry his hands. What was his mistake?

By wiping his hands on his apron, Jose contaminated his hands. Jose should have restocked the paper towels and rewashed his hands.

Slide 37
***Salmonella* Typhi**
Lisa, a clerk at the local convenience store, was sick with nausea and vomiting. She went to the doctor and was told she had been infected with *Salmonella* Typhi. Lisa reported it to her manager. Thinking she only handled money and prepackaged foods, and needing the help, he had her continue to work as scheduled. What was his mistake?

The manager should have waited to get clearance from her doctor before he let her return to work handling foods or money.

Making it Real

When Should You Wash Your Hands – Slide 31

- There are several products on the market that allow you to demonstrate with the use of a powder, gel, or liquid and an ultraviolet light, the effectiveness of hand washing.
    - Supplies
        - Commercial product such as GlowGerm (www.glowgerm.com)
    - Procedures:
        - **Shake the hand of the learner that used the glow germ (used in a later Making it Real)
        - Follow manufacturer's procedures
    - Teaching Tip – This is a great exercise to do both before and after you view the slide.

Proper Handwashing Procedures – Slide 32

- Field Trip!!! Take the class to the bathroom and have them show you how to wash their hands properly!
    - Supplies
        - Hand sink
        - Soap
        - Paper towel
        - Garbage can
    - Procedures (watch for all of the following steps to be followed):
      1. Wet hands.
      2. Apply soap.
      3. Briskly rub hands for 20 seconds.
      4. Scrub fingertips and between fingers.
      5. Scrub forearm to just below the elbow.
      6. Rinse forearms and hands.
      7. Dry hands and forearms.
      8. Turn off water using towel.
      9. Turn doorknob and open door using paper towel.
      10. Discard towel.
    - Teaching Tip – This is a great exercise to do both before and after you view the slide.

Disposable Gloves – Slide 33

- Have the students practice proper procedures for glove usage.
    - Supplies
        - Box of single-use gloves
        - Hand sink
        - Soap
        - Paper towel
        - Garbage can
    - Procedures (watch for the all of the following steps to be followed):
      1. Wash hands.
      2. Select gloves.
      3. Put on gloves.
      4. Engage in ONE (1) task.
      5. Discard gloves.
      6. Wash hands properly.

- Have people put on disposable gloves. Show how germs from dirty hands can be transferred to gloves.
- Teaching Tip – Ask the students
    1. What do you do if you tear your glove?
    2. When do you change your gloves?
    3. When is it inappropriate to wear a glove?

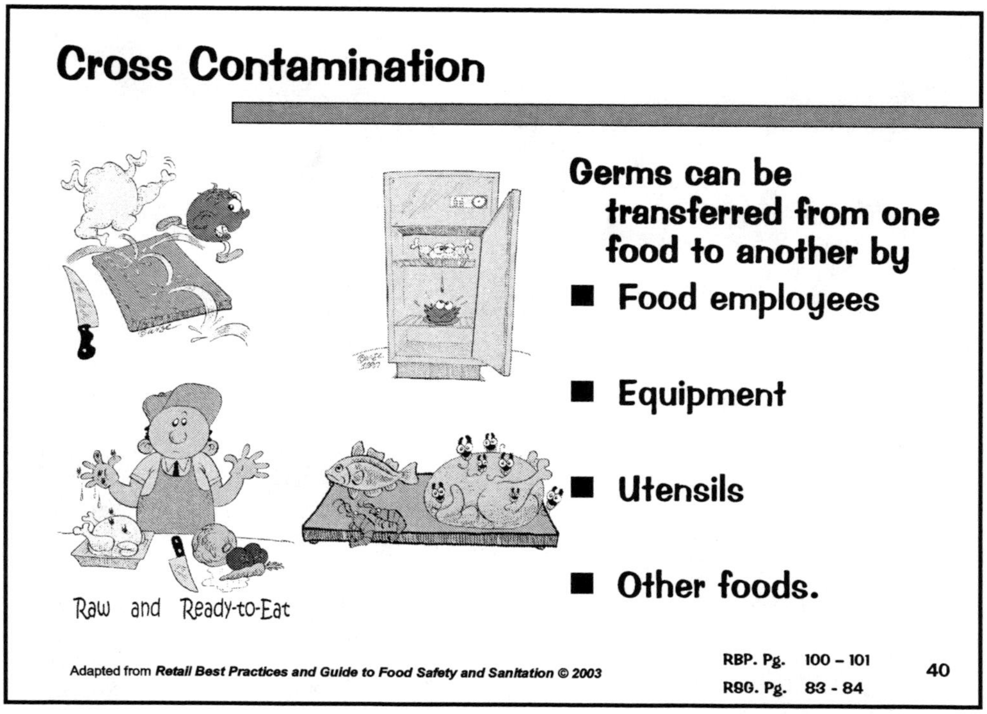

Slide 40

Contaminated food contains germs or harmful substances that can cause foodborne illness.

Cross contamination is the transfer of harmful organisms from one food item to another, most commonly:

- When germs from raw foods are transferred to a cooked or ready-to-eat food via contaminated hands, equipment or utensils.

Germs can be transferred by:

- Food employees
- Equipment
- Utensils
- Other foods.

Common sources of cross contamination include:

- Cutting boards
- Improper glove usage and hand washing
- Improper storage of foods
    - Juices dripping from raw foods onto cooked or ready-to-eat foods.

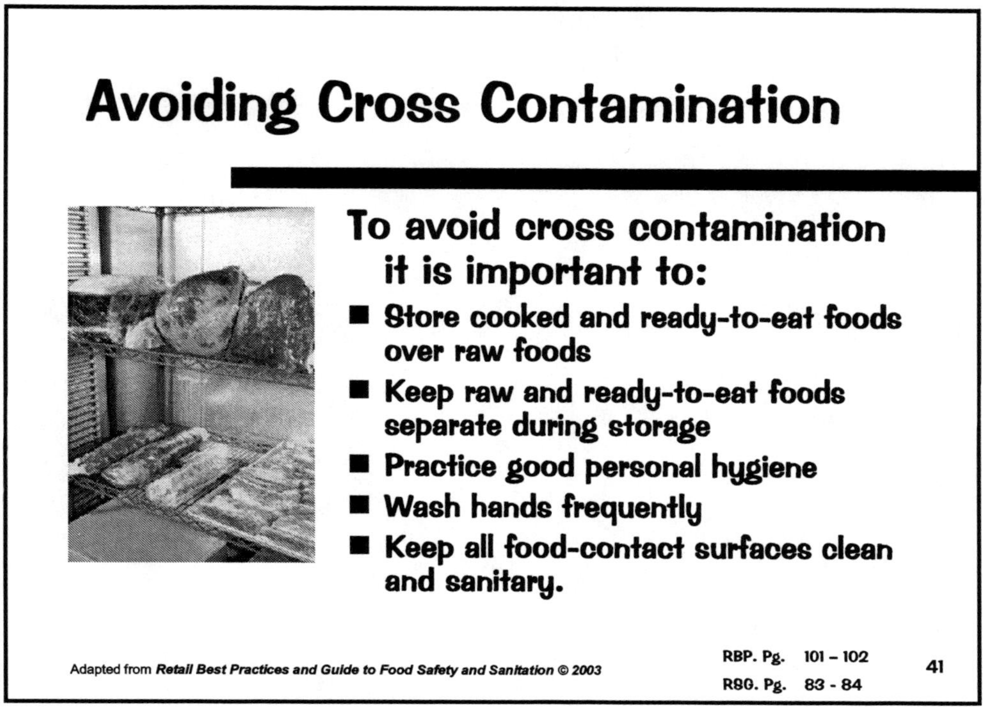

Slide 41

To avoid cross contamination it is important to:

- Store cooked and ready-to-eat foods over raw foods
- Keep raw and ready-to-eat foods separate during storage
- Practice good personal hygiene
- Wash hands frequently
- Keep all food-contact surfaces clean and sanitary.

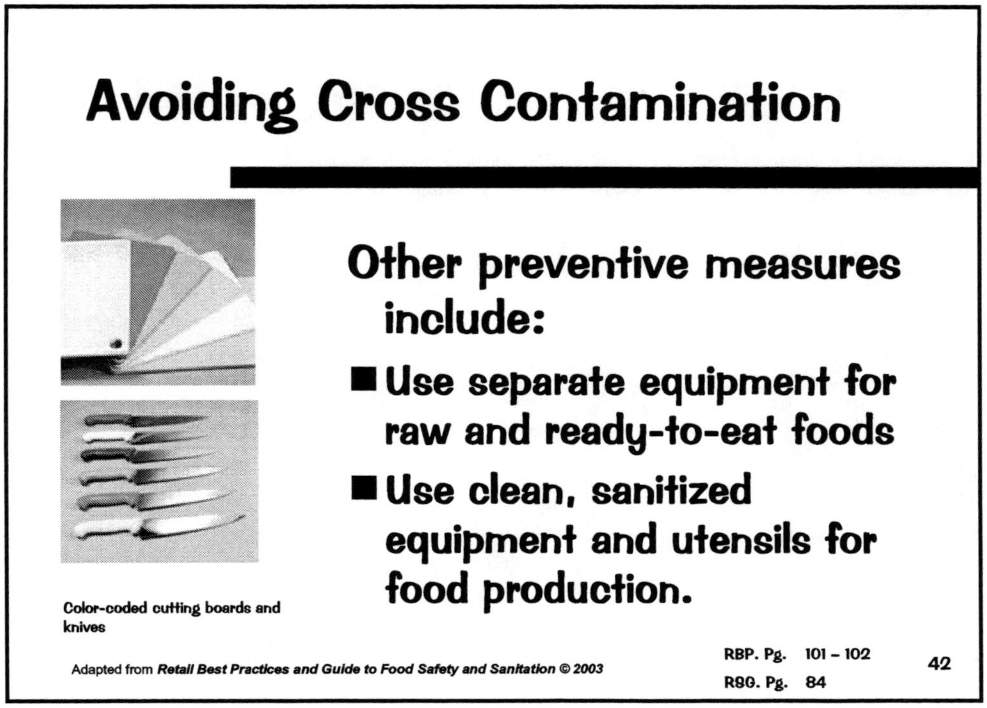

Slide 42

Other preventive measures include:

- Use separate equipment for raw and ready-to-eat foods
- Use clean, sanitized equipment and utensils for food production.

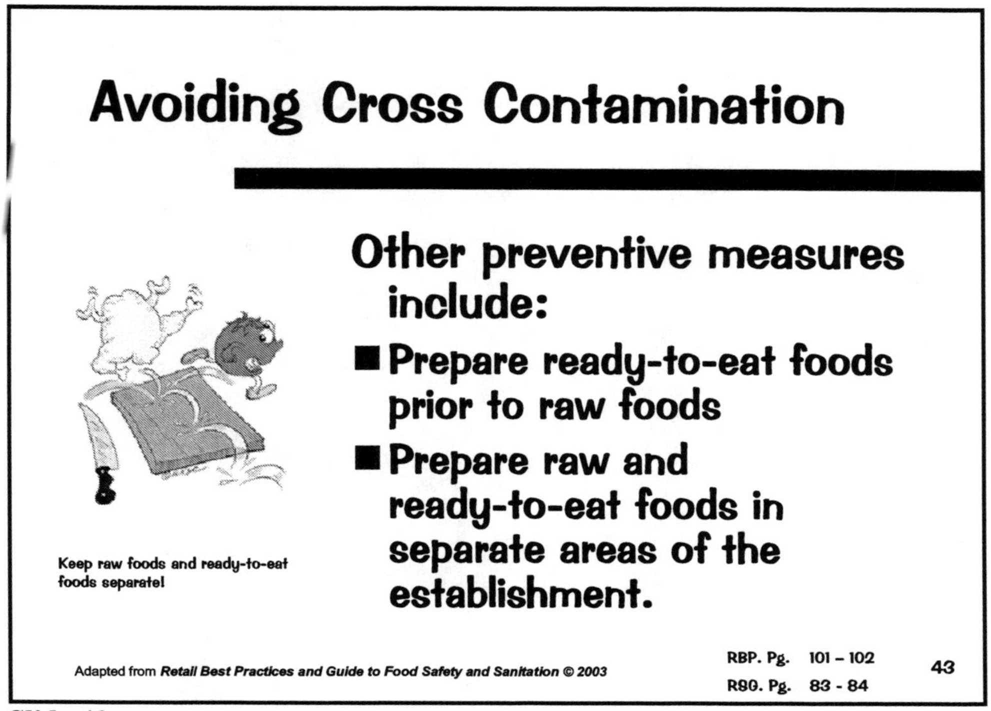

Slide 43

Other preventive measures include:

- Prepare ready-to-eat foods prior to raw foods
- Prepare raw and ready-to-eat foods in separate areas of the establishment.

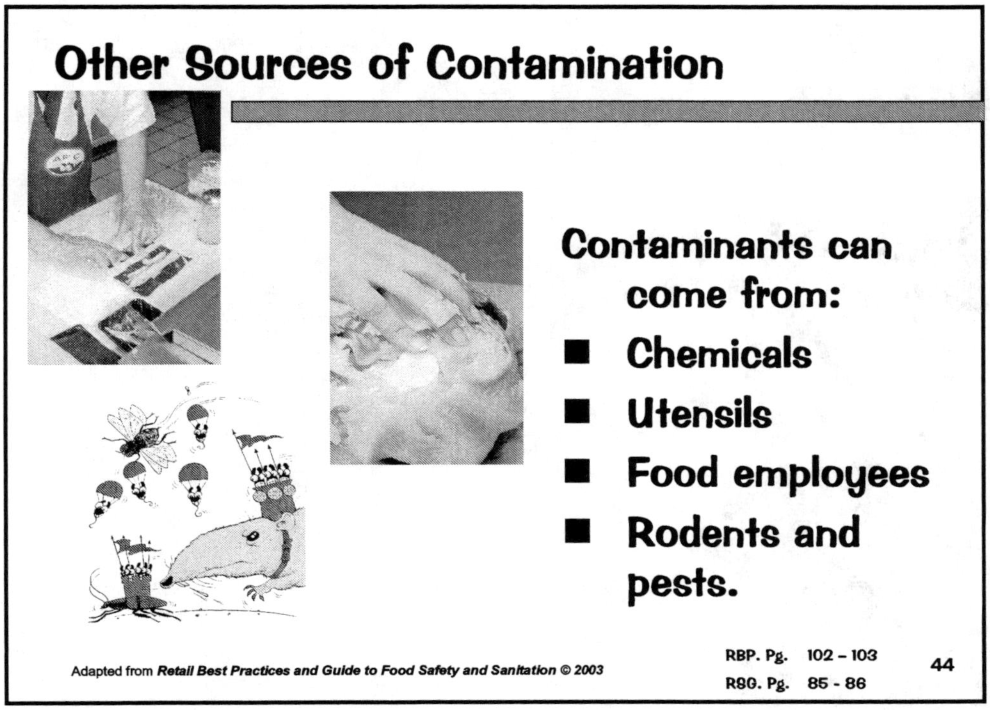

Slide 44

Other sources of contamination include:

- Food employees
- Chemicals
    - Store chemicals away from food items
    - Wash raw foods (such as fruits and vegetables) prior to processing or before use to remove pesticide residue
- Serving/dispensing utensils
    - Handle utensils properly
    - Utensils on hot- or cold-holding tables must be stored in food
- Scoops and tongs
    - Should be kept in a clean area away from possible contaminants
    - Use different utensils for different items to avoid possible allergens
- Animals
    - Wash your hands after touching live animals, such as live lobsters, shellfish, or fish
- Tasting
    - Use proper tasting techniques
- Rodents and pests
    - Ensure a proper pest control system is in place.

Slide 45

Keep anything that comes in contact with food clean and sanitary, including:

- Human hands
- Equipment
- Utensils
- Storage and holding areas
- Self-service areas (salad bars, bagel centers, coffee centers).

The goal of cleaning is to:

- Remove visible soil.

The goal of sanitizing is to:

- Reduce the number of harmful microbes that may be present on a clean surface to a safe level.

Slide 46

Safety Stop! 3.5

Sample Test Questions

1. Harmful microorganisms can be transferred by:
    A. Food employees.
    B. Equipment.
    C. Other foods.
    D. All of the above.

2. To avoid cross contamination, it is important to:
    A. Store raw foods over cooked foods.
    B. Practice good personal hygiene.
    C. Keep raw and ready-to-eat foods together during storage.
    D. Clean, but not necessarily sanitize, food-contact surfaces.

3. The serving portion of utensils being used for service of hot- or cold-held food products must be:
    A. Changed after each use.
    B. Washed after each use.
    C. Stored outside the food product.
    D. Stored completely in the food product.

4. A piece of equipment or a utensil is considered to be sanitized when:

   A. You wipe it on your apron.
   B. You can't see any dirt.
   C. You rinse it in warm water.
   D. The harmful microbes have been reduced to a safe level.

Real-life

Slide 40
Display
Bob was stocking the bagels in the self-service bagel station. He used his hands to fill the bins with bagels. What was his mistake?

Bob should have used clean, sanitized tongs, utensils or deli tissues to transfer the bagels into the bins, not his bare hands.

Slide 41
Storage
Jason was putting away the day's seafood deliveries. He put the pre-cooked shrimp for the seafood salad on the bottom shelf in the walk-in to thaw. Just above it, he placed a bag of raw clams. What was his mistake?

Jason should have stored the ready-to-eat shrimp above the raw clams.

Slide 44
Hot-holding
Trisha was cleaning up the salad bar after the lunch rush. She carefully wiped down the salad bar. She then pulled the serving utensils out of the salad products and placed them on a plate next to each product. What was her mistake?

Trisha should have ensured the serving utensils were completely inserted into the foods on the salad bar, not set beside them.

Slide 45
Cleaning and Sanitizing
Larry was cleaning out the meat display case. He used a clean cloth and hot water with a detergent in it to wipe the case clean. Larry then restocked the meat display case. What was his mistake?

Larry should have properly sanitized the display case before he restocked it.

Making it Real

Contamination (linked to previous Making it Real with slide Number 31) – Slide 40-41

- Remember shaking the hand of the learner that used the glow germ. Use the ultraviolet light to now show how the germs transferred from their hand to yours or to other surfaces.
    - Supplies
        - Commercial product such as GlowGerm (www.glowgerm.com)
    - Follow manufacturer's procedures but be sure to shake hands with the student that is performing the exercise.

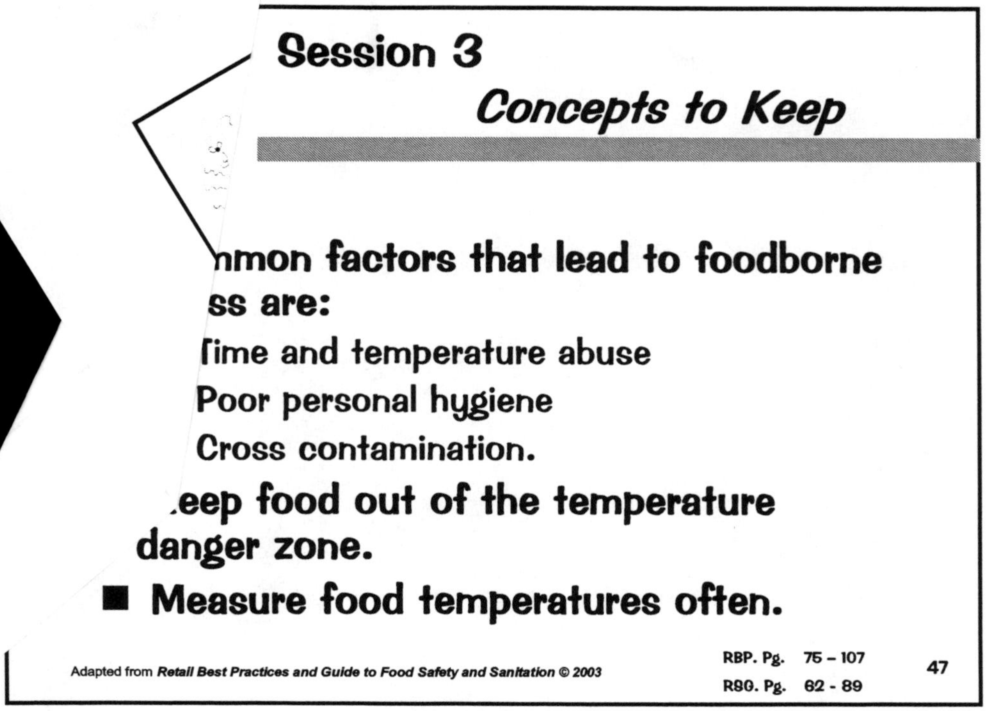

Slide 47

Concepts to Keep

- Common factors that lead to foodborne illness include:
  - Time and temperature abuse
  - Poor personal hygiene
  - Cross contamination
  - Cleaning and sanitation.
- Keep food out of the temperature danger zone.
- Measure food temperatures often.

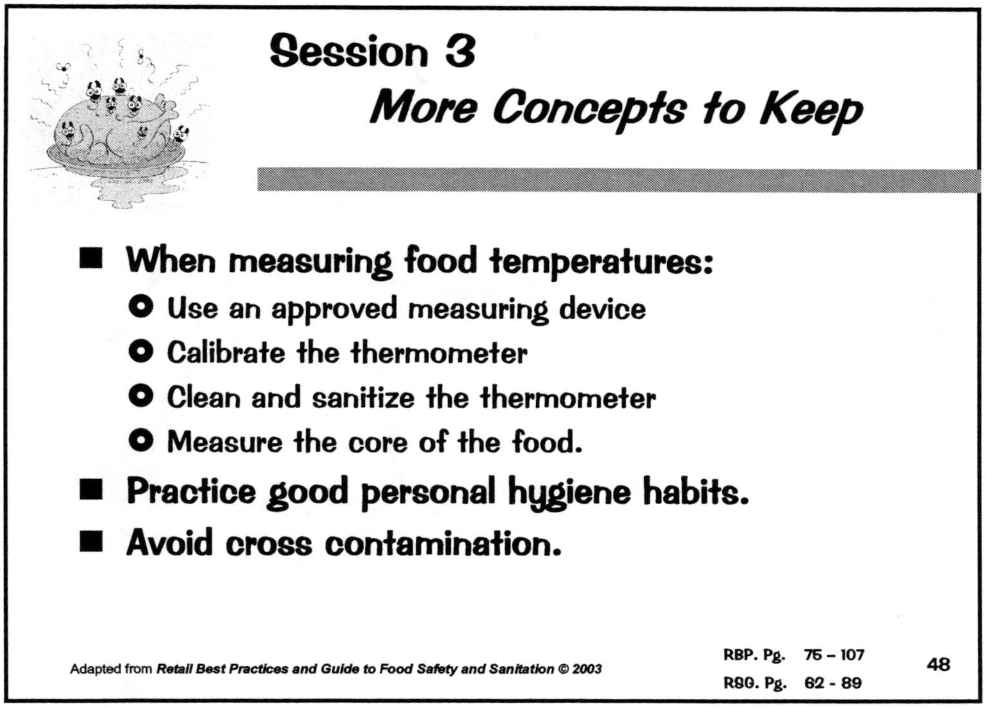

Slide 48

<u>More Concepts to Keep</u>

- When measuring food temperatures:
  - Use an approved measuring device
  - Calibrate the thermometer
  - Clean and sanitize the thermometer
  - Measure the core of the food.
- Practice good personal hygiene habits.
- Avoid cross contamination.

"Take This Back To your Team!"

Importance of:

- Temperature-measuring devices.
- Calibrating thermometers.
- Accurate measuring of temperatures.
- Handwashing procedures.
- Proper disposable glove usage.
- Personal health and proper hygiene.
- How to prevent cross contamination.

Adapted from *Retail Best Practices and Guide to Food Safety and Sanitation* © 2003

RBP. Pg. 75 – 107
RGG. Pg. 62 - 89

49

Slide 49

Take This Back to Your Team!

Importance of:

- Temperature-measuring devices.
- Calibrating thermometers.
- Accurate measuring of temperatures.
- Handwashing procedures.
- Proper disposable glove usage.
- Personal health and proper hygiene.
- How to prevent cross contamination.

Session 4

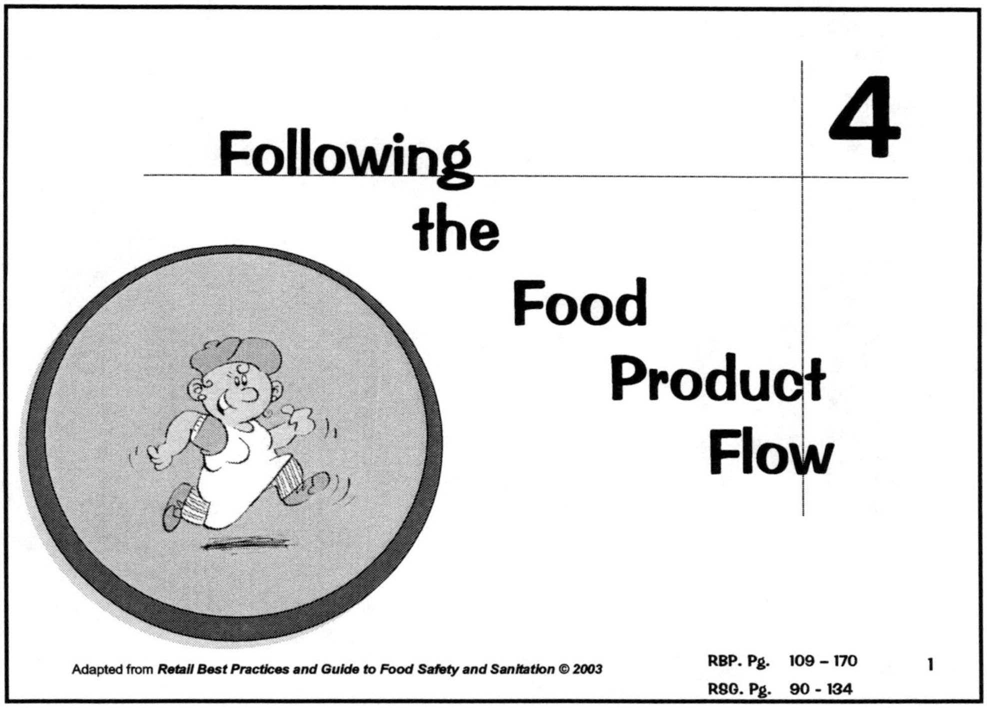

Following the Food Product Flow – Session 4 Key Points

Slide 1

Training Tip:

The content of each slide provides the most essential information in the session. The written text below each slide is provided to help you understand and better teach the essential information. The written text may also be used to answer questions that may be posed during the session. Depending on the length of your training session, you may not be able to address all the information contained beneath each slide.

Overview:

Session 4 takes the learners through the steps of the flow of food within retail food establishments – from the moment the purchase is made to bring the food into the retail establishment to the moment the food leaves the establishment. One of the biggest impacts this session can make on a learner is the number of people that can affect the safety of a single food item.

This session reinforces the importance of:

- Moving food quickly through delivery into storage
- Avoiding temperature abuse
- Proper receiving and storage procedures
- The steps in food preparation including
  - Proper thawing of food products for preparation
  - Proper cooking times and temperatures
  - Proper hot- and cold-holding of products for display and service
  - Proper cooling of foods that have been cooked and/or held hot
  - Proper reheating of foods for hot-holding
- Following procedures at the checkout stand for safe food handling.

Suggested Timeline:
8-hour course: 1 hour 15 minutes using the PowerPoint© slides only.

- 46 ***Slides.***

16-hour course: 2 hours using the 51 PowerPoint© slides and the 5 ***Safety Stop!*** slides containing:

- 28 ***Sample Test Questions***
- 17 ***Real-life*** scenarios
- 7 ***Making it Real*** interactive exercises.

Video suggestions and supplemental resources:

Videos:
The Invisible Challenge, Food Marketing Institute, 1-301-843-3084.

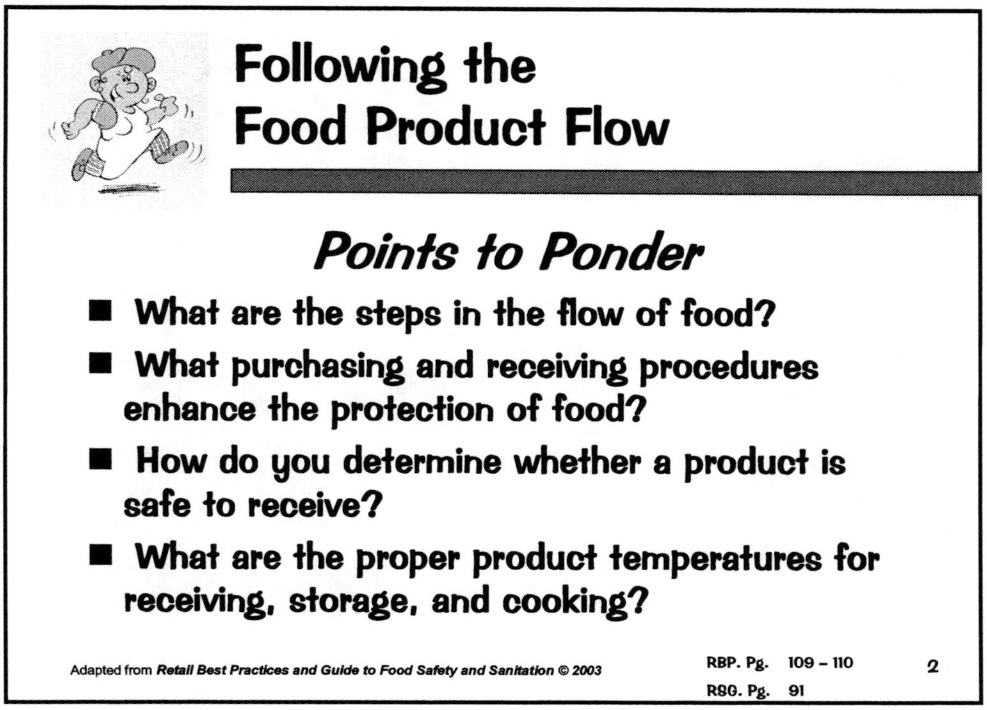

Slide 2

Points to Ponder

- What are the steps in the flow of food?

- What purchasing and receiving procedures enhance the protection of food?

- How do you determine whether a product is safe to receive?

- What are the proper product temperatures for receiving, storage, and cooking?

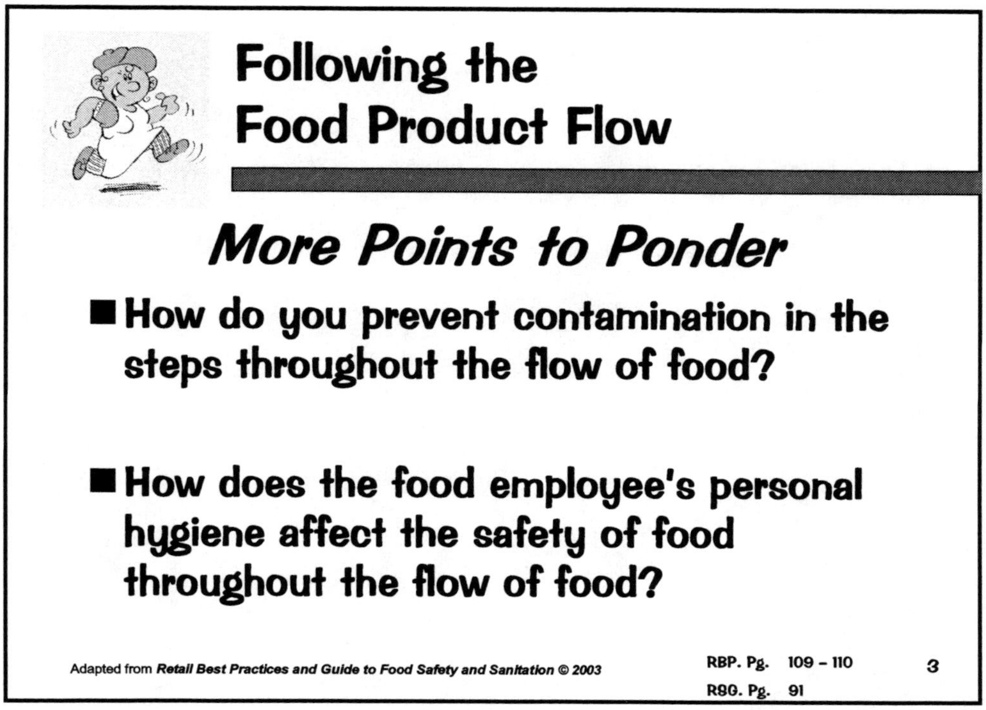

Slide 3

More Points to Ponder

- How do you prevent contamination in the steps throughout the flow of food?

- How does the food employee's personal hygiene affect the safety of food throughout the flow of food?

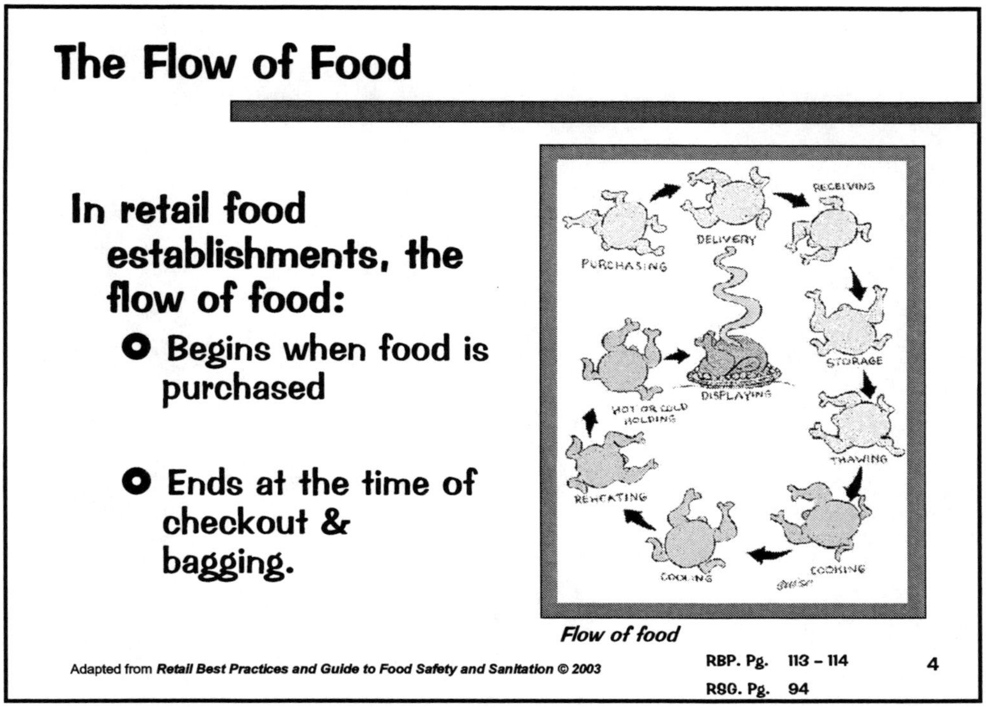

Slide 4

The flow of food begins at the time food is harvested and ends when it is displayed and sold.

The steps that can be impacted by retail food employees in the flow of food include:

1. Purchasing.
2. Delivery.
3. Receiving.
4. Storage including:
    a. Frozen.
    b. Refrigerated.
    c. Dry.
5. Preparation including:
    a. Thawing.
    b. Cooking.
    c. Cooling.
    d. Reheating.
6. Hot- or cold-holding and display.
7. Checkout and bagging.

Purchasing

Purchasing product specifications include:

- Quality grade
- Weight
- Count
- Contents
- Packaging type.

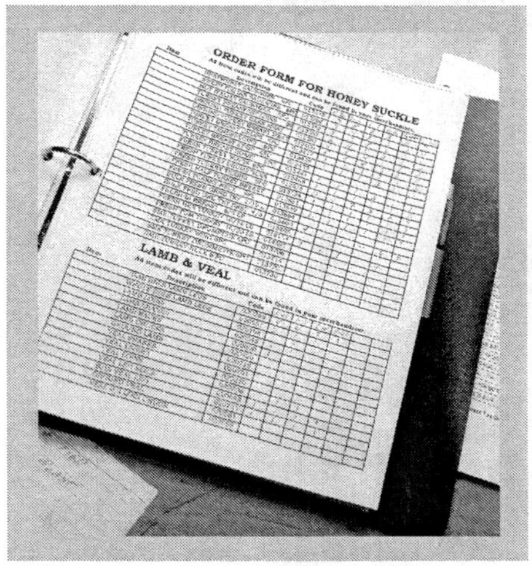

Purchase from reputable sources only.

Adapted from ***Retail Best Practices and Guide to Food Safety and Sanitation* © 2003**

RBP. Pg. 110 – 111 5

RSG. Pg. 92

Slide 5

Purchasing consists of buying the products needed for your establishment. Purchasing:

- Requires knowledge of products and market activity.

Objectives of an effective purchasing plan include:

- Buy the product best suited for the job
- Buy the proper quantity of the product
- Buy from only reputable and dependable suppliers
- Pay the right price for the product.

Purchasing product specifications:

- Quality grade
- Weight
- Count
- Contents
- Packaging type.

Purchase from approved sources only:

- Food from an approved source will be inspected to assure it is wholesome
- **Wholesome** food is safe to eat and promotes the consumers' health
- Approved sources comply with all local, state, and federal food laws
- A private home is not an approved source. Some local regulatory agencies may provide exceptions.

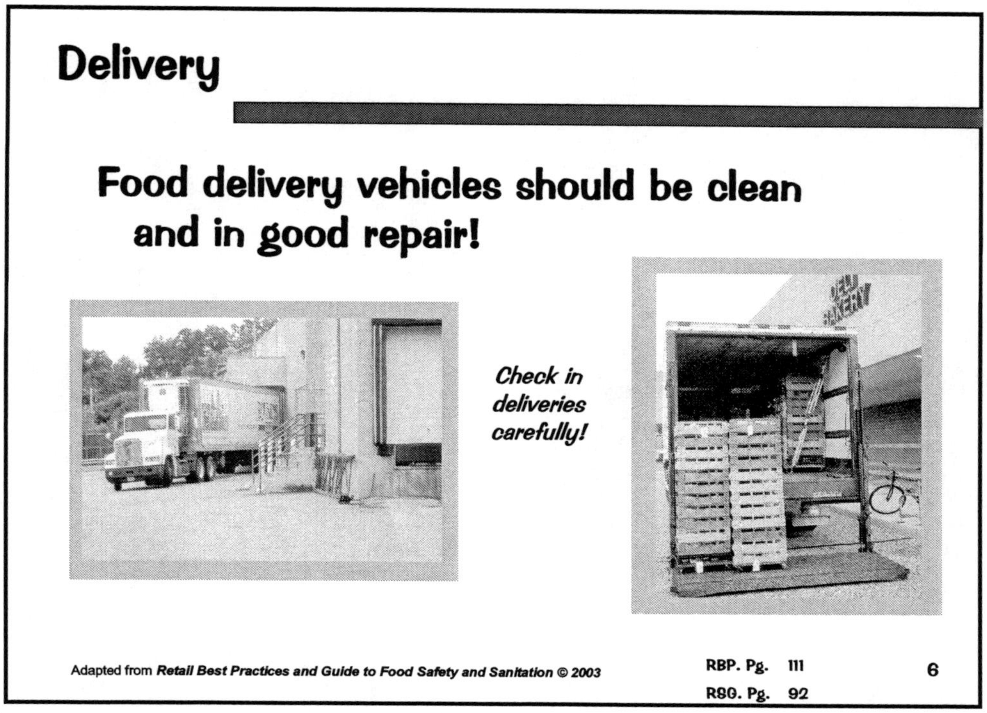

Slide 6

Reputable suppliers will ensure:

- Foods are delivered in vehicles that are clean and in good repair
- Perishable and potentially hazardous foods are maintained at safe temperatures during transport
- Products are loaded in a manner that separates food from non-food items
- Food packages are loaded so they are protected from being damaged during transit.

Reputable suppliers deliver food products to your retail food establishment in vehicles that are clean and in good repair.

Check the delivery vehicle for:

- Cleanliness of the cargo area
- Proper temperatures in refrigerated and frozen sections
- Proper separation of food and non-food items
- Signs of insect, rodent, or bird infestation
- Damaged packages that might result in contamination of food items
- Foods packages handled properly to keep them from being damaged.

Slide 7

Safety Stop! 4.1

Sample Test Questions

1. Which of the following should you purchase products from?
   A. Unlicensed street vendors.
   B. Home canners.
   C. Approved sources.
   D. One supplier.

2. Food should be delivered:
   A. Early in the morning.
   B. In clean vehicles that are in good repair.
   C. Once a week.
   D. Frozen.

3. The flow of food ends at the time the food is:
   A. Cooked.
   B. Sold.
   C. Received.
   D. Frozen.

Real-life

<u>Slide 4</u>
<u>Flow of Food</u>
Josie was in charge of setting up and refilling the salad bar for the day. She wanted to save herself some time later in the day, so she overfilled all the salads. What was her mistake and which step in the flow of food did she compromise?

By overfilling the salad ingredients, Josie compromised the cold-holding step in the flow of food. The product that was overfilled was above the cooling line and was being held inside the temperature danger zone.

<u>Slide 6</u>
<u>Delivery Vehicles</u>
Wayne was in charge of receiving the day's frozen food order. When the delivery came, the driver told him the freezer had just stopped working in the truck. Wayne believed him and accepted the frozen food order. What should Wayne have done?

Wayne should have taken the temperature of the products he was receiving to be sure they were not being held in the temperature danger zone, and more preferably, below freezing. Wayne should have also checked the product quality. If there were any issues with the temperature the food was being delivered at or the quality of the product, he should have refused the delivery. We will learn later Wayne could have also looked for large ice crystals on the surface of products and other signs of thawing and refreezing.

Making it Real

The Flow of Food – Slide 4

- Understanding the flow of food within retail food establishments and how many times any given product can be put in jeopardy can be difficult. Reinforce this concept by going around the room and having the learners explain the steps in the appropriate order and how the product can be impacted.
    - Supplies
        - Class
        - Food product of your choice (optional)
    - Procedure
        - Hand the first student the food product (plastic food products work well for this exercise). Have the student tell you the first step in the flow of food, how that product's food safety could be impacted, and how they could keep it safe.
        - Allow the student to hand the product to another student who will continue with the flow of food.
        - Continue until you have reached the final step in the flow of food – checkout and bagging.
    - Teaching Tip – Have each of the students remain standing until the end of the exercise. Allow them to really see how many people touch any given product throughout its flow through the establishment.

Receiving:
Inspection

Determine food quality by using your senses –

- Sight
- Touch
- Smell.

Determine food quality before you accept it!

Adapted from *Retail Best Practices and Guide to Food Safety and Sanitation* © 2003

RBP. Pg. 112
RSG. Pg. 93

8

Slide 8

<u>Sensory evaluation</u> uses the senses of smell, touch, sight, and taste to help determine food quality.

Steps in determining food quality:

Observe the appearance of the food using your sight. Look for:

- Color
- Texture
- Evidence of spoilage such as
  - Slime
  - Mold growth
  - Discoloration
- Package defects
  - Tears
  - Punctures
  - Dents
  - Bulges.

Observe the flavor of the food using your taste. Flavor is a combination of smell and taste. Negative flavors to observe include:

- Smell of ammonia, hydrogen sulfide (smell of rotten eggs), and alcohol.

Taste, although not recommended, is another sense sometimes used to determine wholesomeness of food.

- Taste of acid, alcohol, or a bitter taste.

Touch the product as well. Feel the product and avoid textures that are:

- Slimy
- Mushy.

These are all signs of spoilage, not food safety. A food that shows no signs of spoilage is not necessarily safe to eat.

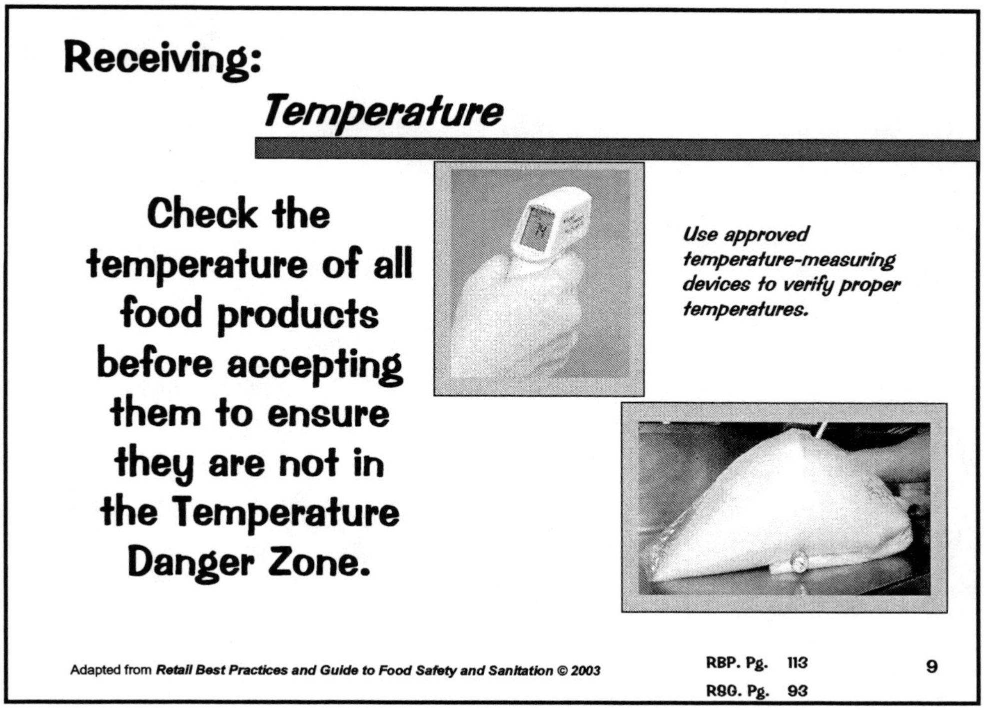

Slide 9

Holding foods at proper temperatures during delivery is essential to food safety.

Check the temperature of foods before accepting them:

- Packaged foods:
    - Insert the probe of the thermometer between two packages or fold the package over the probe of the thermometer
    - Ensure good contact is made with the probe
    - Infrared thermometers can be used to measure the surface temperature of foods
- Unpackaged foods:
    - Insert the probe of the thermometer into the thickest part of the food
    - Clean, sanitize, and air-dry the probe between products.

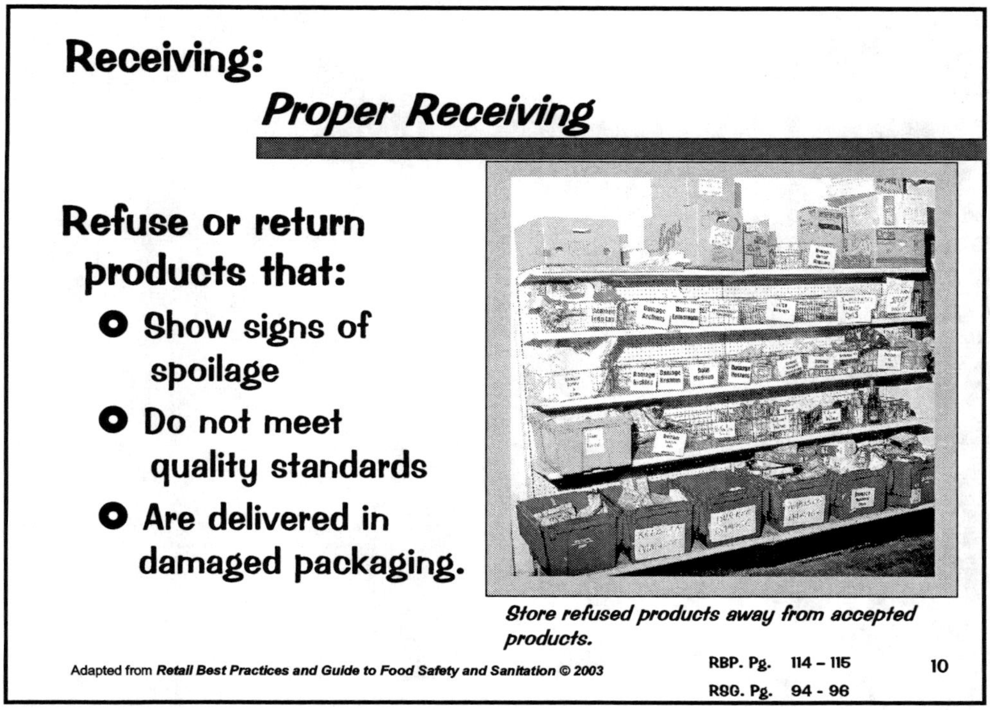

Slide 10

Receiving requires:

- Prompt handling
- Exacting quality control procedures
- Trained staff that can interpret:
    - Product specifications
    - Coding
    - Temperature measurement
    - Proper handling of distressed merchandise.

Refuse or return products that:

- Show signs of spoilage
- Do not meet quality standards
- Are delivered in damaged packaging that may threaten the safety of the food.

Poor receiving procedures increase the chance of:

- Theft
- Acceptance of underweight merchandise
- Contamination
- Shrink
- Acceptance of products that do not meet specifications.

It is important to dispose of, or properly store, distressed products away from wholesome foods.

Receiving:

Packaged Goods – Reduced Oxygen Packaging

Vacuum packed vegetables

Reduced Oxygen Packaging (ROP)

- Packaging that has had oxygen removed or replaced by another gas
- Increases shelf life of foods.

Adapted from ***Retail Best Practices and Guide to Food Safety and Sanitation* © 2003**

RBP. Pg. 116, 151 – 154 11
RSG. Pg. 96 - 100

Slide 11

<u>**Reduced Oxygen Packaging (ROP)**</u> is any packaging method that reduces the amount of oxygen in a sealed package. ROP:

- Has increased in the past decade in some retail food establishments to increase shelf life
- Must contain a "Use By" date
- Should not have broken seals.

Types of Reduced Oxygen Packaging include:

- Cook-chill – Plastic bag filled with hot food, air is removed, then quickly chilled and sealed
- Controlled Atmosphere Packaging – maintains the desired atmosphere within a package by using agents to bind or scavenge oxygen
- Modified Atmosphere Packaging – a gas flushing with carbon dioxide or nitrogen and sealing practice.
- <u>**Sous Vide**</u> – product is vacuum sealed in a plastic container, cooked at low temperatures, then quickly cooled or frozen
- Vacuum Packaging – all air is removed and the package is hermetically sealed (more on next slide).

Reduced Oxygen Packaging benefits:

- Creates a largely oxygen-free environment that prevents the growth of aerobic bacteria, yeast, and molds
- Prevents chemical reactions that can produce off odors and colors
- Reduces product shrinkage by preventing water loss.

Reduced Oxygen Packaging safety concerns:

- ROP creates an anaerobic environment that creates the potential for microbial growth.
- *Clostridium botulinum* and *Listeria monocytogenes* grow in anaerobic conditions.

The *FDA Food Code* provides very specific requirements for supermarkets that desire to use ROP technology. A summary of the requirements are presented below:

- The supermarket must have a HACCP plan in place for the ROP operation that details how the seven principles (see Session 10) are incorporated into the operations food safety management system.

- Because *C. botulinum* is a significant health hazard, the establishment is required to have two microbial growth barriers. Time/temperature are commonly used as one barrier, but they need to be coupled with the use of pH, A_w, or product formulation to assure *C. botulinum* will not grow inside the package.

- All potentially hazardous foods in RPO which rely on refrigeration as a barrier to microbial growth must bear the statement Important - Must be kept refrigerated at 41°F (5°C) or Important - Must be kept frozen in the case of foods which rely on freezing as a primary safety barrier.

- Each container must bear a Use-By date. This date cannot exceed 14 days from retail packaging or repackaging without a variance granted by the regulatory authority.

- Employees responsible for the ROP process must receive training that will enable them to understand the key components of the process, the equipment that is used, and the specific procedures that must be followed to ensure critical limits identified in the HACCP plan have been met. The training program must also outline the employees responsibilities for monitoring and documenting the process and describe what corrective actions they must take when critical limits are not met.

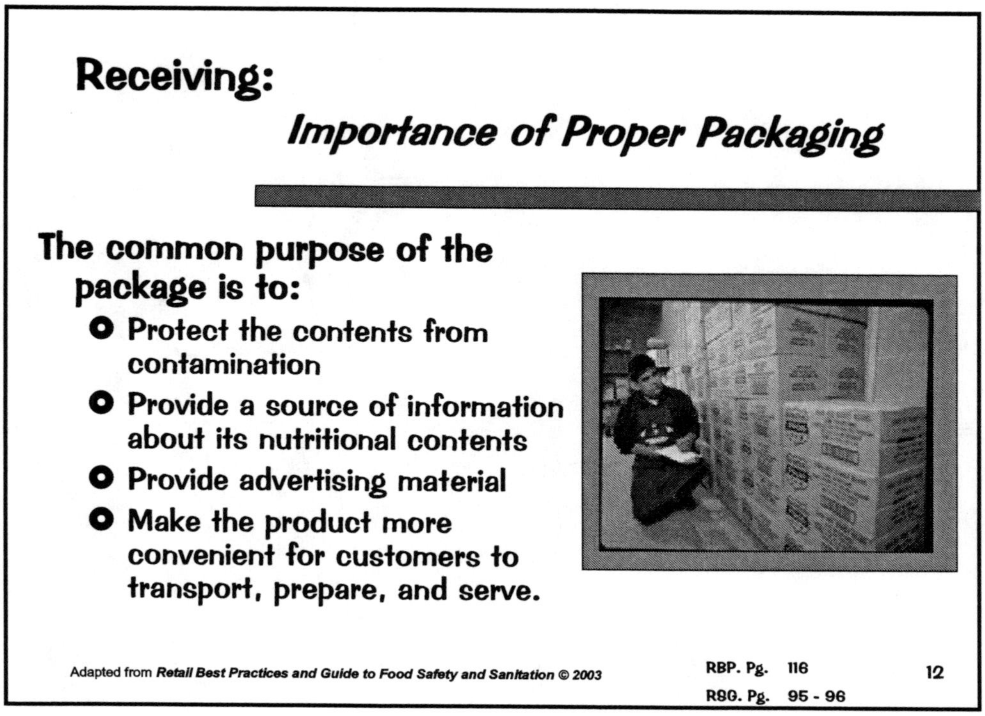

Slide 12

The foods in a retail food establishment come in many different types of packages. Examples of food packaging include:

- Cans
- Bottles
- Jars
- Pouches
- Tubs
- Trays
- Bags
- Boxes.

The common purpose of the package is to:

- Protect the contents from contamination
- Provide a source of information about its nutritional contents
- Provide advertising material
- Make the product more convenient for customers to transport, prepare, and serve.

Dry foods such as flour, sugar, rice, and beans are not potentially hazardous foods and can be safely stored at room temperature.

- Packages of dry food must be inspected for tears and other signs of damage and contamination by chemicals and other substances that could cause foodborne illness.

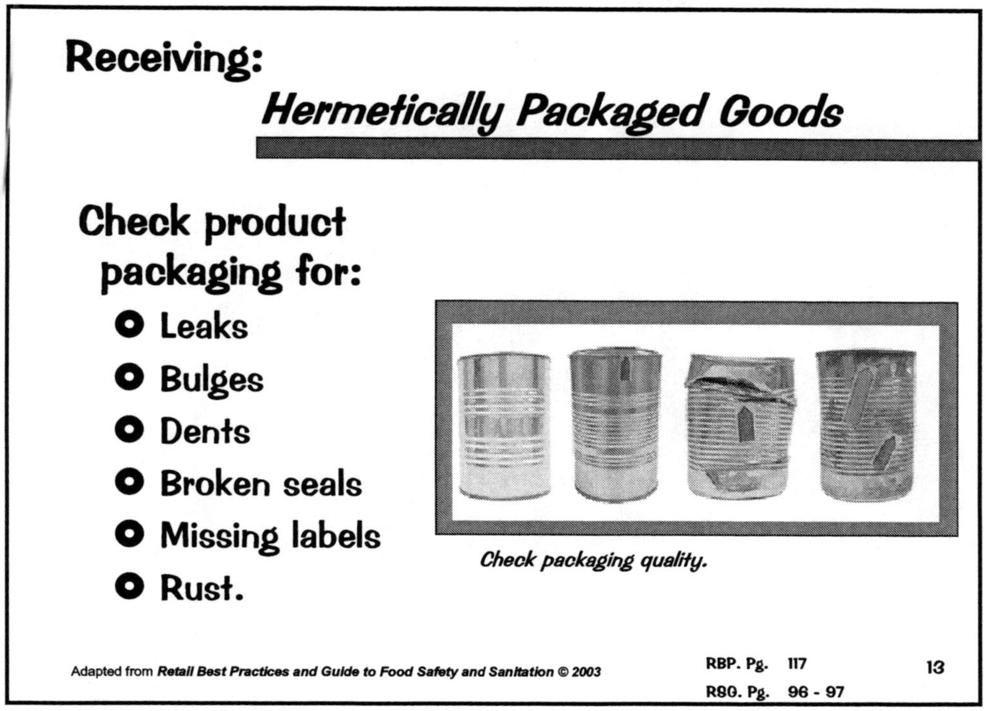

Slide 13

Hermetic packaging
Product is completely sealed in a container to prevent the loss and entry of gases.

Common hermetically sealed containers include:
- Vacuum packages
- Metal cans
- Glass jars.

Check for, and refuse, hermetically sealed containers if they:
- Leak
- Bulge
- Have dents
- Have broken seals
- Have damage along seams
- Have penetrating rust
- Are missing labels.

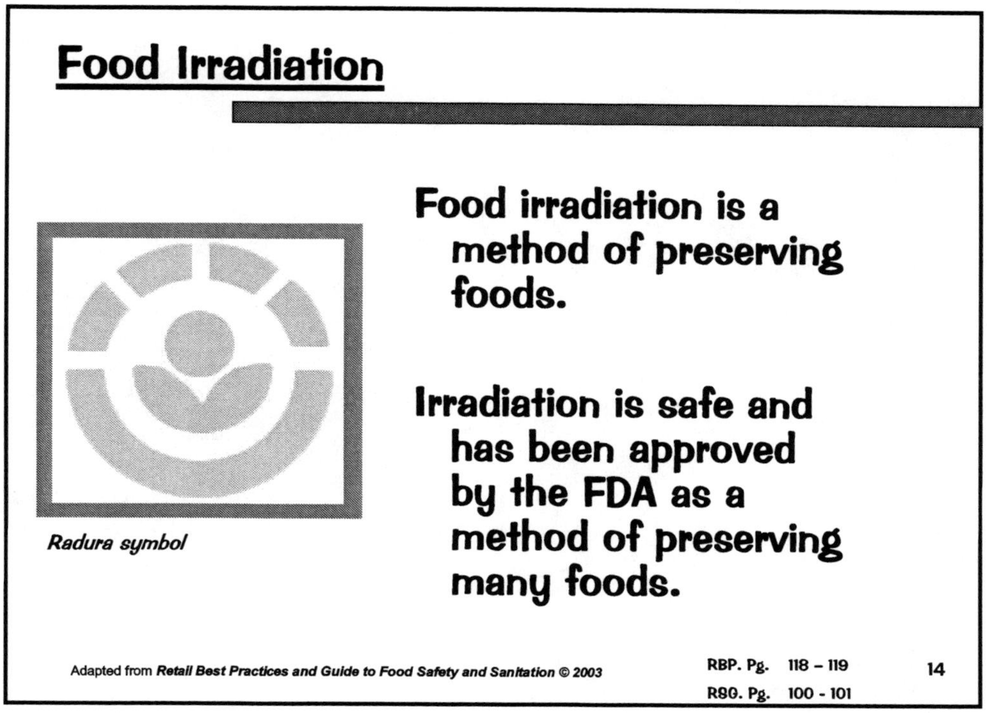

Slide 14

Food **irradiation** is a method of preserving foods where food is exposed to ionizing radiation to destroy disease-causing microbes and delay spoilage

Food irradiation is FDA approved for:

- Fruits
- Vegetables
- Grains
- Spices
- Poultry
- Pork
- Lamb
- Ground beef.

Food irradiation does not pose a risk to consumers.

All irradiated foods must be marked with the ***Radura*** symbol accompanied by the words "Treated with Irradiation" or "Treated with Radiation."

Irradiation of food can effectively reduce or eliminate pathogens and spoilage microbes while maintaining the quality of most foods.

Slide 15

Safety Stop! 4.2

Sample Test Questions

1 When receiving food products, you should:
   A. Just sign for the product and put it away.
   B. Ask someone else to sign for the product.
   C. Inspect the product using a thermometer and sensory evaluation.
   D. Leave the product on the floor until you can get to it.

2. Common hermetically sealed containers you may find in a food establishment can include:
   A. Styrofoam containers.
   B. Paper bags.
   C. Cardboard boxes.
   D. Metal cans.

3. Reduced Oxygen Packages (ROP) must contain:
   A. A "Best Eaten By" Date.
   B. Moisture.
   C. A safety seal.
   D. A "use-by and sell-by date".

4. When taking the temperature of packaged food products with a stemmed thermometer, it is important to:
   A. Use an infrared thermometer.
   B. Insert the thermometer into the thickest part of the product.
   C. Insert the probe of the thermometer between the two packages and ensure good contact is being made with the probe.
   D. Touch the product to make sure it is cold.

5. Refuse or return products that:
   A. Show signs of spoilage.
   B. Do not meet quality standards.
   C. Are delivered in severely damaged packaging.
   D. All of the above.

6. Food irradiation is a method of:
   A. Preserving foods.
   B. Adding flavor to foods.
   C. Receiving foods.
   D. Cooking foods.

Real-life

Slide 8 – 13
Receiving
Brian was receiving the meat order for the day. He was very busy and was just told he had a phone call when the delivery came in. He told the driver to put the meat in the cooler and he would sign for it. What was his mistake?

Brian should have had the driver wait until he was ready, or taken the time to check the product carefully before he signed for it.

Slide 10
Storage of Spoiled Product
After receiving a delivery, Gerald came across a package of spoiled cheese in the order. He wanted to save the cheese so he could return it for credit. He wrote "Do Not Use" on the package and put it behind the rest of the cheese. What was his mistake?

Gerald should have removed the spoiled product and stored it away from any good food product. This would help prevent cross contamination from the spoiled food to good food product.

Slide 13
Missing Label
Tom was receiving the dry goods delivery. The driver gave him a case of #10 cans with no label on them and told him the cans had tomato sauce in them. Tom dated them and put them on the shelf behind the other cans of tomato sauce. What was Tom's mistake?

Tom should have refused the cans. He had no way of knowing what the product really was, and without the label, he was missing valuable product information such as ingredients and production dates.

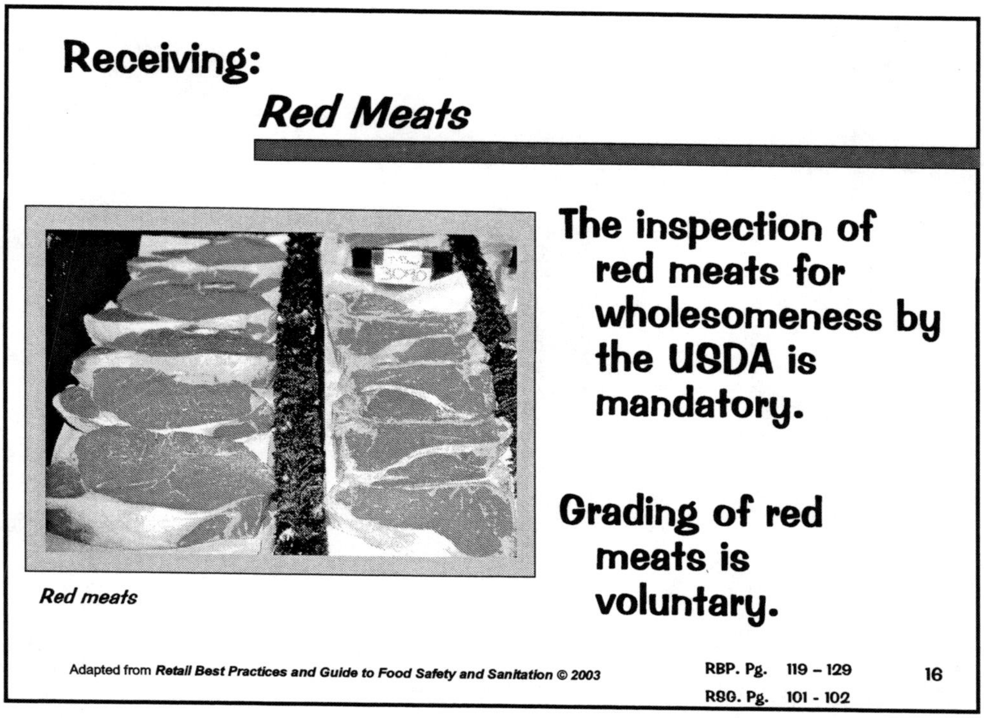

Slide 16

Red meats are inspected and graded by the USDA.

Inspection for wholesomeness – animals are inspected prior to slaughter for disease and defects
- Inspections are mandatory.

Grading – measures the quality or palatability of the product
- Grading is voluntary.

Meats and meat products come in a variety of forms including:
- Frozen
- Fresh
- Cured
- Smoked
- Dried
- Canned.

Fresh red meats can be stored up to three weeks between temperatures of 28°F (-2°C) and 32°F (0°C). They must:
- Be received at temperatures at or below 41°F (5°C)
- Have the appropriate color:
  - Beef – bright red
  - Lamb – light red
  - Pork – light pink
- Have the appropriate texture:
  - Firm flesh
  - Flesh elastic to the touch
  - Not slimy or sticky
- Smell fresh, with no off smelling odors.

Frozen red meats can be stored for several months when held below 0°F (-18°C) and:

- Must be received solidly frozen
- Should not show signs of freezing, thawing and refreezing such as:
  - Blood juices frozen to the bottom of container
  - Large ice crystals on the product.

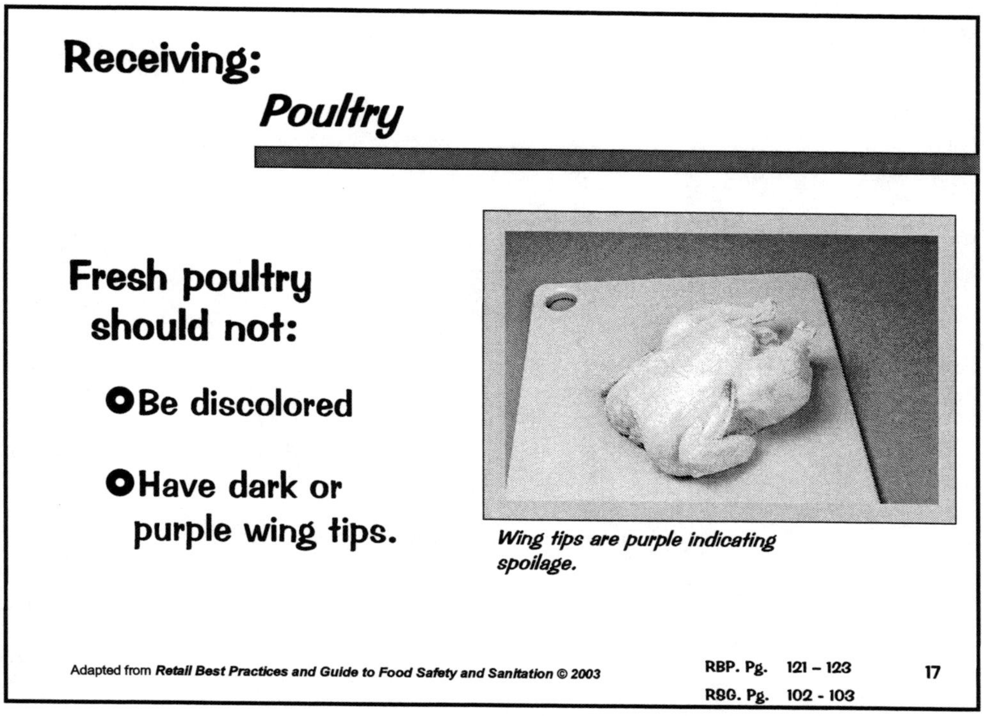

Slide 17

Poultry must be inspected for wholesomeness and graded by the USDA or state inspectors.

Fresh poultry can be stored for short periods of time between temperatures of 28°F (-2°C) and 32°F (0°C). They must:

- Be received at temperatures at or below 41°F (5°C)
- Have the appropriate color:
    - No discoloration
    - Wing tips not dark or purple
- Have the appropriate texture:
    - Firm flesh
    - Flesh elastic to the touch
    - Not slimy or sticky, especially under wings
- Smell fresh, with no off smelling odors.

Frozen poultry can be stored for four to six months when held below 0°F (-18°C) and::

- Must be solidly frozen
- Must not show signs of freezing, thawing and refreezing such as:
    - Blood juices frozen to the bottom of container
    - Large ice crystals.

Receiving:
Eggs and Egg Products

Eggs are a common source of *Salmonella enteritidis* bacteria.

Eggs and egg products
Adapted from ***Retail Best Practices and Guide to Food Safety and Sanitation* © 2003**

RBP. Pg. 123 - 124
RSG. Pg. 103 - 105

18

Slide 18

Eggs are inspected and graded by the USDA

- Eggs are the leading source of *Salmonella enteritidis* bacteria.

Raw shell eggs should be stored and displayed in equipment that maintains an ambient temperature of 45°F (7°C) or below. It is recommended they be stored at 34°F (1°C) to 38°F (3°C) to maintain optimal quality.

Whole fresh eggs must:

- Be received at ambient temperatures at or below 45°F (7°C)
- Be clean and free of cracks
- Have a firm yolk with whites clinging to yolk
- Be free of off odors.

Egg products (an egg without its shell) must be:

- Pasteurized – **<u>pasteurization</u>** destroys all disease-causing microorganisms and reduces the number of bacteria, thus extending the shelf life
- Received in sealed containers
- Received and stored at 41°F (5°C) or below.

Frozen egg products must be received and stored at or below 0°F (-18°C).

Dried eggs may be stored at room temperature, however, once they are reconstituted

- They are considered to be potentially hazardous
- They must be stored out of the temperature danger zone.

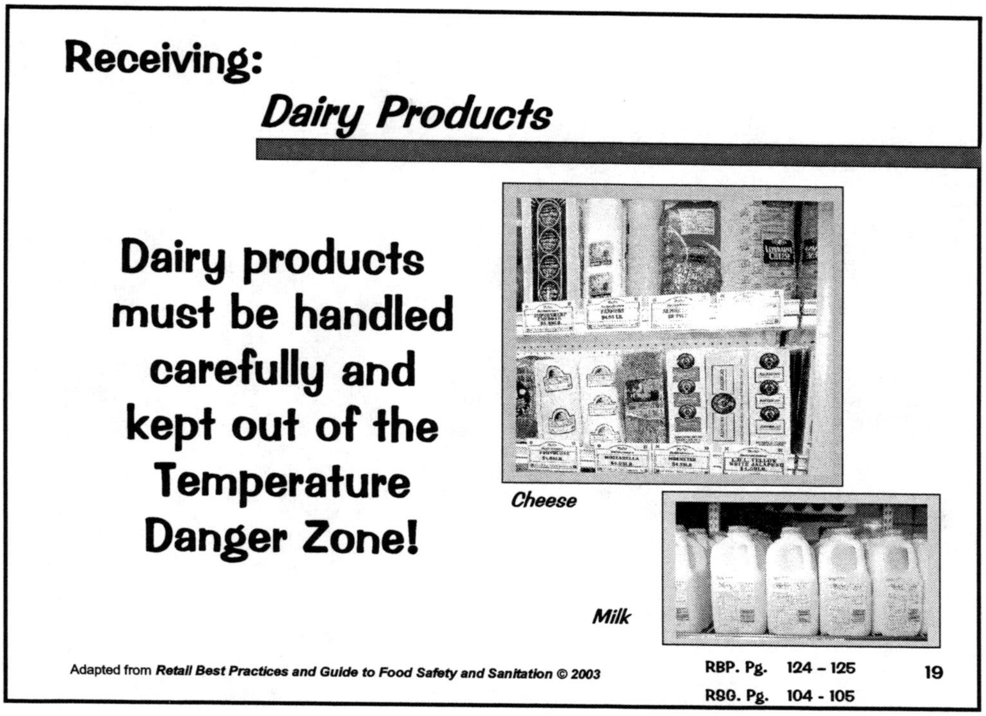

Slide 19

Milk, cheese, and butter

Fluid milk must be:

- Received at temperatures at or below 41°F (5°C)
- Clearly marked with expiration date
- Pasteurized.

Fluid milk is best stored at temperatures between 34°F (1°C) to 38°F (3°C) for up to ten days or longer to maintain optimal quality.

Milk marked with a "UHT" has been pasteurized at ultra high temperatures

- UHT milk can be stored under refrigeration for several weeks and at room temperature for several days
- UHT milk is **packaged aseptically** or free of pathogenic microorganisms.

Cheese must be:

- Received at temperatures at or below 41°F (5°C)
- Inspected for proper
    - Color
    - Flavor
    - Moisture content
- Rejected if it contains mold that is not normally a part of the cheese.

Butter must be:

- Received at temperatures at or below 41°F (5°C)
- Made from pasteurized cream
- Firm in texture
- Uniform in color
- Free of mold
- Delivered in a package that will protect it.

Slide 20

Fish and shellfish

For optimal quality and freshness, fish and shellfish receiving temperatures often range between 32°F (0°C) to 34°F (1°C).

Fish should be received at a temperature at or below 41°F (5°C) and have:

- A mild pleasant odor
- Bright, shiny skin with tightly attached scales
- Clear, bulging eyes
- Bright red, moist gills
- Firm flesh that is elastic to the touch.

Molluscan shellfish (oysters, clams, and mussels) should be received at or below 45°F (7°C) and must be:

- Delivered with shells closed tightly
- Harvested from unpolluted waters
- Free of mud
- Purchased by FDA approved sources
- Accompanied by legible shellfish identification tags.

Shellfish identification tags must be attached to the container until it is empty and kept on file within the retail food establishment for 90 days. These tags must have:

- The harvester's identification number
- The harvesting date
- The harvesting site
- The shellfish type and quantity
- Verbiage about keeping the tag on the container until the shellfish is used
- Instructions to keep the tag on-site for 90 days.

Crustaceans (lobster and crabs) must be:

- Delivered alive.

Slide 21

Fruits and vegetables:

- Can spoil rapidly
- Keep a high quality for only a few days so they should be inspected for freshness
- Must be purchased from approved sources
- Must be washed thoroughly before they are:
    - Cut
    - Combined with other ingredients
    - Cooked
    - Served
    - Offered for human consumption in ready-to-eat form.

Fresh fruits and vegetables are best stored between 34°F (1°C) and 41°F (5°C) and shaded from the light.

Fruits and vegetables that will be washed by the consumer prior to consumption do not have to be washed at the establishment before they are sold.

Fruits and vegetables are not typically considered potentially hazardous, but may be associated with foodborne illness.

Microbes that can be infective with only a few cells and have been found on fruits and vegetables include:

- Shiga toxin-producing *Escherichia coli*
- Hepatitis A virus
- *Cyclospora spp.*
- *Salmonella spp.*

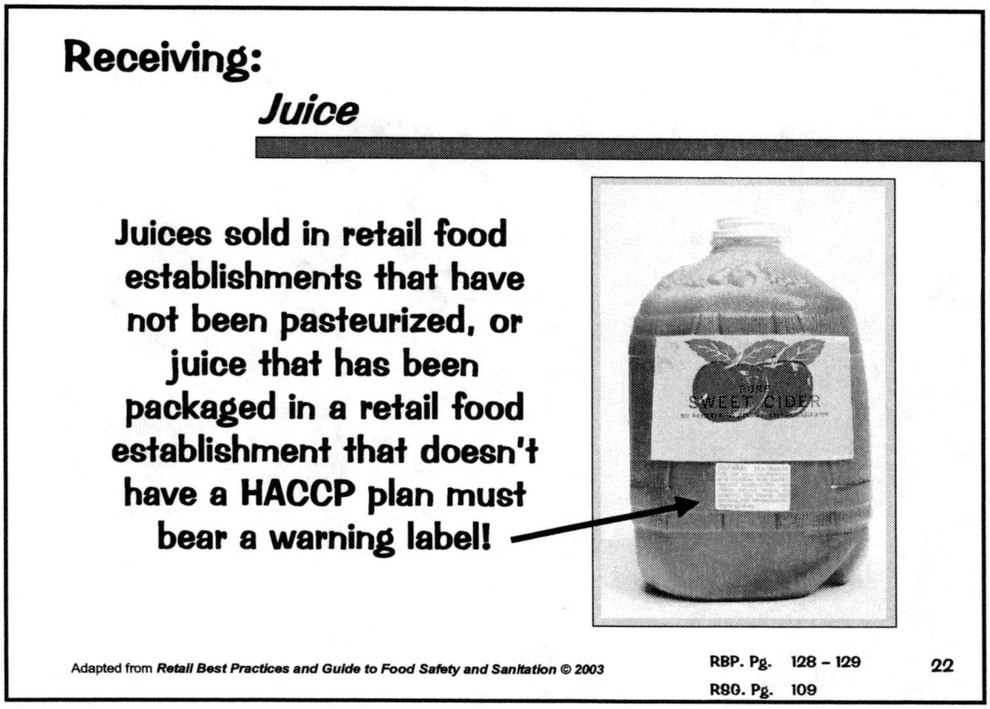

Slide 22

The category of juice includes:

- The liquid extracted from one or more fruits or vegetables
- Purees of the edible portions of one or more fruits or vegetables
- Any concentrates of such liquid or puree.

According to the *FDA Food Code* the juice group includes:

- Juice as a whole beverage
- Juice as an ingredient of a beverage
- A puree as an ingredient of a beverage.

Most juices sold in retail food establishments are obtained from processors in a prepackaged form.

- These juices have been pasteurized to attain 99.999% reduction of the most resistant disease-causing microorganisms
- Juices that have not been pasteurized must have a warning label stating "Warning: This product has not been pasteurized and, therefore, may contain harmful bacteria that can cause serious illness in children, the elderly, and persons with weakened immune systems."

Juice packaged in a retail food establishment must:

- Be treated under a HACCP plan to attain 99.999% reduction of the most resistant disease-causing microorganisms, or
- Bear a warning label stating "Warning: This product has not been pasteurized and, therefore, may contain harmful bacteria that can cause serious illness in children, the elderly, and persons with weakened immune systems."

Slide 23

Safety Stop! 4.3

Sample Test Questions

1. When receiving red meats, the meat should:
   A. Smell fresh.
   B. Be elastic to the touch.
   C. Have a bright color.
   D. All of the above.

2. A key indicator that fresh poultry has spoiled is:
   A. Temperature indicates it is below 41°F (5°C).
   B. The wing tips are purple in color.
   C. The flesh is firm and elastic to the touch.
   D. There is a fresh odor.

3. According to *FDA Food Code*, fresh shell eggs must be received:
   A. At or below 45°F (7°C) internal temperature.
   B. At or below 41°F (5°C) internal temperature.
   C. At or below 41°F (5°C) ambient temperature.
   D. At or below 45°F (7°C) ambient temperature.

4. All fluid milk must:
   A. Be white in color.
   B. Contain at least 5% milk fats.
   C. Be clearly marked with an expiration date.
   D. Be held at or below 45°F (7°C).

5. Shellfish tags must be kept on premise for a minimum of:
   A. 28 days.
   B. 30 days.
   C. 90 days.
   D. 120 days.

6. Shellfish tags must
   A. Be removed from the original packaging and filed immediately upon receipt.
   B. Remain attached to the original packaging until the product is completely used, and then removed and filed.
   C. Be removed, photocopied and returned to the original packaging upon receipt.
   D. Be discarded after the product is completely used.

7. Fruits and vegetables must be thoroughly washed:
   A. Before they are prepared or offered for sale in ready-to-eat form.
   B. Before they are displayed.
   C. Before they are delivered.
   D. When they are received.

8. Juices that have not been heat pasteurized must
   A. Have a safety seal.
   B. Be packaged in a clear container.
   C. Bear a warning label.
   D. Be stored at temperatures below 35°F (2°C).

Real-life

Slide 16
Meat Receiving
Sarah was receiving the meat order for the day. The frozen beef liver had large ice crystals on it and there appeared to be blood juices frozen to the bottom of the package. What should this have indicated to Sarah?

Sarah should have realized these were signs the meat products had been thawed and then refrozen. She should refuse the delivery of meat.

Slide 17
Poultry Receiving
Jonathan was receiving the poultry order for the week. The fresh, whole chicken came in a pool of ice water. What should Jonathan have done?

Jonathan should have refused the product. Whole fresh chicken must be stored on a bed of ice that allows the water to drain.

Slide 18
Egg Receiving
Steven was receiving the dairy order for the week. He properly checked the temperature of the pre-whipped egg products that came in sealed boxes using the proper packaged product temperature-taking procedures. The egg product was delivered at 44°F (7°C). Steven accepted the product and stored them properly. What was his mistake?

Although fresh shell eggs can be delivered at or below 45°F (7°C) ambient temperature, raw egg products need to be received at or below 41°F (5°C).

Slide 19
Dairy Receiving
Nick was receiving the dairy order for the week. One of the containers of whipped margarine had a hole in the lid. He took it directly to the deli area so it could be used first. What did Nick do wrong?

Nick should have refused the product that came in the damaged package. He did not know what had come into contact with the unprotected food product.

Slide 20
Shellfish Receiving
Jason was receiving the seafood delivery for the day. He was checking in the oysters and noticed there was no shellfish tag. When he asked the delivery person where it was, he pulled a tag out of his jacket pocket and said that it fell off when he was bringing in the delivery. Jason inspected it, and threw it away and stored the oysters in the cooler. What did Jason do wrong?

Jason should have returned the tag to the delivery person and refused the oysters. Shellfish tags must be attached to the container upon delivery and kept with the container until it is empty. The tag then must be kept on file for an additional 90 days.

Slide 21
Fruit and Vegetable Receiving
Brenda was preparing the fruit salad for the day for the salad bar in the deli. She pulled the cantaloupe from the box it was delivered in and cut off the rind. She sliced it, seeded it, and cut it into chunks for the salad. What did Brenda do wrong?

Brenda should have washed the fruit prior to preparing it in order to remove any contaminants that may have been on the rind so they did not cross contaminate the edible fruit.

Slide 22
Juice Receiving
Roberto was receiving an order of unpasteurized apple cider on which his local market was running a Fall special. He checked the case and none of the gallon jugs had a warning label. What should Roberto do?

Roberto should refuse the product. Unpasteurized apple cider must bear a label stating "Warning: This product has not been pasteurized and, therefore, may contain harmful bacteria that can cause serious illness in children, the elderly, and persons with weakened immune systems."

Making it Real

Washing Fruit – Slide 21

- Bacteria on fruit that is being cut or prepared can get onto the knife or into the fruit if it is not properly washed prior to preparation. Using food coloring as your hypothetical bacteria, show the students how that bacteria can be transferred onto the knife and into the fruit.
    - Supplies
        - Food coloring drops
        - Fruit of choice such as a cantaloupe
        - Knife
    - Procedure
        - Place a few drops of food coloring onto the fruit.
        - Using the knife, cut through the drops you had placed on the fruit. The food coloring will get onto the knife and onto the edible portion of the fruit you just cut. Make another cut through the fruit to visually show how the bacteria can be transferred from the rind of the fruit onto the edible surface of the fruit.
    - Teaching Tip – This exercise is best done with a fleshy fruit.

Storage:

FIFO

Store foods properly.

To ensure product freshness and quality, use the First In, First Out inventory method!

Adapted from ***Retail Best Practices and Guide to Food Safety and Sanitation* © 2003**

RBP. Pg. 129 – 130

RSG. Pg. 110

24

Slide 24

Storage principles:

- Utilize **<u>FIFO (First In, First Out)</u>** storage method
- Store older products in front of new products
- Product containers should be marked with the date they were received
- Store foods a minimum of six (6) inches off the floor
- Store raw products under cooked and ready-to-eat foods
- Store species of animals in the following order from bottom to top:
    1. Poultry.
    2. Ground beef and pork.
    3. Fish, eggs, and other red meats.
    4. Cooked and ready-to-eat foods.

Type of Storage: Refrigerator

Refrigeration

- Slows microbial growth
- Used to store potentially hazardous and perishable foods for a short period of time
- Store potentially hazardous foods at 41°F (5°C) or below
- Space products to allow cold air to circulate around them
- Refrigerators must have a temperature-measuring device located to measure the air temperature in the warmest part of the unit.

Adapted from *Retail Best Practices and Guide to Food Safety and Sanitation* © 2003 RBP. Pg. 130 – 131 RSG. Pg. 111 - 112 25

Slide 25

Type of storage: Refrigeration

- Slows microbial growth
- Used to store potentially hazardous and perishable foods for a short period of time
- Store potentially hazardous foods at 41°F (5°C) or below
- Space products to allow cold air to circulate around them
- Refrigerators must have a temperature-measuring device located to measure the air temperature in the warmest part of the unit.

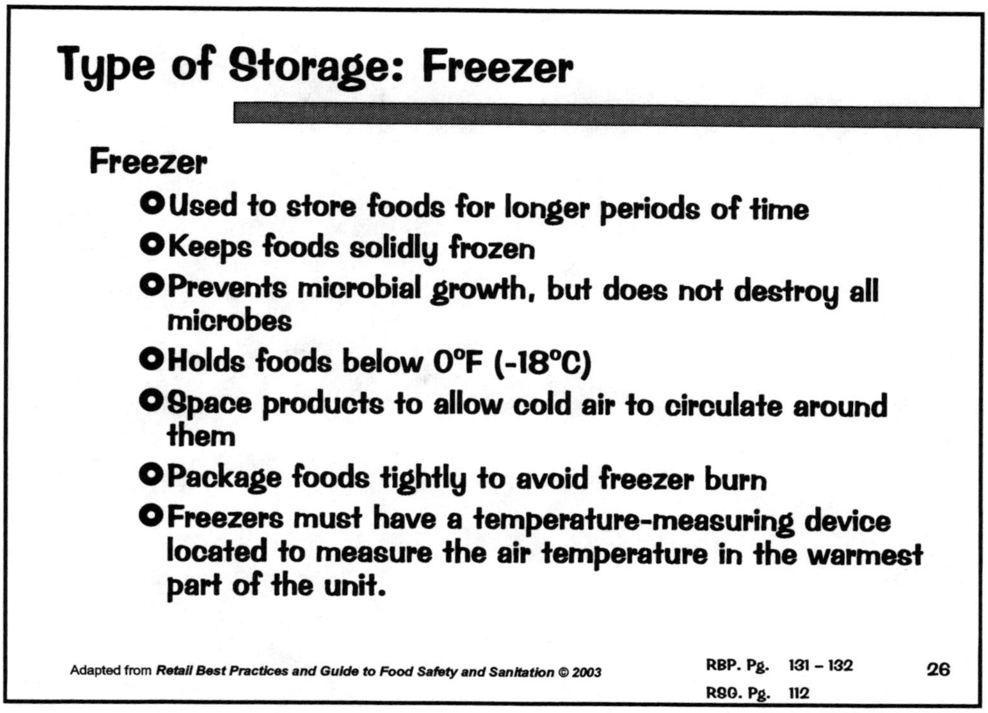

Slide 26

Type of storage: Freezer

- Used to store foods for longer periods of time
- Keeps foods solidly frozen
- Prevents microbial growth, but does not destroy all microbes
- Holds foods below 0°F (-18°C)
- Space products to allow cold air to circulate around them
- Package foods tightly to avoid freezer burn
- Freezers must have a temperature-measuring device located to measure the air temperature in the warmest part of the unit.

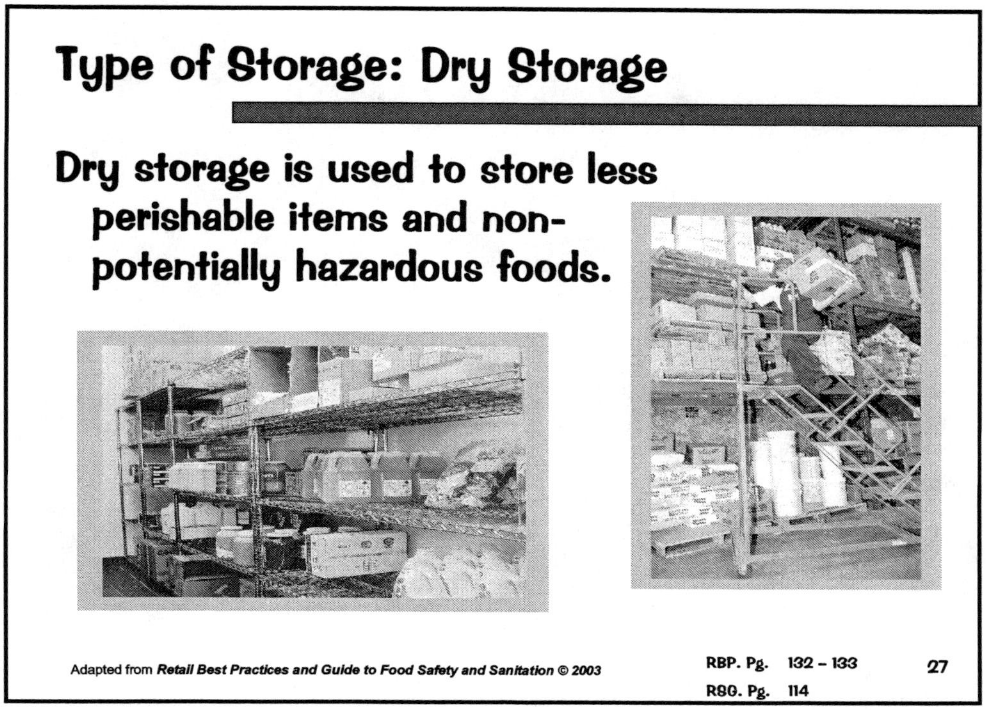

Slide 27

Type of storage: Dry

- Used to store less perishable items and nonpotentially hazardous foods
- Should be room temperature 50°F (10°C) to 70°F (21°C)
- Should have a relative humidity of 50 to 60%
- Sunlight deteriorates the quality of some foods so windows are not recommended in dry grocery storage areas
- Foods should be stored on slatted shelves or skids.

If it is necessary to transfer bulk food items such as flour, sugar, and grain foods into other containers, the containers must be:

- Constructed of food-grade materials
- Equipped with tight-fitting lids
- Properly coded, dated and labeled with the common name of the food.

Containers holding commonly recognized foods such as dry pasta do not need to be labeled.

Scoops to remove food from bulk food containers must:

- Be constructed of materials that are:
    - Nontoxic
    - Nonabsorbent
    - Easily cleanable
- Be equipped with handles
- Be stored in a manner that will allow food employees to grip the handle of the utensil without touching the food.

Storage Don'ts

Store foods in proper places.

Do Not Store Food:

- In toilet rooms
- In locker areas
- In mechanical rooms
- Under piping
- Under sewage lines
- Among chemicals
- With personal items.

Adapted from ***Retail Best Practices and Guide to Food Safety and Sanitation* © 2003**

RBP. Pg. 133 – 135
RSG. Pg. 114

28

Slide 28

Do not store food, single-service items, utensils, equipment, or paper goods:

- In toilet rooms
- In locker areas
- In mechanical rooms
- Under piping
- Under water and sewage lines unless the lines are shielded to interfere with potential drops
- Among chemicals
- With personal items.

Chemical Storage

Most chemicals are poisonous and must be properly labeled and stored away from:

- Food items
- Single-service items
- Utensils
- Equipment
- Paper goods.

Adapted from ***Retail Best Practices and Guide to Food Safety and Sanitation* © 2003**

RBP. Pg. 134 – 135

RSG. Pg. 113 - 114

29

Slide 29

Toxic chemicals that are commonly sold and used in retail food establishments include:

- Cleaners
- Sanitizers
- Pesticides.

Chemicals should be properly labeled with information on:

- Directions on how to properly use the chemical
- Chemical identification
- Instructions on first aid in case of an accident.

Slide 30

Safety Stop! 4.4

Sample Test Questions

1. Store species of animals in this order from bottom to top.
   A. Cooked and ready-to-eat; fish, eggs, and other red meats; ground beef and pork; poultry.
   B. **Poultry; ground beef and ground pork; fish, eggs, and other red meats; cooked and ready-to-eat.**
   C. Fish, eggs, and other red meats; cooked and ready-to-eat; ground beef and pork; poultry.
   D. Cooked and ready-to-eat; poultry; fish, eggs, and other red meats; ground beef and pork.

2. Freezing foods will:
   A. Kill all harmful microorganisms.
   B. Slow microbial growth.
   C. Not affect microbial growth.
   D. Increase microbial growth.

3. All food must be stored at least:
   A. 4 inches off the floor.
   B. 6 inches off the floor.
   C. 8 inches off the floor.
   D. 12 inches off the floor.

4. Chemicals must be stored

A. **Away from foods and food products.**
B. Above foods and food products.
C. In the freezer.
D. In a locked room.

Real-life

Slide 24
FIFO

Ron was putting away pallets of dry cereal he had just received. He put new cereal in front of the old cereal. What was his mistake?

Ron should have used the FIFO inventory method and put the old cereal in front of the new cereal.

Slide 29
Chemical Storage

Jennifer was cleaning a display case that held raw red meats. She followed the cleaning steps in detail and cleaned out the part of the display case that held the meat. She returned the meat to the display case and realized she had neglected to clean the back wall of the display case and the mirrors above the case. She proceeded to clean the back wall and the mirrors. Did Jennifer do anything incorrectly?

Jennifer should not have used a cleaning solution around the meats she had restocked. She should have removed the meats and then cleaned the back wall and mirrors.

Making it Real

First In, First Out – Slide 24

- FIFO storage method sounds very easy, however it can be confusing to some. This exercise will allow the learners to better understand the inventory process and the steps involved in it.
    - Supplies
        - 9 to 12 metal cans
        - A black magic marker
        - Shelf
    - Procedure
        - On the top of three of the metal cans write yesterday's date
        - On the top of three more cans, write the date of three days prior
        - Leave the remaining cans with no date
        - Place the dated cans onto the shelf in the appropriate manner with the older dated cans behind the newer dated cans
        - Give the cans with no dates to the student and have them stock the cans in the appropriate manner on the shelf
    - Teaching Tip – Seems too easy, right? Check to see if the student put today's date on the top of the new can. To make it slightly more difficult, choose a shelf that is slightly higher than eye level.

Freezing

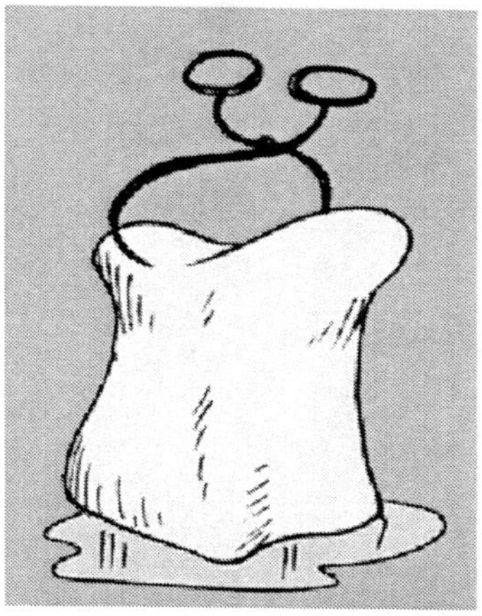

Freezing foods helps to maintain quality and freshness for longer periods of time as long as the food remains solidly frozen.

Adapted from ***Retail Best Practices and Guide to Food Safety and Sanitation* © 2003**

RBP. Pg. 129, 139
RGG. Pg. 109 - 110

31

Slide 31

Freezing foods helps to maintain quality and freshness for longer periods of time as long as the food remains solidly frozen.

- Most disease-causing microorganisms do not grow at temperatures below 41°F (5°C)
- Some spoilage microorganisms can grow at temperatures well below 41°F (5°C).

To properly destroy parasites in raw-marinated and marinated, partially cooked fish by freezing:

Food should be frozen throughout to -4° F (-20°C) and held for 7 days in a freezer,

or

Food should be frozen throughout to -31°F (-35°C) using a blast chiller, and held at that temperature for 15 hours.

Certain species of fish can be served or sold in a raw, raw-marinated or partially cooked ready-to-eat form without freezing. These include certain species of tuna such as:

- Yellowfin
- Bigeye
- Bluefin-Northern
- Bluefin-Southern.

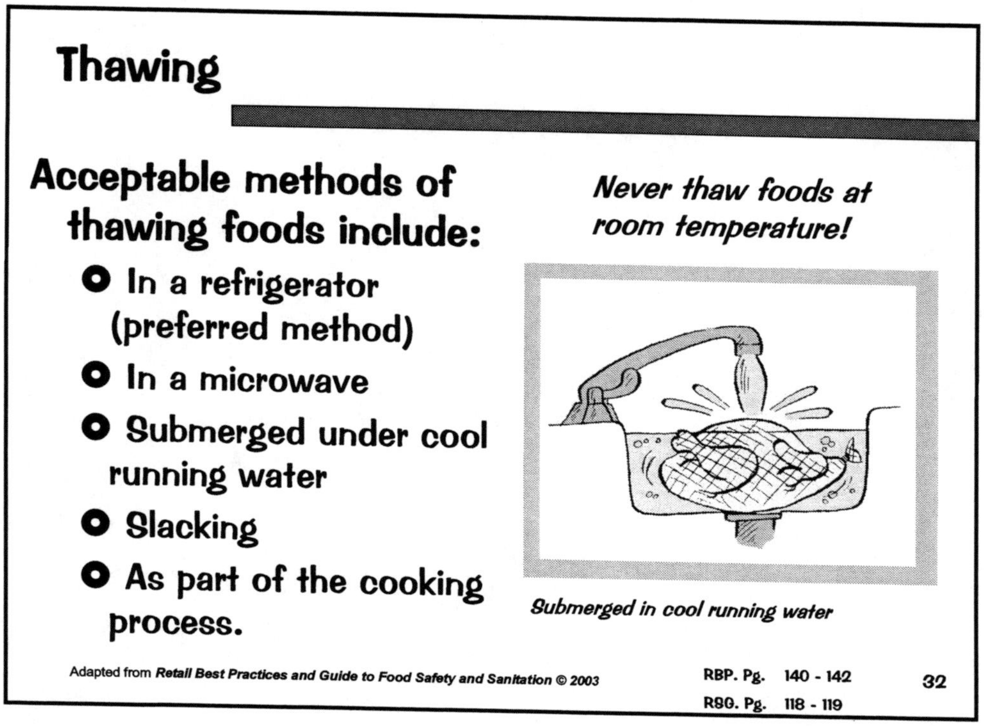

Slide 32

Thawing

Most common, and acceptable thawing methods:

1. In a refrigerator
   - Preferred method because food never enters the temperature danger zone.
2. In a microwave oven
   - Food thawed in a microwave must be immediately cooked.
   - Do not thaw foods in the microwave and then hold them in the refrigerator before cooking.
3. Completely submerged in cool [70°F (21°C)] running water
   - Properly clean and sanitize the sink prior to thawing using this method
   - Thaw raw animal foods in less than 4 hours [this time includes cooking and preparation time or the time to lower the food temperature to 41°F(5°C) under refrigeration].
   - Ready-to-eat foods must never exceed 41°F (5°C).
4. Slacking (defrosting)
   - Gradual increase of food temperature prior to deep fat frying.
   - Allows for even heat penetration during cooking.
   - Maintain food under 25°F (-4°C).
5. As part of the cooking process

**Never thaw foods at room temperature.

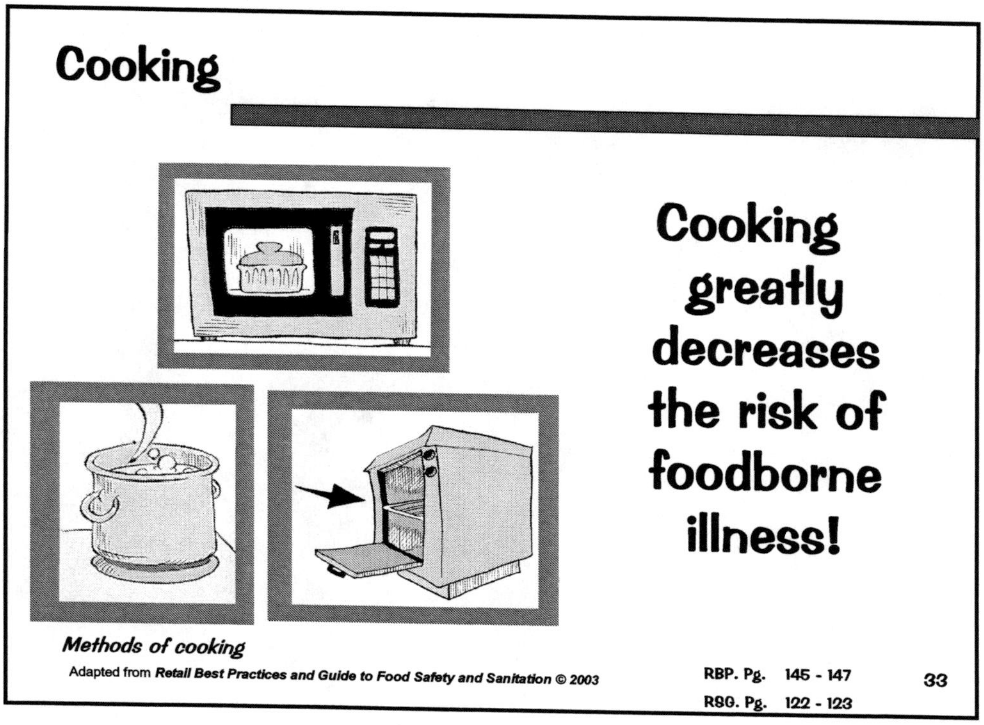

Slide 33

Cooking

The following foods should not be served raw:

- Meat
- Poultry
- Fish
- Seafood
- Eggs
- Unpasteurized milk.

Food establishments that choose to serve these foods raw increase their risk of causing a foodborne illness.

<u>Cooking</u>:

- Makes food more palatable
- Decreases the risk of foodborne illness
- Is a relationship of time and temperature (next slide)
- Requirements for cooking a particular food item should include:
    - A final temperature, and
    - A prescribed length of time it must be held at that temperature.

Most cooking is done with:

- Stoves
- Conventional ovens
- Microwave ovens
    - Heat distribution in a microwave is uneven

- *FDA Food Code* requires raw animal foods cooked in a microwave to reach 165°F (74°C) throughout the food and allowed to stand at this temperature for at least 2 minutes.

Cooking is extremely important for the preparation of safe foods.

Cooking foods to the proper temperature will ensure harmful organisms are destroyed.

Cooking:

Time and Temperature Guidelines

- **Beef roast (rare) –**
  - 130°F (54°C) for 112 minutes or
  - 140°F (60°C) for 12 minutes
- **Eggs, beef/pork, fish –**
  - 145°F (63°C) for 15 seconds
- **Ground beef, ground pork, and ground game animals –**
  - 155°F (68°C) for 15 seconds
- **Beef roast (medium), pork roast, ham –**
  - 145°F (63°C) for 4 minutes
- **Poultry, stuffed meats –**
  - 165°F (74°C) for 15 seconds.

Adapted from ***Retail Best Practices and Guide to Food Safety and Sanitation*** © 2003

RBP. Pg. 145 – 147
RSG. Pg. 122 - 123

34

Slide 34

Cooking

To destroy harmful microorganisms, potentially hazardous foods must be heated to minimum temperatures for a minimum amount of time:

- Beef roast (rare) – 130°F (54°C) for 112 minutes or 140°F (60°C) for 12 minutes
- Eggs, beef/pork, fish – 145°F (63°C) for 15 seconds
- Ground beef, ground pork, and ground game animals – 155°F (68°C) for 15 seconds
- Beef roast (medium), pork roast, and ham – 145°F (63°C) for 4 minutes
- Poultry, stuffed meats – 165°F (74°C) for 15 seconds.

Foods, such as casseroles, that contain more than one type animal food must have a final cooking temperature that coincides with the highest risk food.

Cooling

To ensure safe food, hot foods must be:

- **Cooled from 135°F (57°C) to 70°F (21°C) or below within 2 hours**

and

- **From 135°F (57°C) to 41°F (5°C) within six hours.**

Proper cooling techniques

Adapted from ***Retail Best Practices and Guide to Food Safety and Sanitation* © 2003**

RBP. Pg. 147 – 150
RSG. Pg. 123 - 124

35

Slide 35

Cooling

Improper cooling is one of the leading contributors to foodborne illness in retail food establishments.

FDA Food Code recommends after proper cooking, potentially hazardous foods not used for immediate service or hot display, be cooled:

- From 135°F (57°C) to 70°F (21°C) within 2 hours, and
- From 135°F (57°C) to 41°F (5°C) within 6 hours or less.

Potentially hazardous foods prepared from ingredients that have been held at room temperature (i.e., canned tuna) must be cooled from 135°F (57°C) to 41°F (5°C) within 4 hours or less.

Some methods to ensure rapid cooling are:

- Blast chillers
- Containers that facilitate heat transfer (stainless steel)
- Transferring foods into smaller containers
- Cooling foods at a depth of 3 inches or less
- Cutting foods into smaller portions
- Stirring food while cooling
- Utilizing an ice bath
- Utilizing cooling paddles
- Adding ice directly to food as an ingredient.

Reheating

For quality and safety reasons, reheat foods only once.

To reheat foods quickly:

- Reheat in small quantities
- Use preheated ingredients
- Stir foods frequently.

Adapted from *Retail Best Practices and Guide to Food Safety and Sanitation* © 2003

RBP. Pg. 150 – 151
RSG. Pg. 125 - 126

36

Slide 36

Reheating

All potentially hazardous foods meant to be held hot, which have been cooked and then cooled, must be reheated to at least 165°F (74°C) within two hours.

Ready-to-eat foods that are commercially prepared and packaged and from a food processing plant under regulatory inspection should be free of harmful microorganisms.

- Therefore, these foods (i.e., cooked hot dogs) must only be reheated to a temperature of at least 135°F (57°C) for hot-holding [rather than a minimum of 165°F (74°F)].

To reheat **leftovers** (food that has been prepared and held over for service at a future meal period) quickly:

- Reheat in small quantities
- Use preheated ingredients
- Stir foods frequently.

It is good practice to reheat foods only once, and then discard any uneaten product.

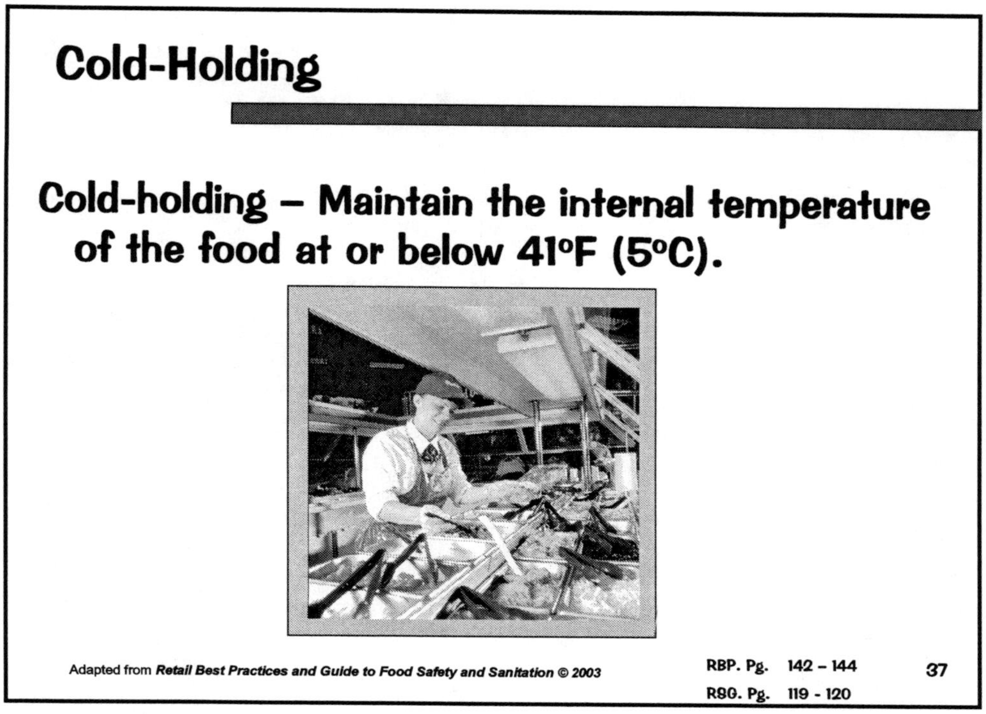

Slide 37

When holding cold foods for service:

- Maintain the internal temperature of the food at or below 41°F (5°C).

Proper holding temperatures slow down, or prevent, the growth of harmful microorganisms.

Refrigerator or display cases are typically set 3°F (2°C) to 5°F (3°C) below the desired cold-holding temperature.

The *FDA Food Code* recommends cold, ready-to-eat, potentially hazardous foods that are prepared and held refrigerated for more than 24 hours:

- Be clearly marked at the time of preparation to indicate "sell by" or "use by" for the customer
- Can be held for up to seven calendar days when held at or below 41°F (5°C) during that time.

Ready-to-eat salads such as chicken salad or seafood salad must be prepared in a way that will help prevent temperature abuse.

- Pre-chill all ingredients (in a refrigerator) before mixing and storing.

Frozen, ready-to-eat, potentially hazardous foods should be sold within 24 hours after thawing whenever possible.

When large amounts of food are removed from the freezer:

- It should be marked to indicate the date by which the food must be sold.

Food removed from the freezer must be used within seven calendar days or less minus the time before freezing that food is held refrigerated, if the food is maintained at 41°F (5°C) or less before freezing.

Hot-Holding

Hot-holding – Maintain the internal temperature of the food at or above 135°F (57°C).

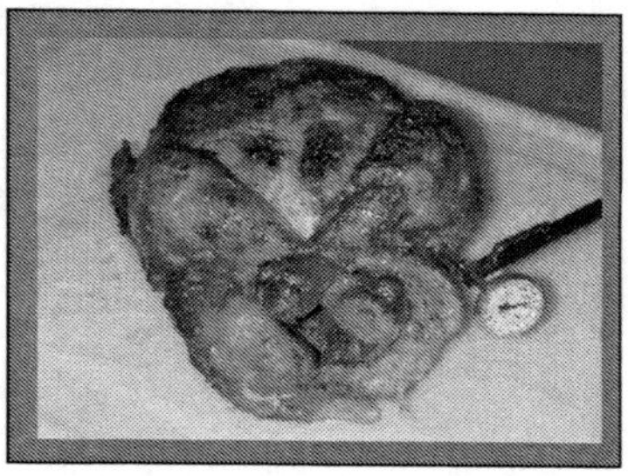

Adapted from ***Retail Best Practices and Guide to Food Safety and Sanitation* © 2003**

RBP. Pg. 150 – 151
RGG. Pg. 125 - 126

38

Slide 38

When holding hot foods for service:

- Maintain the internal temperature of the food at or above 135°F (57°C).

Proper holding temperatures slow down, or prevent, the growth of harmful microorganisms.

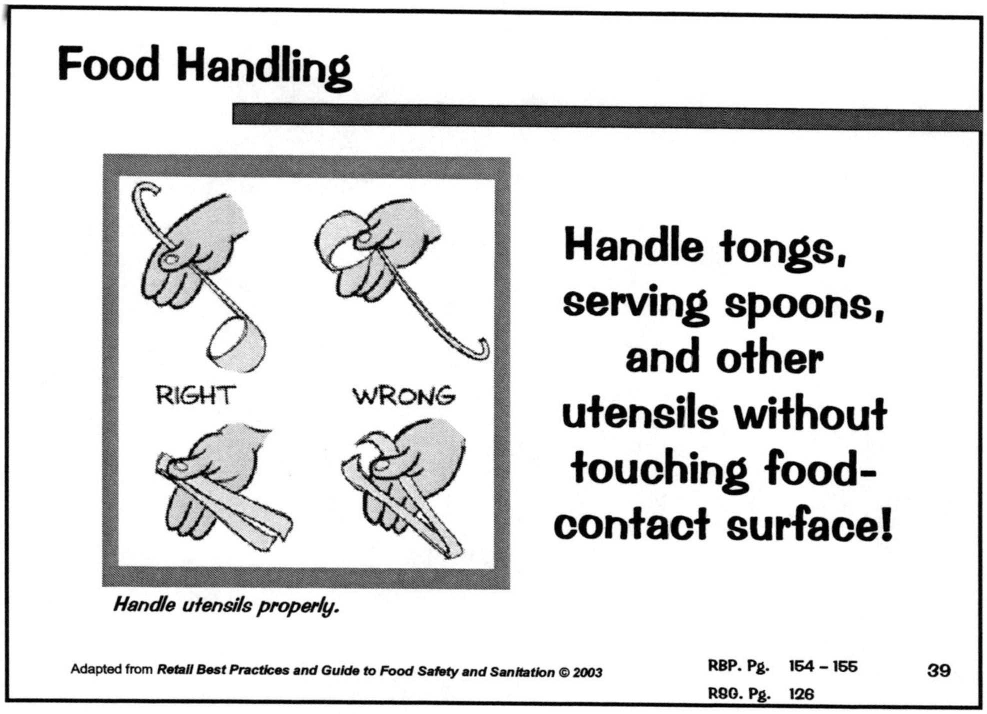

Slide 39

Keeping food safe during food handling and display includes:

- Practice good personal hygiene
    - Wear a clean uniform
    - Wear a hair restraint
    - Wash hands frequently
    - Never dry your hands on your apron
- Food handlers in all departments should avoid touching food with their bare hands
- Hold utensils by their handles only
- Use tongs, serving spoons, disposable gloves or deli tissue to avoid bare hand contact
- Use a single utensil for each food item (stored in the food item between uses)
- Store utensils in a way that allows the employee to grab the handle without touching the food or food-contact surface.

Food Sold or Received

Food that has been served or sold to, and is in the possession of, a customer may not be returned for service or sale with few exceptions

Adapted from ***Retail Best Practices and Guide to Food Safety and Sanitation* © 2003**

RBP. Pg. 155 – 156
RSG. Pg. 127

40

Slide 40

Food that has been served or sold to, and is in the possession of, a customer may not be returned for service or sale with the exception of:

- A container of nonpotentially hazardous food that is dispensed in a way that protects the food from contamination and the container is closed between uses (i.e., a narrow-neck bottle of catsup)
- A nonpotentially hazardous food such as crackers, salt, or pepper that is in an unopened, original package and is maintained in a sound condition.

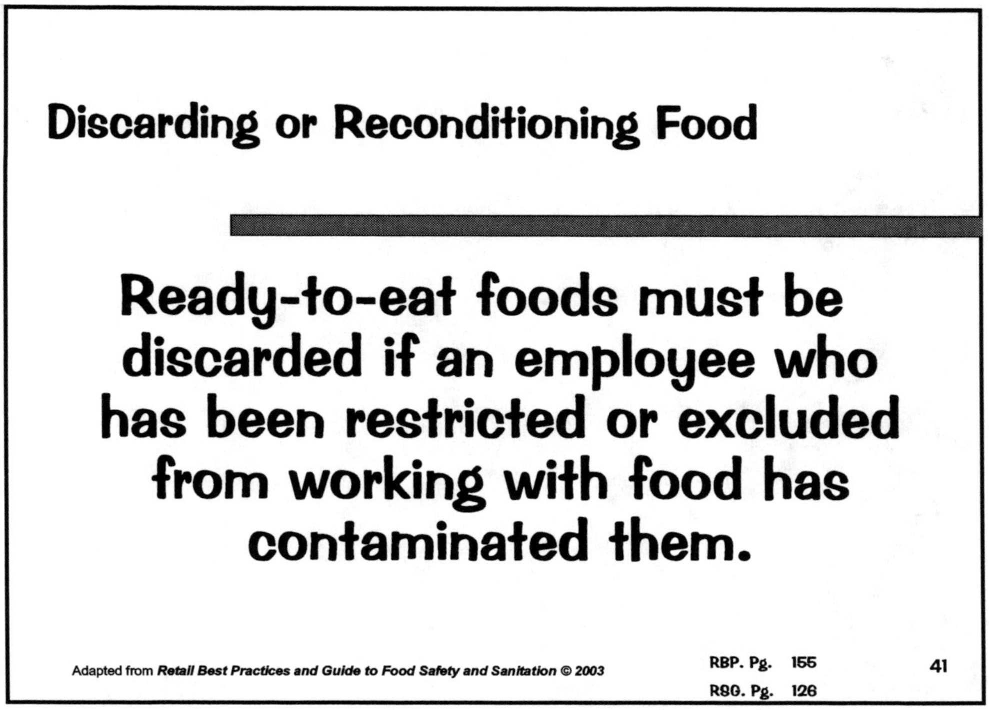

Slide 41

Discarding or reconditioning food:

- A food that is unsafe, adulterated or not honestly presented shall be reworked or reconditioned using a procedure that has been approved by the regulatory authority in the jurisdiction, or it must be discarded
- Food must be discarded if it is not from an approved source or has been contaminated by food employees, consumers, or other persons via soiled hands, bodily discharges (i.e., coughs and sneezes), or other means
- Ready-to-eat foods must be discarded if an employee who has been restricted or excluded from working with food has contaminated them.

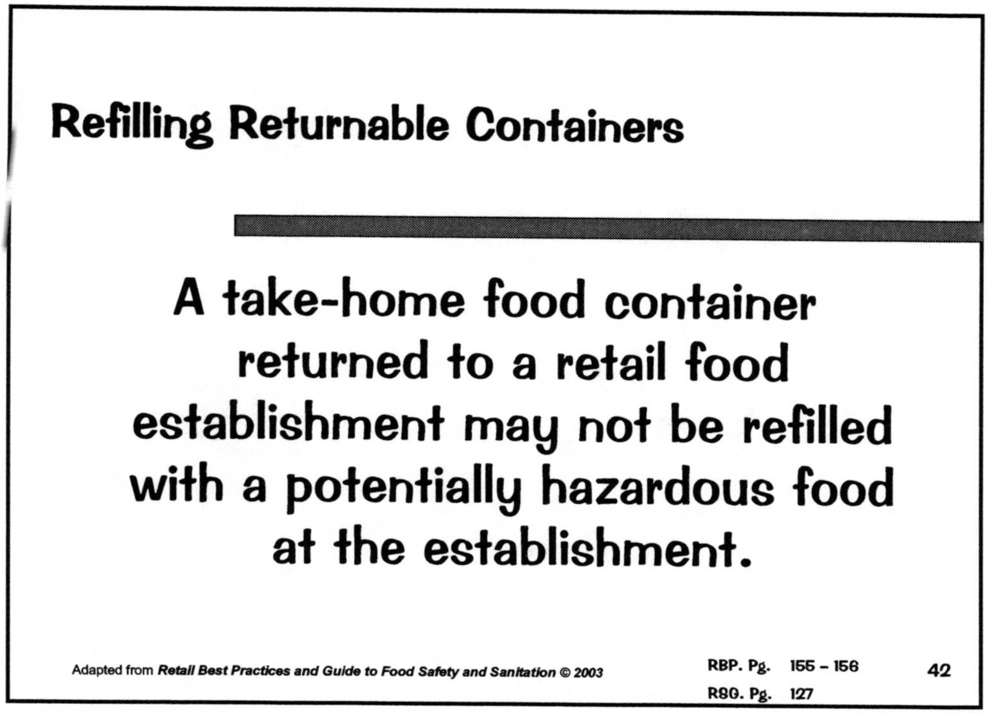

Slide 42

Refilling returnable containers

A take-home food container returned to a retail food establishment may not be refilled with a potentially hazardous food at the establishment. A food-specific container for beverages may be refilled at a food establishment if:

- Only a beverage that is not a potentially hazardous food is dispensed into the container,
- The design of the container and of the rinsing equipment and the nature of the beverage, when considered together, allow effective cleaning of the container at home or in the retail food establishment,
- Facilities for rinsing before refilling returned containers with fresh, hot water that is under pressure and not recirculated are provided as part of the dispensing system,
- The consumer-owned container returned to the retail food establishment for refilling is refilled for sale or service only to the same consumer, and
- An employee of the retail food establishment or the owner of the container refills the container if the beverage system includes a contamination-free transfer process that cannot be bypassed by the container owner.

Personal take-out beverage containers, such as thermally insulated bottles, non-spill coffee cups, and promotional beverage glasses, may be refilled by employees or the consumer if the refilling process will protect the food and food-contact surface of the container from contamination.

Slide 43

Self-service bars can be hot-holding, cold-holding, or a combination of both.

- Maintain the internal temperature of hot foods at or above 135°F (57°C)
- Maintain the internal temperature of cold foods at or below 41°F (5°C).

Self-service bars best practices include:

- Assign properly trained personnel to this area
- Monitor temperatures frequently
- Sneeze guards – keep foods from being contaminated by customers
- Clean and sanitize utensils prior to use
- Store the serving portion of utensils completely in the food when not being used.

Never place oysters and other types of raw animal foods on a self-service bar except for ready-to-eat foods like sushi.

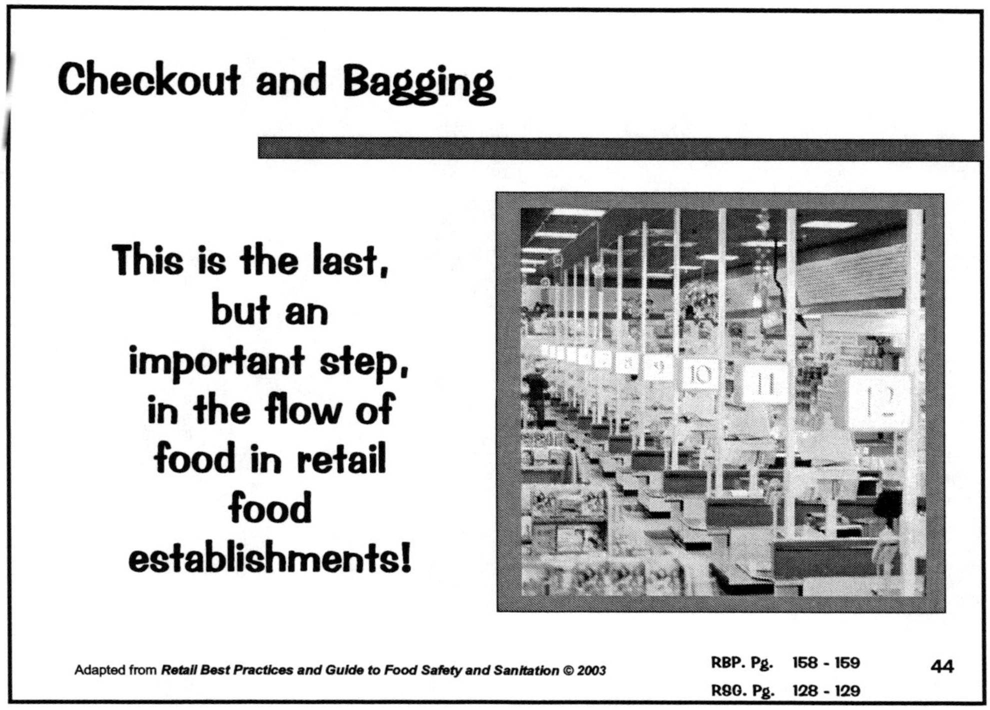

Slide 44

Checkout and bagging
This is the last step in the flow of food within retail food establishments.

Checkers and baggers must work carefully to prevent contamination and cross contamination:

- They must practice good personal hygiene
- They must not allow their hands to transfer contaminants from raw foods to cooked and ready-to-eat foods
- They must wipe juices from raw foods off conveyor belts consistently
- They must wash their hands often
- They must not put raw meat, poultry, and fish in the same bags with fresh fruits and vegetables
- They must not put non-food items such as detergents, pesticides, and light bulbs, in with food items
- They should use freezer bags to separate frozen goods.

Some customers may decide not to purchase a potentially hazardous food while they are at the checkout counter. These products must be returned to proper storage or display environment as quickly as possible.

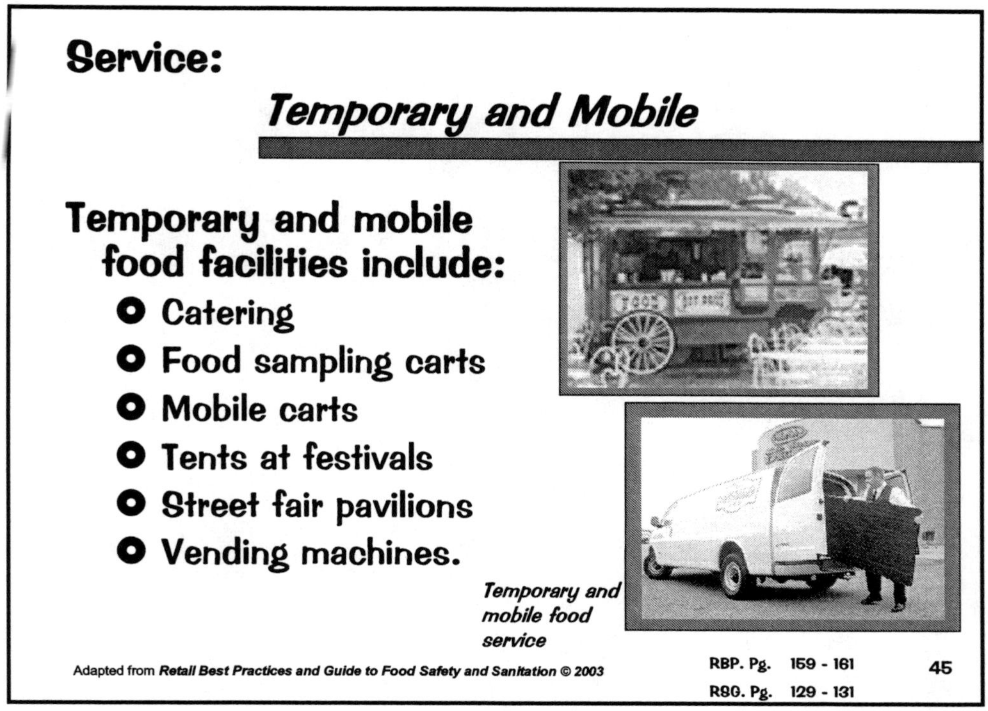

Slide 45

A **temporary food establishment** (TFE) operates for a period of no more than 14 consecutive days in conjunction with a single event.

- They can operate indoors or outdoors
- They often have limited physical and sanitary facilities available.

A variety of foods can be prepared and served from temporary stands including:

- Sandwiches
- Pizza
- Barbecue ribs
- Gyros
- Corn on the cob
- Beverages
- Confections (i.e. cotton candy, funnel cakes, etc.).

Mobile facilities are trucks and trailers used to cater events located away from the establishment.

- The extent of items that can be offered is practically unlimited.

At both temporary food establishments and mobile facilities:

- The same food safety practices must be in place as in other areas of the establishment.
- Cooking equipment must be capable of heating foods to required internal temperatures in appropriate amount of time.
- Cooling equipment must be capable of keeping foods below 41°F (5°C).
- Must have a source of hot and cold potable water for:
    - Hand washing
    - Cleaning
    - Sanitizing.

- Use disposable utensils whenever possible. If nondisposable utensils must be used the following four-step process for cleaning and sanitizing must be followed:
    - Wash in hot water with a detergent
    - Rinse in clean water
    - Chemically sanitize
    - Air-dry.
- Garbage and paper wastes should be placed in containers that are lined with plastic bags and equipped with tight-fitting lids.

Vending Machines:
A vending machine is a self-service device that dispenses individual sized servings of food and beverages.

Vending machines dispense food and beverages in bulk or in individually wrapped packages in the following forms:

- Potentially hazardous foods
- Ready-to-eat foods
- Nonpotentially hazardous foods.

Vending machines that store and dispense potentially hazardous foods must have

- Adequate refrigeration and/or heating units
- Insulation
- Controls to keep cold foods cold and hot foods hot
- An automatic control that prevents the machine from vending food if there is a power failure, mechanical failure or other condition that prevents cold food from being maintained at 41°F (5°C) or below and hot food maintained at 135°F (57°C) or above.

The temperature cut-off requirement does not apply:

- In a cold food machine during a period not to exceed 30 minutes immediately after the machine is filled, serviced or restocked; or
- In a hot food machine during a period not to exceed 120 minutes immediately after the machine is filled, serviced or restocked.

Refrigerated, ready-to-eat, potentially hazardous foods prepared in a retail food establishment and dispensed through a vending machine (with an automatic shutoff control) shall be discarded if it a) exceeds the time and temperature combinations prescribed by the *FDA Food Code* or b) is not correctly date labeled.

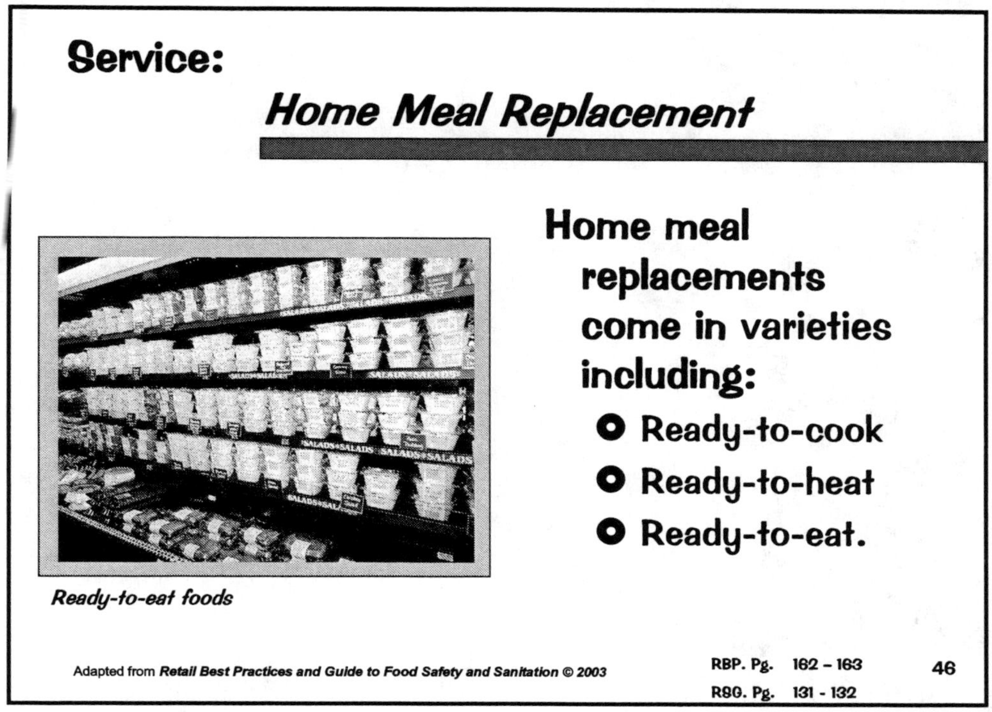

Slide 46

Home meal replacement and meal solutions

- Terms that refer to high-quality meals prepared away from home that are eaten at home.

Home meal replacements come in varieties such as:

- Ready-to-cook (stuffed green peppers)
- Ready-to-heat (fully cooked ham or turkey)
- Ready-to-eat (rotisserie chicken).

Home meal replacements safety guidelines:

- Follow all requirements throughout the flow of food
- Keep foods out of the temperature danger zone while on display and during delivery
- Label home meal replacements so the customer knows how to keep the product safe at home
- Retailers must document procedures that were followed to keep the food safe.

Slide 47

Safety Stop! 4.5

Sample Test Questions

1. The final cooking temperature for poultry is:
   A. 145°F (63°C).
   B. 155°F (68°C).
   C. 165°F (74°C).
   D. 130°F (54°C).

2. All of the following are acceptable cooling methods except:
   A. Cutting large foods into smaller portions.
   B. Setting out at room temperature.
   C. Using an ice bath.
   D. Using ice as a final ingredient.

3. When reheating potentially hazardous foods for hot-holding, you should ensure it reaches:
   A. 145°F (63°C) within 4 hours.
   B. 155°F (68°C) within 3 hours.
   C. 165°F (74°C) within 2 hours.
   D. 180°F (82°C) within 4 hours.

4. Proper hot- and cold-holding temperatures:
   A. Kill harmful microorganisms.
   B. Increase the number of harmful microorganisms.
   C. Reduce the number of harmful microorganisms to safe levels.
   D. Prevent the growth of harmful microorganisms.

5. During checkout and bagging, baggers or cashiers should:
   A. Wipe juices from raw foods off conveyor belts.
   B. Wear open-toed shoes.
   C. Take the temperature of potentially hazardous foods.
   D. Place raw meats and fish in the same bag.

6. A food establishment that operates for a period of no more than 14 consecutive days in conjunction with a single event is called a:
   A. Retail food establishment.
   B. Mobile food establishment.
   C. Temporary food establishment.
   D. Home meal replacement.

7. An example of a home meal replacement in ready-to-eat form is:
   A. Beef kebobs.
   B. Rotisserie chicken.
   C. Frozen pizza.
   D. Stuffed peppers.

Real-life

Slide 34
Cooking temperatures
Leah was preparing a casserole that contained both meat and poultry for the hot line in the deli. She cooked the casserole to 145°F (63°C), ensuring it was at that temperature for at least 4 minutes. What was Leah's mistake?

Leah needed to cook the casserole to 165°F (74°C) and ensure it was held at, or above that temperature for at least 15 seconds because it contained both meat and poultry. Since poultry is a higher risk food, she would need to ensure it was cooked to a high enough temperature to kill the microbes associated with poultry.

Slide 35
Cooling
Levina finished cooking a 5-gallon batch of chili and needed to cool it. She separated the batch into five 1-gallon batches. She left them on the prep table while she took her hour lunch break. She returned to them, covered them, and put them in the cooler. What was her mistake?

Levina should not have allowed the hot chili to sit out at room temperature. She should have ensured the chili cooled from 135°F (57°C) to 70°F (21°C) within two hours and from 135°F (57°C) to 41°F (5°C) within six hours.

Slide 36
Reheating
At 8 a.m. Karl started setting up the soup well for the day. He took the previous night's soups, placed them in soup tureens, and put them in the hot well. At 10:30 a.m. Karl did a temperature check on the hot foods and the soup was at 145°F (63°C). Since that temperature met his hot-holding requirements, he put ladles in them and moved on to another task. What was Karl's mistake?

Karl should have rapidly reheated the soups to 165°F (74°C) within two hours instead of allowing them to slowly warm. Additionally, the temperature of the reheated soup never reached the temperature needed to kill all the harmful organisms.

Making it Real

Utensil Handling – Slide 39

- The proper handling of utensils is a very important concept. Hands are a primary source of food contamination and are a means of transferring harmful microorganisms.
    - Supplies
        - 4 to 6 utensils of choice
        - Half hotel pan to simulate food being held hot
    - Procedure
        - Lay the utensils out on a table
        - Have a student show you the proper way to handle the utensils
        - Have the student explain the proper storage of the utensil
            - When it is being stored after it is clean
            - When it is being used as a service utensil
    - Teaching Tip – Be sure the class understands the importance of storing the utensil completely in the hot food so bacteria will not be allowed to grow.

Bagging – Slide 44

- Bagging foods at the checkout stand is the final step in the flow of food within retail food establishments. We cannot expect our customers to understand the full importance of bagging products appropriately to keep them safe.
    - Supplies
        - Variety of foods
            - Meat
            - Poultry
            - Fish
            - Vegetables
            - Ready-to-eat products
            - Dry goods
            - Frozen goods
        - Plastic or paper bags
        - Freezer bags
    - Procedure
        - If your facilities allow for the holding of a wide variety of foods this exercise will show your class the importance of proper bagging. Have the learners bag the products appropriately according to your procedures.
        - Question the learners on specific reasons why they are bagging different foods into different bags.
    - Teaching Tip – Some class members may never bag food products. These are the best ones to choose to perform the task. It will help them understand the importance of all food handlers, including checkout employees and baggers, in the safety of food. This exercise can be used to reinforce proper cooking time and temperatures, proper storage and receiving, and the different manner in which various foods need to be handled so they remain safe. This is an excellent exercise to review key points that were made throughout the session.

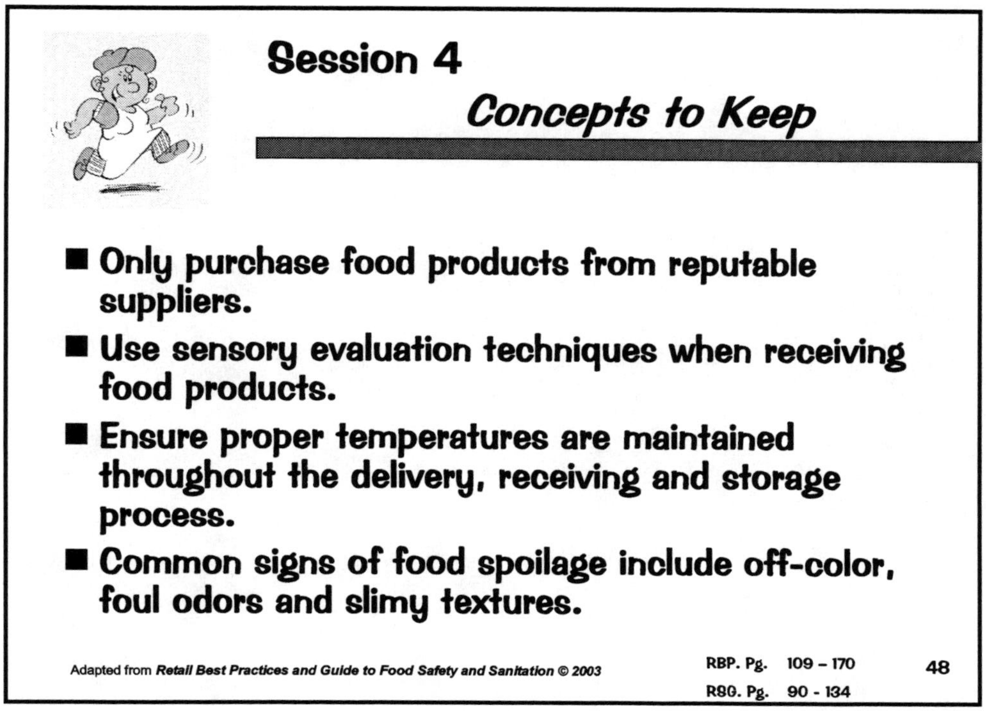

Slide 48

Concepts to Keep

- Only purchase food products from reputable suppliers.
- Use sensory evaluation techniques when receiving food products.
- Ensure proper temperatures are maintained throughout the delivery, receiving and storage process.
- Common signs of food spoilage include off-color, foul odors and slimy textures.

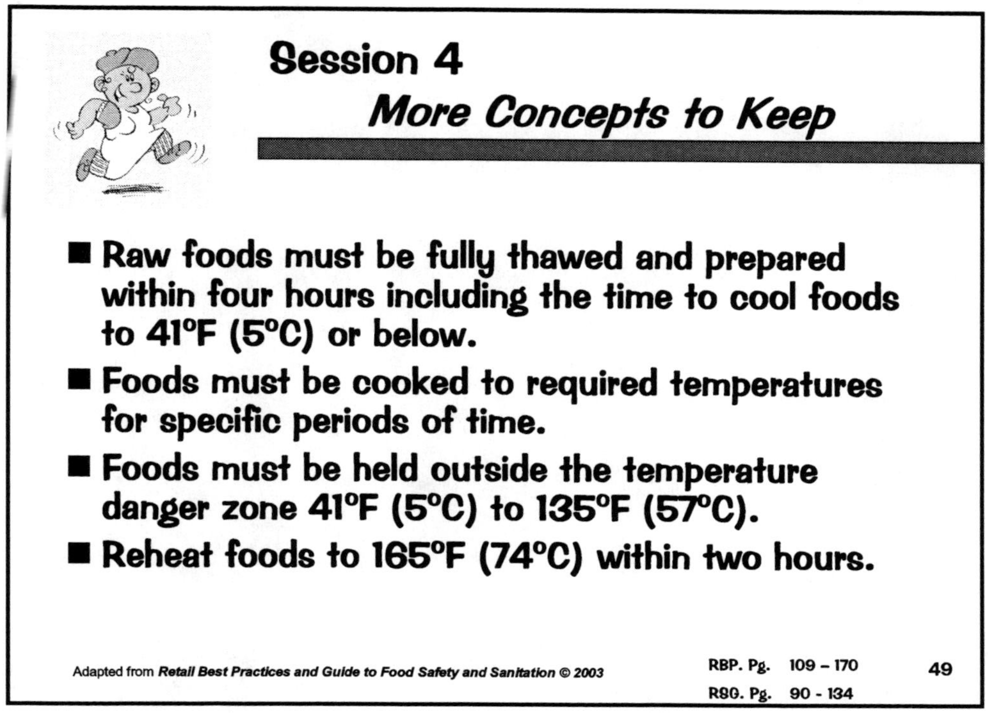

Slide 49

More Concepts to Keep

- Foods must be fully thawed, including the preparation process, within four hours.
- Foods must be cooked to required temperatures for specific periods of time.
- Foods must be held outside the temperature danger zone 41°F (5°C) to 135°F (57°C).
- Reheat foods to 165°F (74°C) within two hours.

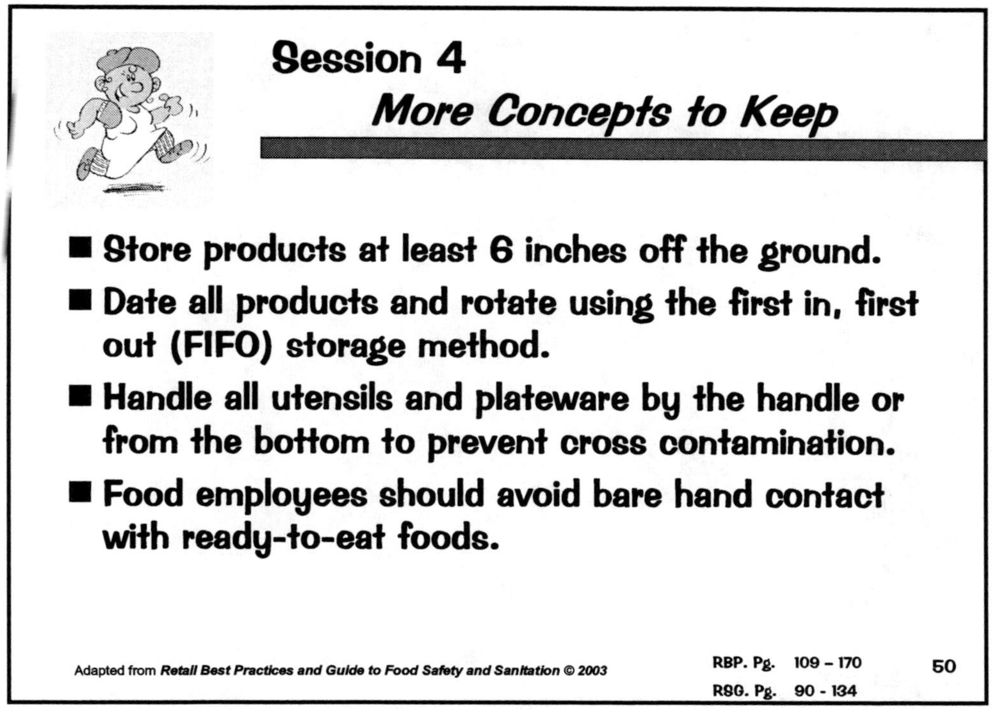

Slide 50

More Concepts to Keep

- Store food products at least 6 inches off the ground.
- Date all products and rotate using the first in, first out storage method.
- Handle all utensils and plateware by the handle or from the bottom to prevent cross contamination.
- Food employees shall avoid bare contact with ready-to-eat foods.

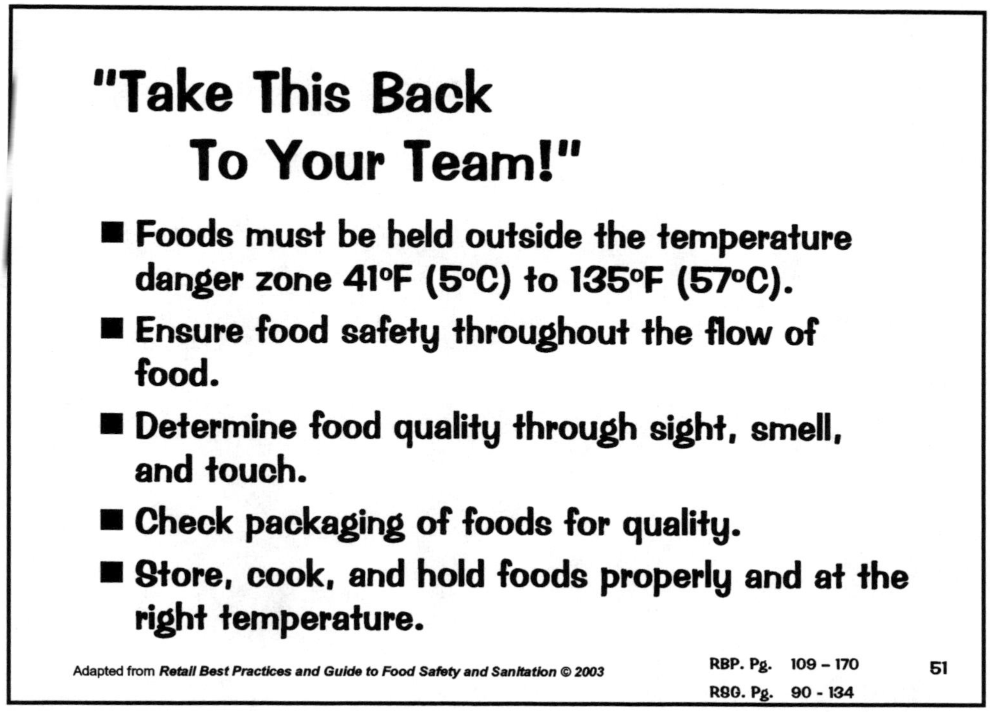

Slide 51

<u>**Take This Back to Your Team!**</u>

- Foods must be held outside the temperature danger zone 41°F (5°C) to 135°F (57°C).
- Ensure food safety throughout the flow of food.
- Determine food quality through sight, smell, and touch.
- Check packaging of foods for quality.
- Store, cook, and hold foods properly and at the right temperature.

Session 5

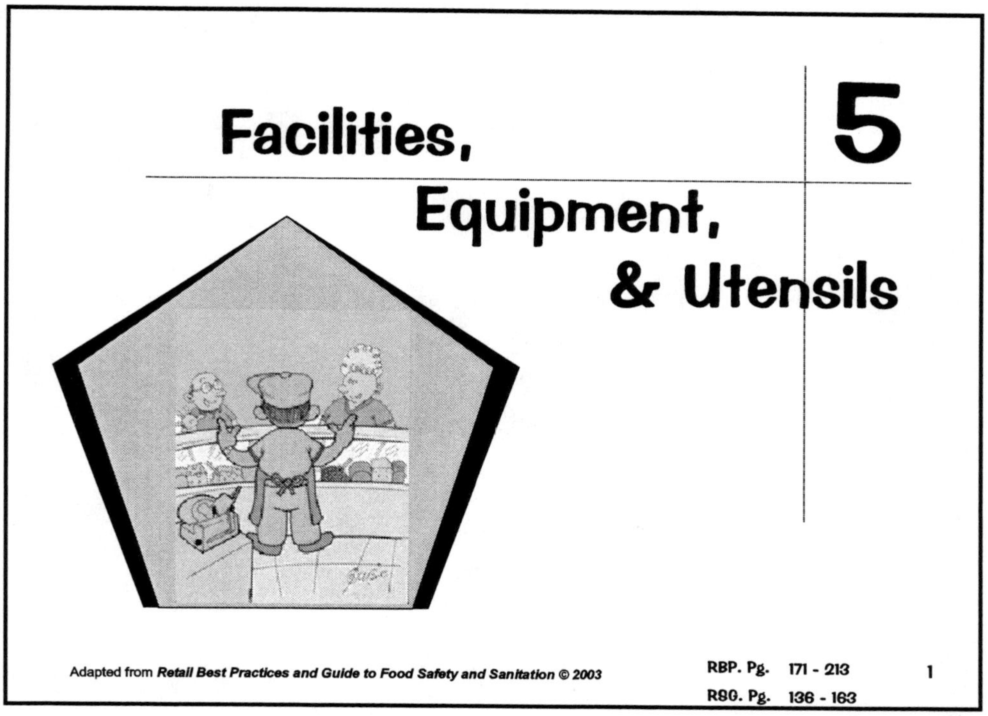

Facilities, Equipment and Utensils – Session 5 Key Points

Slide 1

Training Tip:

The content of each slide provides the most essential information in the session. The written text below each slide is provided to help you understand and better teach the essential information. The written text may also be used to answer questions that may be posed during the session. Depending on the length of your training session, you may not be able to address all the information contained beneath each slide.

Overview:

Retail food establishment managers may or may not be directly involved in the selection, purchasing and installation of equipment. However, it is important to understand the criteria involved in the process. Session 5 will review:

- Proper facilities planning to allow for maximum productivity and efficiency
- Facility design that allows for smooth flow of work
- Factors involved in equipment selection, purchasing and installation
- Materials used to manufacture equipment and utensils, and how they can affect food safety
- Lighting and HVAC regulations and requirements.

Get your walking shoes on for this session. The best way to understand facilities design is to see the facility.

Suggested Timeline:

8-hour course: 30 minutes using the PowerPoint© slides only.

- 22 ***Slides.***

16-hour course: 1 hour using the 36 PowerPoint© slides and the 3 ***Safety Stop!*** slides containing:

- 19 ***Sample Test Questions***
- 8 ***Real-life*** scenarios
- 3 ***Making it Real*** interactive exercises.

Video suggestions and supplemental resources:

Videos
Invisible Challenge, Food Marketing Institute, 1-301-843-3084
Food Safety Success, Food Marketing Institute, 1-301-843-3084

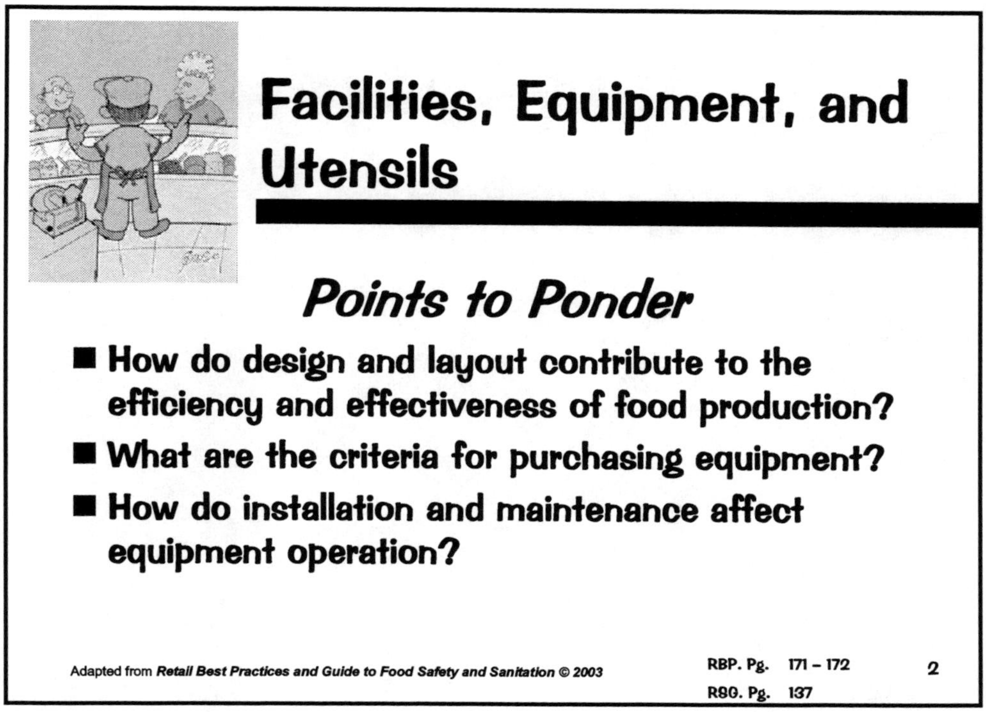

Slide 2

Points to Ponder

- How do design and layout contribute to the efficiency and effectiveness of food production?

- What are the criteria for purchasing equipment?

- How do installation and maintenance affect equipment operation?

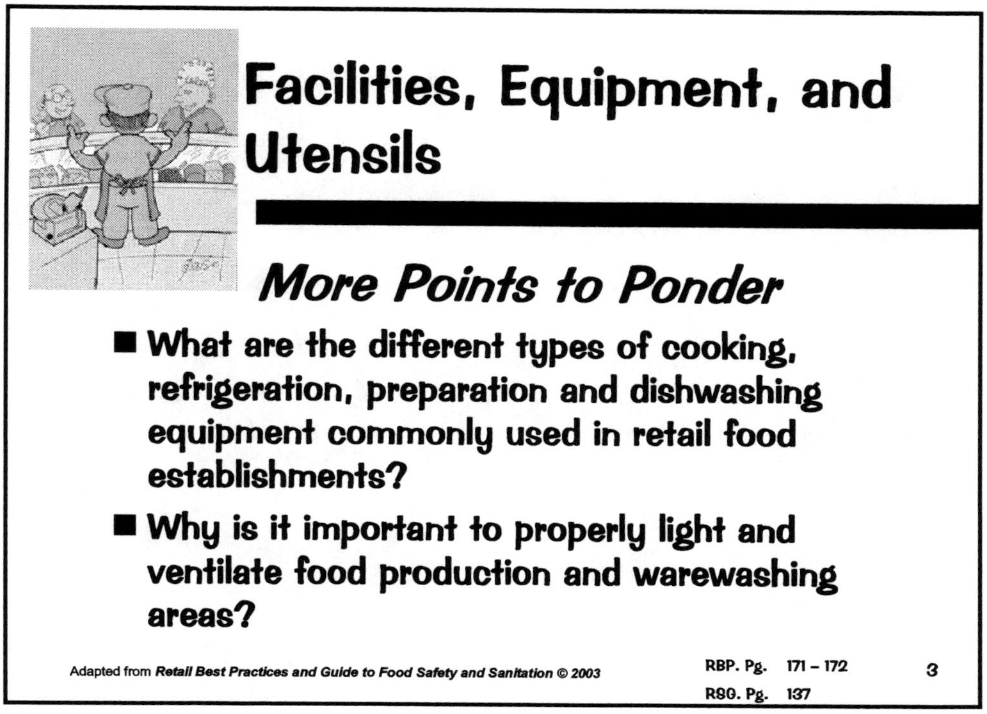

Slide 3

More Points to Ponder

- What are the different types of cooking, refrigeration, preparation and dishwashing equipment commonly used in retail food establishments?

- Why is it important to properly light and ventilate food production and warewashing areas?

Facilities, Design, and Layout

Retail food establishment facilities may include areas for:

- Receiving and delivery
- Storage
- Preparation and holding
- Service
- Warewashing
- Refuse storage and pickup
- Food display area or dining room
- Housekeeping
- Toilets.

Adapted from ***Retail Best Practices and Guide to Food Safety and Sanitation* © 2003**

RBP. Pg. 172 - 173
RSG. Pg. 138 - 139

4

Slide 4

Well-designed facilities:

- Are based on the types of foods being sold
- Anticipate future technological advances and needs
- Enable the use of multiple sources of energy (gas and electric)
- Include input from architects, engineers, regulators, and end users
- Consider each function within the facility
- Create a usable flow throughout the facility.

Well-designed facilities will have a positive influence on:

- Employee safety and productivity
- Labor and energy costs
- Customer satisfaction.

General areas to consider when designing a retail food establishment include:

- Delivery and receiving
- Storage
- Preparation
- Hot- and cold-holding
- Display
- Handling and service
- Warewashing, cleaning and sanitizing
- Waste storage and pickup
- Food display area

- Housekeeping
- Toilet facilities.

All facilities must comply with codes related to:

- Health
- Safety
- Building
- Fire
- Zoning
- Environment.

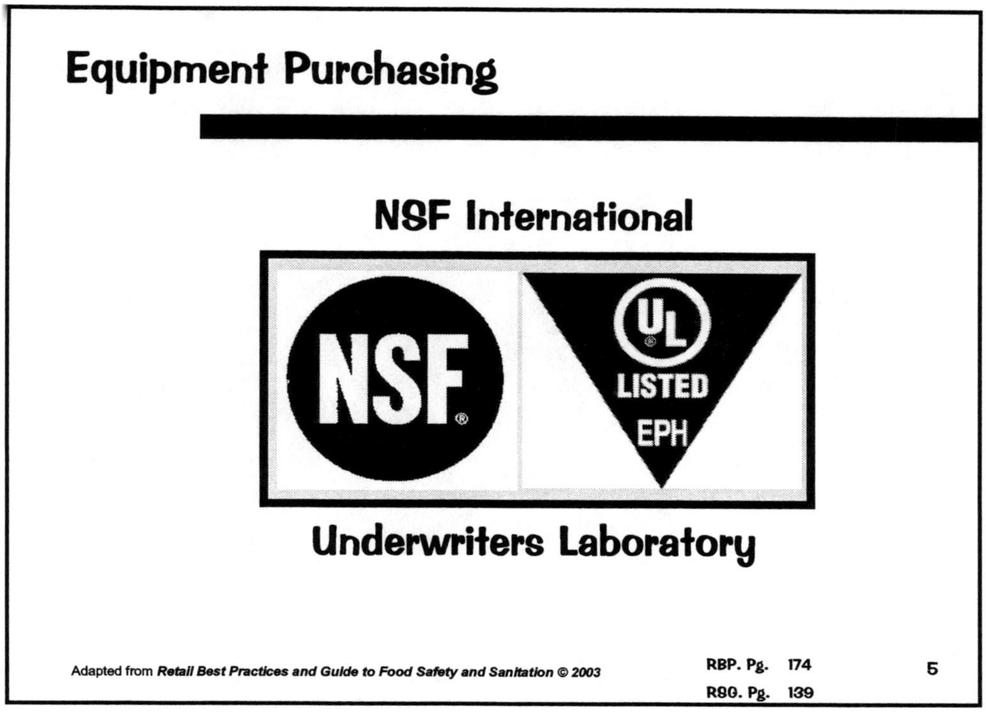

Slide 5

Equipment includes appliances and storage containers (i.e., refrigerators) used in food establishments.

- Equipment used in retail food establishments:
  - Should meet American National Standards Institute (ANSI) standards
  - Should bear a stamp of approval from recognized third-party certification organizations including:
    - NSF International (NSF)
    - Underwriters Laboratories, Inc. (UL)
    - American Gas Association (AGA)
- Equipment evaluated and approved by these third-party organizations will:
  - Assure quality materials were used in construction
  - Assure they were designed and constructed to meet accepted food sanitation criteria.

Equipment installation must conform to local codes and will be monitored by:

- The local health department
- Building officials
- Fire protection agencies.

Utensils are tools or containers that contact food, such as knives, spoons, forks, and tongs.

Work Center Planning

Well-designed work center

A properly designed work center provides adequate facilities for:

- Efficient production
- Fast service
- A pleasant environment
- Effective cleanup.

Adapted from ***Retail Best Practices and Guide to Food Safety and Sanitation* © 2003**

RBP. Pg. 174 – 175
RGG. Pg. 140

6

Slide 6

<u>Work center</u> – Areas where an individual or group of individuals perform common tasks.

When planning work centers:

- Consider the total space needed
- Consider arrangement of equipment in that space.

A properly designed work center provides adequate facilities and space for:

- Efficient production
- Fast handling and service
- A pleasant environment
- Effective cleanup.

Work centers should:

- Allow employees to complete tasks efficiently
- Be large enough to do the job yet compact enough to conserve energy and time.

Equipment Selection

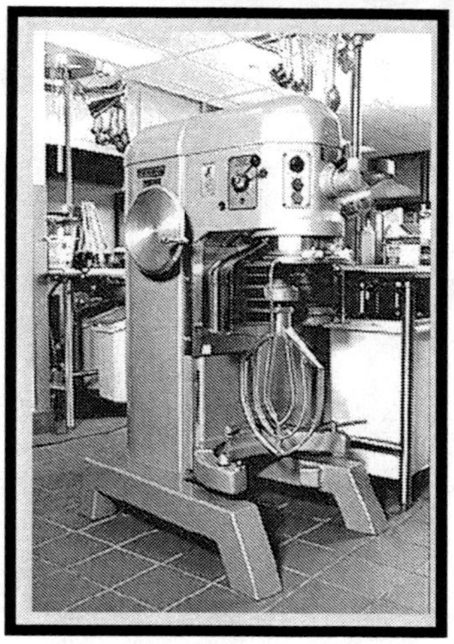

Equipment features include:

- Design
- Construction
- Durability
- Ability to clean easily
- Size
- Cost
- Safety
- Ability to do the job.

Review equipment features prior to purchasing.

Adapted from ***Retail Best Practices and Guide to Food Safety and Sanitation* © 2003**

RBP. Pg. 175 - 178

RSG. Pg. 140 - 141

7

Slide 7

Factors to consider when purchasing equipment include:

- Needs of the establishment
  - Will the equipment:
    - Improve the quality of food
    - Reduce costs
    - Improve sanitation
    - Contribute to the bottom line of the establishment
  - Prioritize the establishment's equipment needs
- Size and design
  - Legs of floor equipment should provide a minimum of 6-inch clearance so the floor is easier to keep clean and it discourages pest infestation
  - Legs of table-mounted equipment should provide a minimum of 4-inch clearance for ease of cleaning under and around the equipment
  - Equipment that is not **easily moveable** (on wheels) or on legs should be sealed to the table or floor
  - Display shelving units, display refrigeration units and display freezer units located in customer areas do not have to be elevated or sealed to the floor, however, the floor under these units must be kept clean and sanitary
- Construction
- Ability to do the job
- Durability
- Ability to clean easily
- Safety
- Cost.

Major costs associated with purchase of equipment are:

- Purchase price
- Installation cost
- Operating cost
- Maintenance cost
- Finance charges.

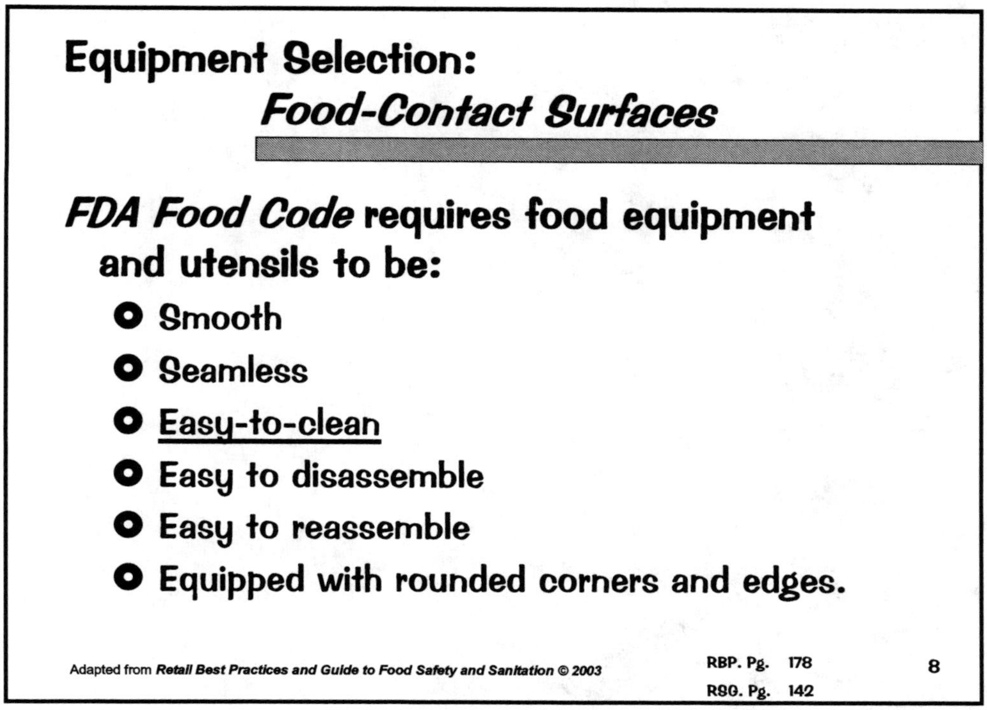

Slide 8

Food-contact surfaces are surfaces of equipment with which food is likely to come into contact.

Non-food-contact surfaces are the parts of the equipment and surrounding area with which food is not intended to come into contact.

These surfaces may contact food through:

- Touch
- Drain
- Drip
- Splash and spill making it a **splash-contact surface**.

FDA Food Code requires food equipment and utensils to be:

- **Smooth** (free of irregularities)
- Seamless
- **Easy-to-clean** (materials which facilitate cleaning)
- Easy to disassemble
- Easy to reassemble
- Equipped with rounded corners and edges.

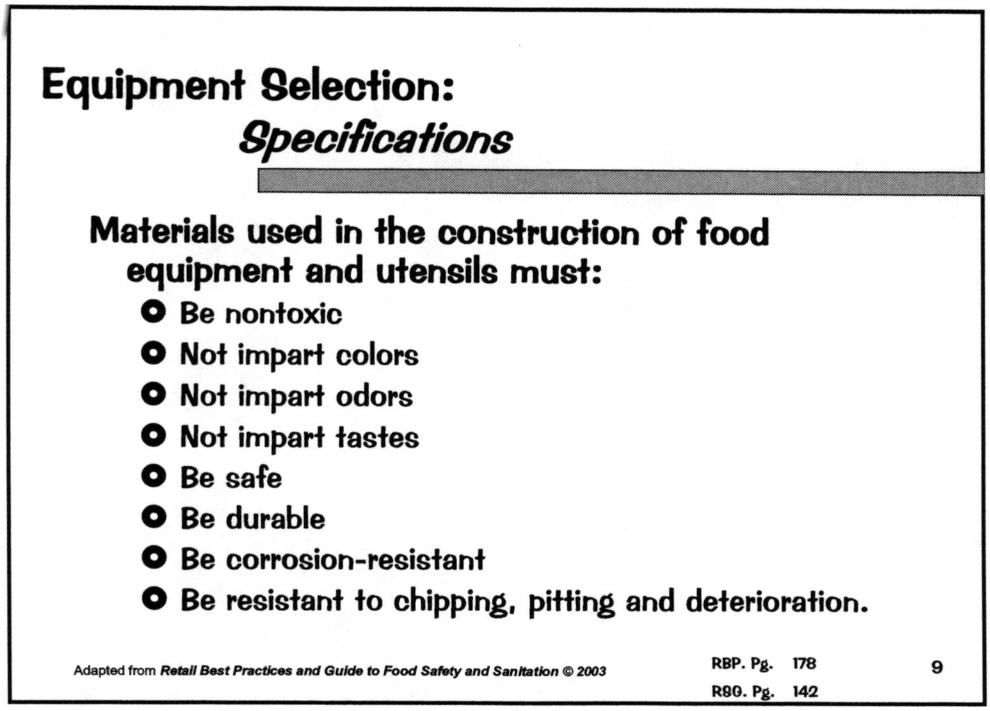

Slide 9

Materials used in the construction of food equipment and utensils must:

- Be nontoxic
- Not impart colors
- Not impart odors
- Not impart tastes
- Be safe
- Be durable
- Be corrosion-resistant
- Be resistant to chipping, pitting and deterioration.

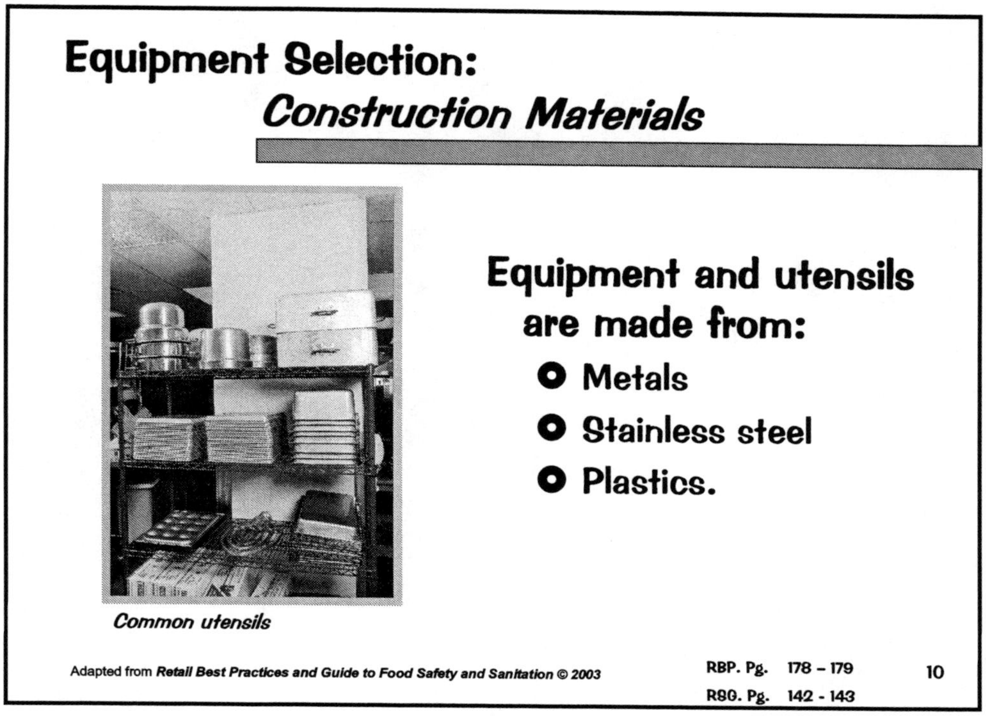

Slide 10

Preferred construction materials for equipment are:

- Metals
  - Chromium is the most preferred metal for its high luster finish
  - Steel, and alloys of iron, nickel are also used
  - Lead, brass, copper, cadmium and galvanized metals are not acceptable

- Stainless steel
  - Most preferred
  - Used for food containers, table tops, sinks, hood systems, etc.
  - Highly durable
  - Resists high temperatures
  - One factor that influences the cost of stainless steel is the extent of polishing desired:
    - The polish of the stainless steel is indicated by numbers 1-8; the higher the number the higher the polish
    - Finish #4 is most commonly found in food preparation areas, higher finishes are found in display areas
  - The gauge, or thickness, of stainless steel can also affect the cost:
    - The higher the number of gauge, the thinner the steel
    - Common gauges include 14, 16, 18, and 22
    - The gauge you use should be determined by the particular use (i.e. 14 and 16 gauge for worktables or sinks while 20 is ideal for cabinet doors)
  - Avoid abrasive cleaners; scratches in stainless steel create "germ farms"

- Plastics
    - Frequently used because they are durable, inexpensive and can be molded
    - Use only food-grade plastics
    - Examples of plastics used in retail food establishments include:
        - Acrylics (used to make covers for food containers)
        - Fiberglass (used in boxes, bus trays, and trays)
        - Polyethylene (used in storage containers and bowls)
    - Consult NSF Standard No. 51 when selecting plastic materials.

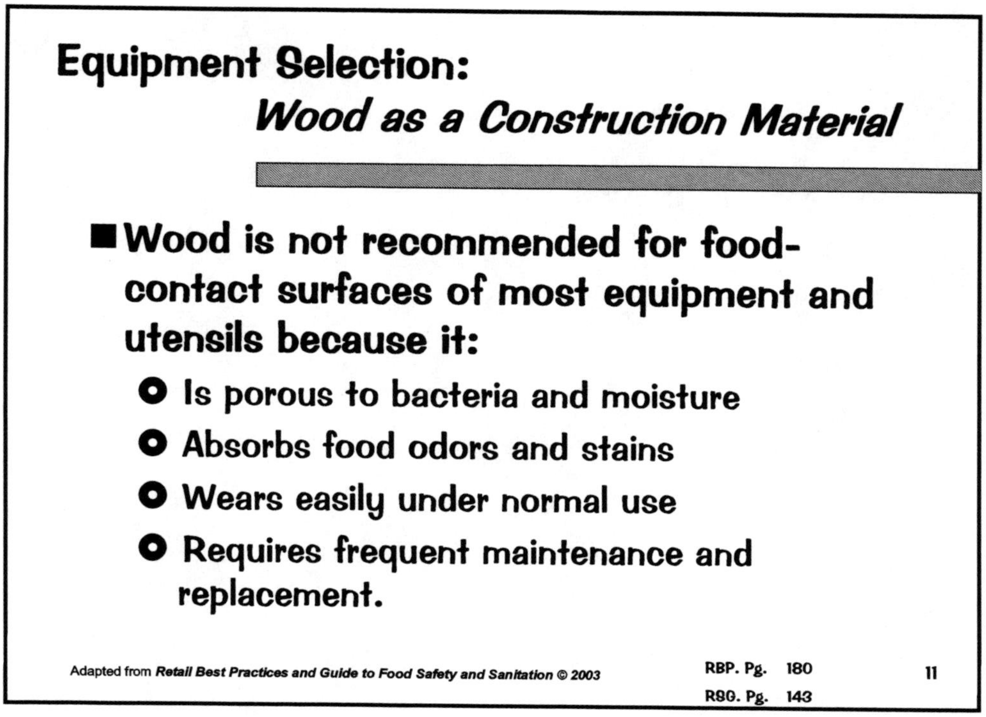

Slide 11

- <u>**Wood**</u> is not recommended for food-contact surfaces of most equipment and utensils because it:
    - Is porous to bacteria and moisture
    - Absorbs food odors and stains
    - Wears easily under normal use
    - Requires frequent maintenance and replacement.

- Wood is approved by the FDA for use as:
    - Paddles in candy kitchens
    - Paddles for pizza ovens
    - Cutting boards if they are made of hard maple (or equally hard close-grained wood).

Slide 12

Safety Stop! 5.1

Sample Test Questions

1. Which of the following organizations does not routinely evaluate the design and construction of food equipment?
   A. Underwriters Laboratories, Inc.
   B. NSF International.
   C. The American Gas Council.
   D. The U.S. Food and Drug Administration.

2. When buying equipment, retail food establishment managers should look for equipment that:
   A. Can meet the need and demands of the operation.
   B. Can be used for a long time.
   C. Does not require excessive repair and upkeep.
   D. All of the above.

3. The *FDA Food Code* requires food equipment and utensils be:
   A. Smooth and free of irregularities.
   B. Easy-to-clean.
   C. Seamless.
   D. All of the above.

4. Areas where a group of common tasks are performed by individuals are called:
   A. Work centers.
   B. Jobs.
   C. Task areas.
   D. Departments.

5. Legs of table-mounted equipment should provide a minimum of _____ -inch clearance.
   A. 2.
   B. 4.
   C. 6.
   D. 8.

6. Legs of floor-mounted equipment should provide a minimum of ____-inch clearance.
   A. 3.
   B. 4.
   C. 5.
   D. 6.

7. The most preferred construction material for food-contact surfaces is
   A. Metal.
   B. Plastic.
   C. Stainless steel.
   D. Wood.

Real-life

Slides 4 – 6
Planning facilities
XYZ Inc., a large grocery store chain, was in a hurry to beat its competition to market in a growing city. They rushed through the planning and design phase of the building. What are some possible problems they may encounter in the future?

There are several potential issues that may arise through poor planning including:

- Work center space may be too small or too large
- Under or overestimate of volume
- Poor energy use
- Improper equipment purchases
- Poor future planning/anticipation of future needs
- Plumbing, or site problems like drainage, etc.
- Flow of food within the facility.

Making it Real

Utensils and Equipment – Slides 10 – 11

- Equipment and utensils are made of a variety of construction materials. It is important to understand the differences in the materials and some potential hazards with them.
    - Supplies
        - Utensils
            - Spring-loaded tongs
            - Wood cutting board
            - Plastic cutting board
            - Knives
            - Spoons
            - Ladles
            - Plastic containers
            - Microwave safe containers
    - Procedure
        - Have the students
            - Identify what materials the utensils are made from
            - Identify potential hazards with the equipment
            - Demonstrate how to use the equipment.

Facilities design – Slides 4 – 11

- Take a tour! If you are performing this training within a retail food establishment, take your class for a tour around the facility.
    - Supplies
        - A retail food establishment
        - Measuring tape
    - Procedure
        - During your tour of the facility, have the learners point out:
            - Various work centers
            - Examples of floor-mounted equipment
            - Examples of table-mounted equipment
            - Examples of various construction material
                - Plastic
                - Metal
                - Stainless steel
        - Have the learners discuss what they see
            - Were the work centers built to the appropriate size?
            - Did they recognize any errors in planning?
        - Have the learners point out:
            - Food-contact surfaces
            - Non-food-contact surfaces
            - Splash-contact surfaces
    - Teaching Tip – Measure the space on table- and floor-mounted equipment. Do they all meet the guidelines discussed in this session?

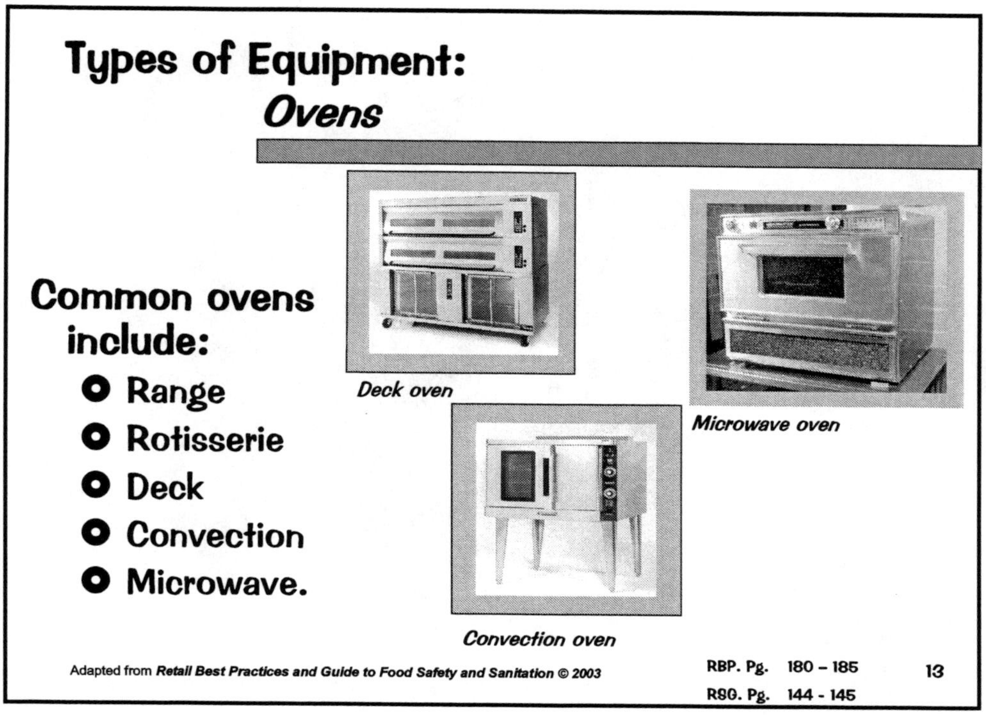

Slide 13

Ovens are commonly found in the deli and bakery. The heat in an oven is distributed by:

- **Conduction** – Heated air circulated around the outside of the heating chamber
- **Convection** – High-speed fan circulates heated air
- **Radiation** – Microwaves (electromagnetic waves of energy).

A good oven should rise to 450°F (232°C) within 20 minutes and be able to cool rapidly. Common ovens include:

- **Range ovens**
  - Most common oven in smaller retail food establishments
  - Have a cooking surface above the oven
  - Must have good insulation/construction
- **Rotisserie ovens**
  - Commonly used to cook chickens but can also accommodate turkeys, roasts, etc.
  - Use gas, electricity or wood as the heat source
  - Can be used to roast products or to hold products hot
- **Deck ovens** (pizza oven is a common deck oven)
  - Stationary with separate units stacked on top of one another
  - The heated air is circulated around the outside of the heating chamber
- **Convection ovens**
  - Use a fan to circulate heated air around foods
  - Reduce the amount of cooking time (⅓ to ¾ the amount of time)
  - Multiple racks increase the amount of cooking that can be done at one time

- **Microwave ovens**
    - Commonly used for thawing, heating, and reheating
    - Electromagnetic waves are absorbed by foods that, in turn, heat the product
    - Do not use metal or aluminum foil in microwaves
    - Cooks small amounts of foods efficiently.

Other types of ovens include rotary, conveyor, roll-in, revolving, and infrared.

Types of Equipment:
Refrigeration and Low Temperature Storage Equipment - Display

Refrigeration equipment should maintain temperatures between 32°F (0°C) and 41°F (5°C) and can be used to store, cool or thaw foods.

Adapted from ***Retail Best Practices and Guide to Food Safety and Sanitation*** © 2003

RBP. Pg. 185 – 187
RSG. Pg. 145 - 149

14

Slide 14

All types of refrigeration units:

- Are used to transport and store perishable foods
- Help to extend the shelf life of perishable foods
- Are used to hold foods between 32°F (0°C) and 41°F (5°C)
- Vary in size and should be purchased based on individual needs
- Can be used to help cool foods from 135°F (57°C) to 70°F (21°C) within two hours and from 135°F (57°C) to 41°F (5°C) within six hours
- Can be used to thaw products from the frozen state
- Must have proper air circulation both inside and outside the cabinet.

It is important not to exceed the maximum capacity of refrigeration units.

- Will cause improper cooling and holding temperatures.

Efficient operation of a refrigeration unit depends on factors such as:

- Design
- Construction
- Capacity of the equipment.

Foods should be stored:

- In shallow containers
- On slatted shelves (do not foil line shelving)
- Properly spaced
- Below maximum load levels or safe load lines
- Using the first-in–first-out (FIFO) inventory method.

Maintenance:

- Replace worn gaskets
- Clean regularly to avoid odors
- Fan grates and condensers should be cleaned frequently.

Types of refrigeration units include:

- Walk-ins
    - Store large quantities of product
    - May have the door opening covered with 4-inch wide plastic strips called strip curtains used to keep the cool air in when the door is open
- Reach-ins
    - Upright
    - Under the counter
    - Mobile
- Display cases
    - Coffin
    - Upright/vertical
    - Reach-in.

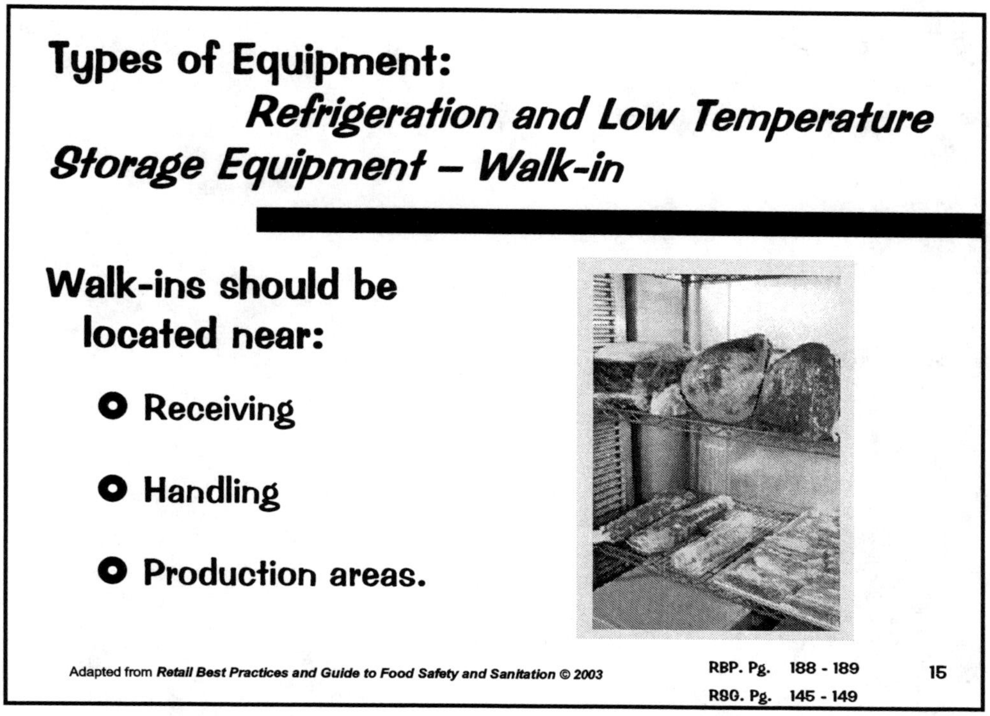

Slide 15

Walk-in refrigeration units:

- Store large quantities of perishable products between 32°F (0°C) to 41°F (5°C)
- Can be used to safely thaw frozen products
- Can be combined with reach-in dairy/deli display cases that are loaded from within the refrigerator
- Should be located at a site easy to get to from areas of:
    - Receiving
    - Production
    - Handling.

Walk-ins should have:

- Floors made of strong, durable, and easily cleanable materials
- Walls with a washable finish that will resist the effects of moisture
- An outside wall-mounted thermometer with the sensing part of the thermometer in the warmest part of the unit
- A safety device for opening the door from the inside if it is locked on the outside.

Some walk-in door openings are covered with 4-inch wide plastic strips called strip curtains that help:

- Reduce cold air loss
- Conserve energy.

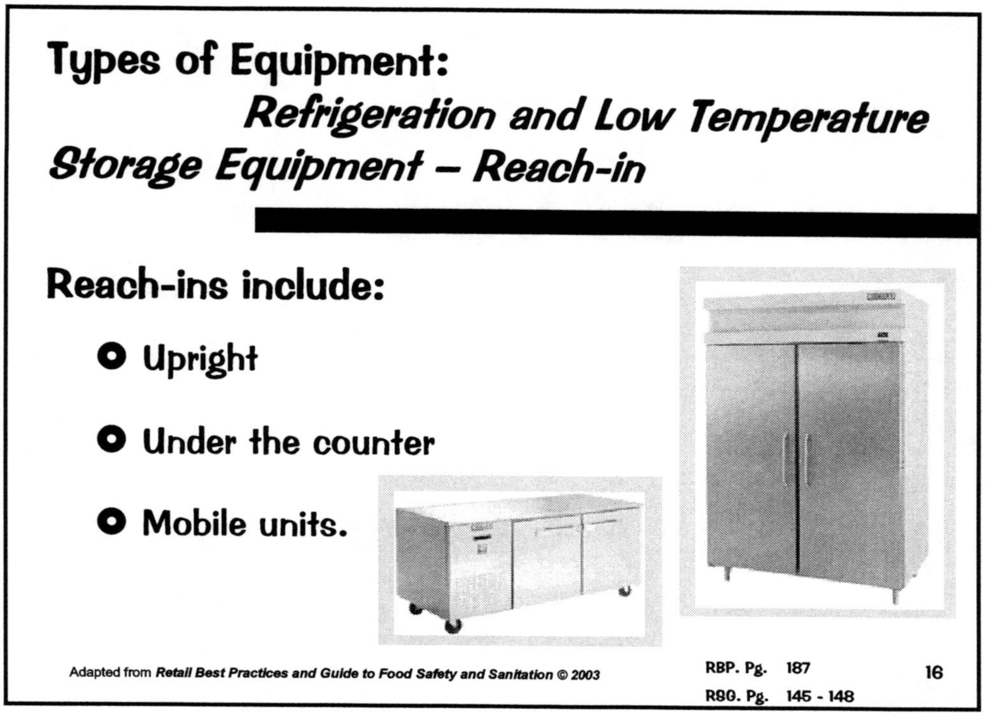

Slide 16

Reach-in refrigeration units:

- Store perishable food products between 32°F (0°C) to 41°F (5°C)
- Can be combined with walk-ins and be used for dairy/deli display cases that are loaded from within the walk-in refrigerator.

Reach-in refrigeration units are commonly found in the:

- Deli
- Bakery
- Meat and seafood departments
- Dairy department.

Reach-in refrigeration unit capacities vary greatly:

- Upright can be as large as 90 cubic feet.

Under the counter typically have only a few cubic feet of storage space.

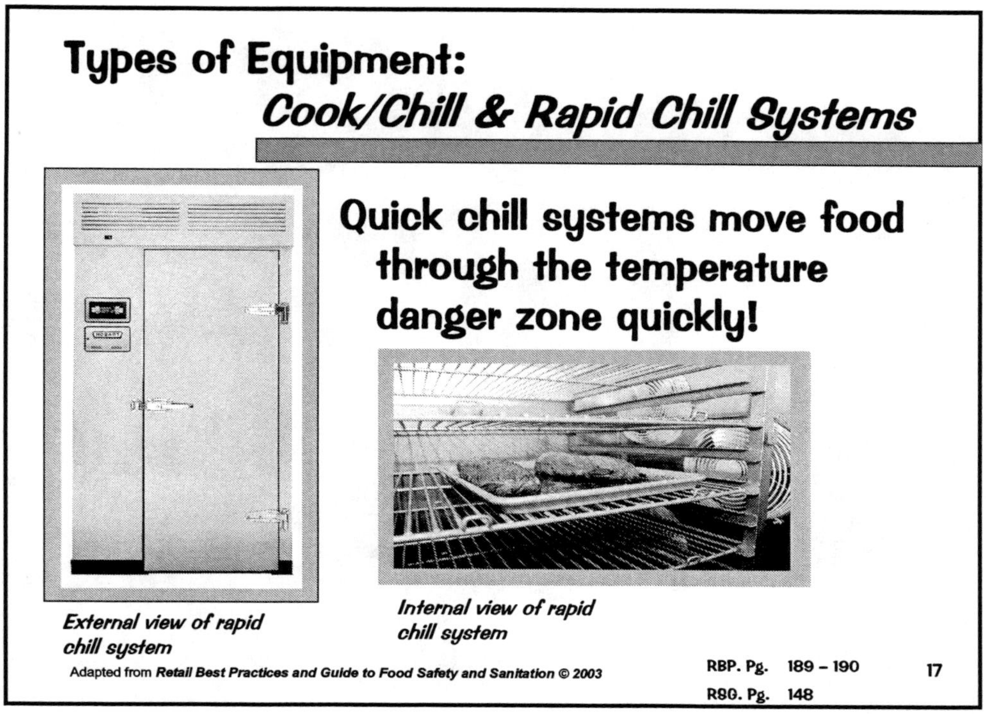

Slide 17

Cook/chill is a system in which food is:

- Cooked using conventional methods
- Placed in plastic pouches
- Rapidly chilled [to 37°F (4°C) within 90 minutes]
- Stored between 33°F (1°C) to 38°F (4°C) for up to five days
- Reheated prior to service.

A major advantage of cook/chill methods is the stability of food production and consistent availability of foods.

Rapid chill systems:

- Cool cooked foods very quickly
- Can move hundreds of pounds of hot food through the Temperature Danger Zone (TDZ) within two hours
- Food is usually **sealed** (closed tightly) in plastic pouches fastened with a metal clip (the clip is an example of a physical hazard you need to be aware of).

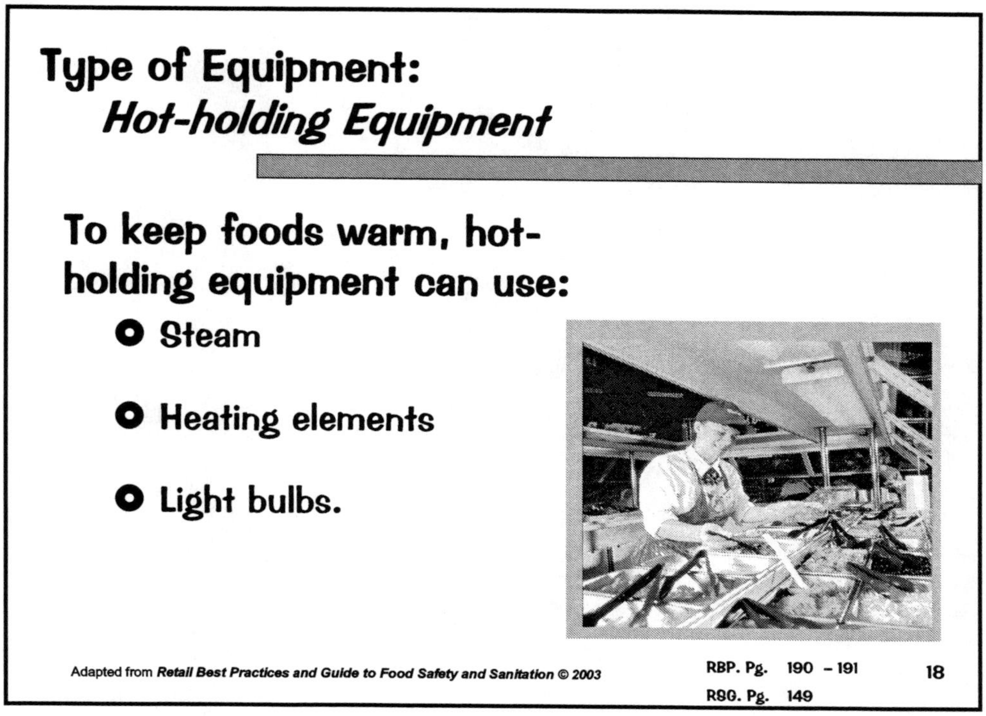

Slide 18

Hot-holding equipment

- Used to hold cooked foods at a temperature of 135°F (57°C)
- Will not raise the temperature of food product much more than 135°F (57°C)
- Not meant to be used as a method of cooking, foods should be cooked to proper temperature prior to being placed in hot-holding
- To keep foods warm, hot-holding equipment can use
    - Steam
    - Heating elements
    - Light bulbs
- Check temperatures regularly to ensure equipment is working properly
- For product quality reasons, use the FIFO rotation method.

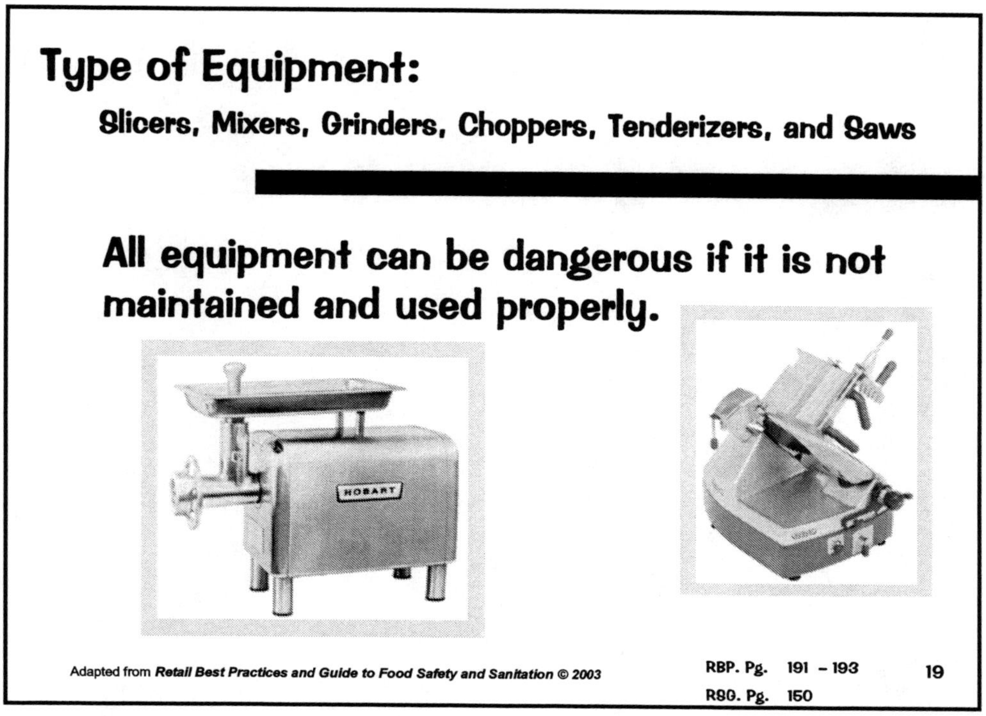

Slide 19

Slicers

- Can be operated manually or automatically, depending upon the model
- Can be dangerous if not used properly; follow proper operating instructions.

Mixers, grinders, choppers, tenderizers and saws

- Can be floor models or table models, depending upon size
- Mixers
  - Can be used to mix, shred, and grind depending upon the accessories and attachments used. Floor model mixers have three standard attachments:
    - Paddle beater for general mixing
    - Whip to incorporate air into products
    - Dough hook to mix and knead dough
- Mechanical tenderizers
  - Commonly used to improve the edible quality of meat by cutting the connective tissue in meats and poultry
  - Amount of tenderization is controlled by the depth and spacing of the blades and the frequency with which the meat is penetrated
- Grinders and choppers
  - Work well with a variety of fresh meats
  - Contain
    - A mixer (paddles or blades)
    - Auger
    - Grinder disk.

Type of Equipment:
Band Saws

Band saws must be:

- Durable
- Easily cleanable
- Safe to use.

Adapted from ***Retail Best Practices and Guide to Food Safety and Sanitation* © 2003**

RBP. Pg. 193 – 194 20

RSG. Pg. 151

Slide 20

Band saws

- Commonly used in meat and fish processing departments
- Must be durable, easily cleanable, and safe to use
- Must be made of nontoxic materials
- Must be properly guarded or have a "dead man's" switch (a switch that activates equipment when depressed and stops the equipment if the pressure is relieved) to protect the operator from the blade
- Heavy duty saws are:
  - Designed to cut meat and bone
  - Designed to hold up under heavy use
- Tabletop models are:
  - Commonly used in small delis, meat shops and small establishments
  - Designed to cut boneless meats and poultry.

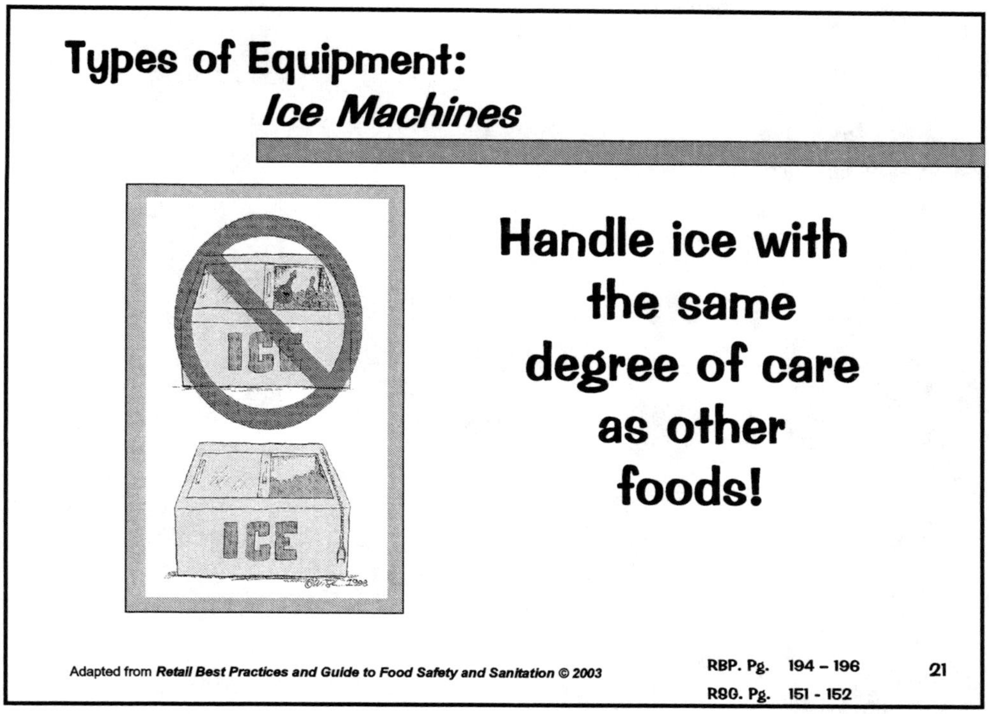

Slide 21

Ice machines:

- Ice is important in retail food establishments because it is used:
    - To chill beverages
    - To preserve the freshness of fish and some produce items
    - For display appearance and temperature control
- Ice is food, and be handled with the same degree of care as other food items
- Ice must be made from potable water
- Ice machines must use potable water and protect the ice during production and storage
- Parts of the ice machine that touch the ice are considered food-contact surfaces and must be:
    - Smooth
    - Durable
    - Made from nontoxic materials
    - Easily cleanable
    - Cleaned and sanitized regularly
- Drain line from the machine must have an air-gap (more in Session 7)
- Ice dispensing scoops must be stored outside the machine or in a bracket that is mounted to the inside of the machine
- Never use a glass to scoop ice (glass could pose a threat as a physical hazard)
- Never store food or beverages in ice machines.

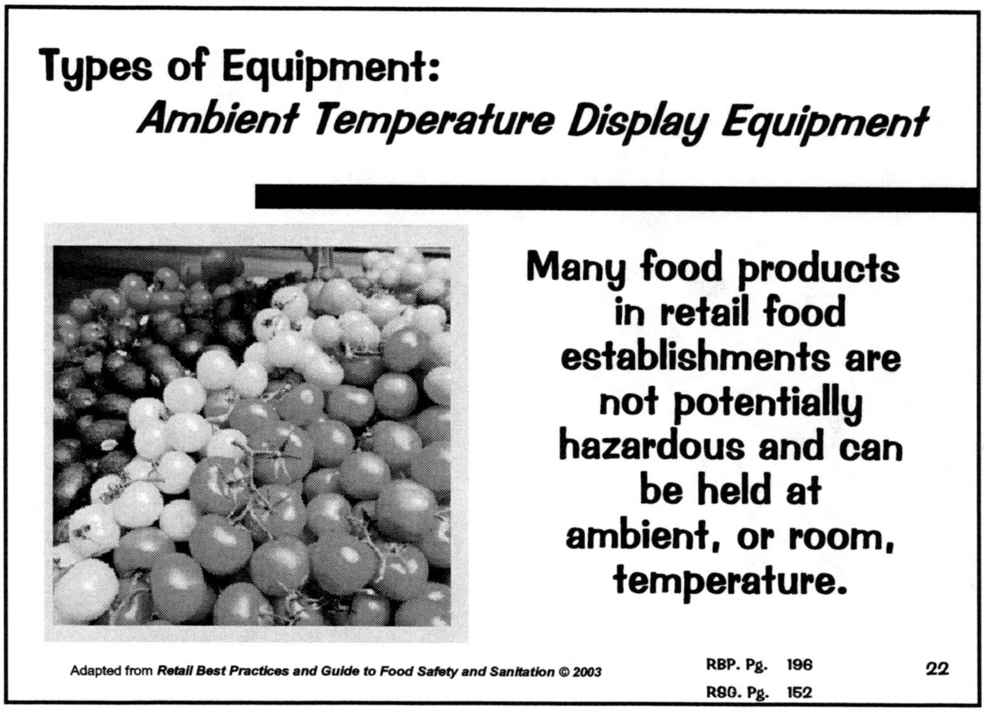

Slide 22

Ambient temperature display equipment:

- Many food products in retail food establishments are not potentially hazardous and can be held at ambient, or room temperature including:
  - Some produce such as bananas, apples, and potatoes
  - Some bakery items such as bagels, cookies, and doughnuts
  - Bulk foods such as candy, nuts, and cereals
- These foods must be held in display equipment and must be protected from contamination by:
  - Customers
  - The general environment
- This display equipment has:
  - Food-contact surfaces that must be cleaned and sanitized regularly to remove contaminants
  - Non-food-contact surfaces that must be cleaned regularly to remove spills and eliminate soil
- Stock should be rotated using the FIFO method
- Utensils should be provided so customers will not contaminate food products.

Types of Equipment:
Ambient Temperature Display Equipment with Misters

System heads must be properly cleaned, sanitized, and maintained to assure:

- Maximum freshness of produce
- Safety of produce
- Wholesomeness of produce.

Adapted from ***Retail Best Practices and Guide to Food Safety and Sanitation*** © 2003

RBP. Pg. 196 – 197
RGG. Pg. 153

23

Slide 23

Ambient display equipment with misters:

- Misters are used to assure freshness of product and prevent wilting
- Misters must be connected to a potable water supply
- System heads must be properly cleaned, sanitized, and maintained to assure:
    - Maximum freshness of produce
    - Safety of produce
    - Wholesomeness of produce.

Types of Equipment:
Live Seafood Display and Holding Tanks

Tanks are used to protect live perishable products such as:

- **Fish**
- **Crustacean shellfish (crabs and lobsters)**
- **Molluscan shellfish (oysters, clams, and mussels).**

Adapted from *Retail Best Practices and Guide to Food Safety and Sanitation* © 2003 RBP. Pg. 198 RSG. Pg. 153 24

Slide 24

Live seafood display and holding tanks:

- Tanks are used to protect live perishable products such as:
  - Fish
  - Crustacean shellfish (crabs and lobsters)
  - Molluscan shellfish (oysters, clams, and mussels)
- Water quality in the tank is critical and can be controlled through proper filtration.
- Important handling and maintenance steps include:
  - Clean and replace the ultraviolet light to assure proper intensity to kill pathogens
  - Cull out and discard dead, cracked and weak molluscan shellfish daily
  - Before adding molluscan shellfish to the tank, ensure they are healthy and clean
  - Never mix molluscan shellfish with crustacean shellfish
  - Never allow molluscan shellfish to be submerged in the system
  - Never cook dead or dying shellfish from live displays to reduce shrink
  - Clean the tank interior to keep it free of algae and prevent slime buildup that may harbor pathogens.

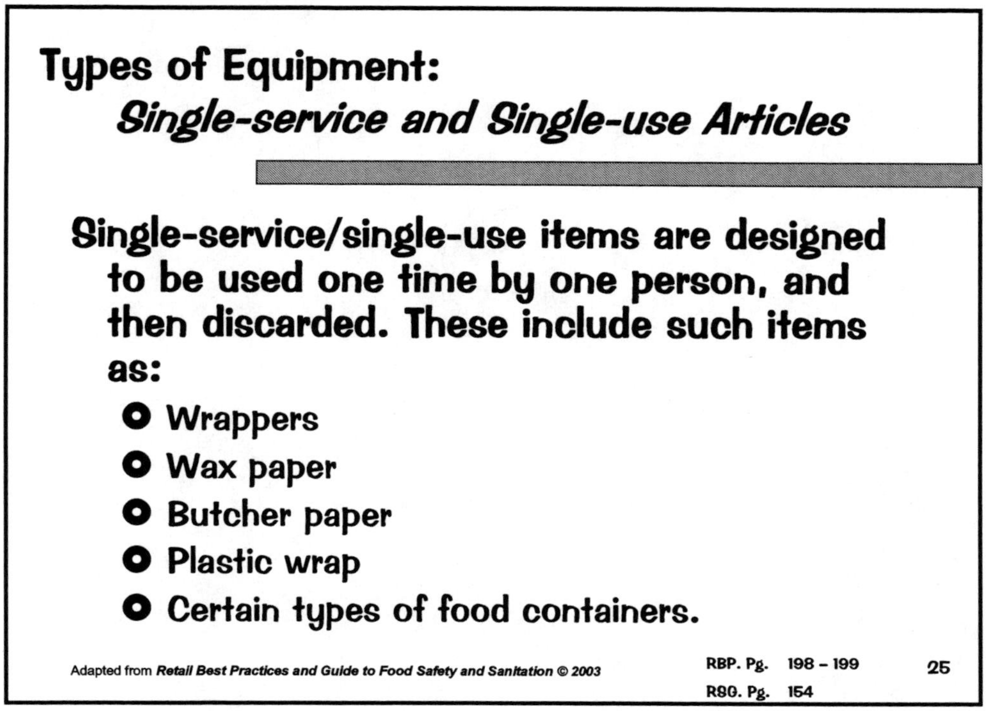

Slide 25

There are unwrapped **single-service** or **single-use** articles. These include items such as:

- Tableware
- Carryout utensils
- Bags
- Containers
- Stirrers
- Straws
- Wrappers
- Wax paper
- Butcher paper
- Plastic wrap
- Certain types of food containers
  - Single-service/single-use items are designed to be used one time by one person, and then discarded
  - Retail food establishments should provide single-service/single-use articles for:
    - Food handlers if they do not have proper facilities for cleaning and sanitizing multi-use kitchenware
    - Customers' use.

Materials used to make single-use items must not permit the transfer of harmful substances, pass on colors, odors, or tastes to food, and must be safe and clean when used in retail food establishments.

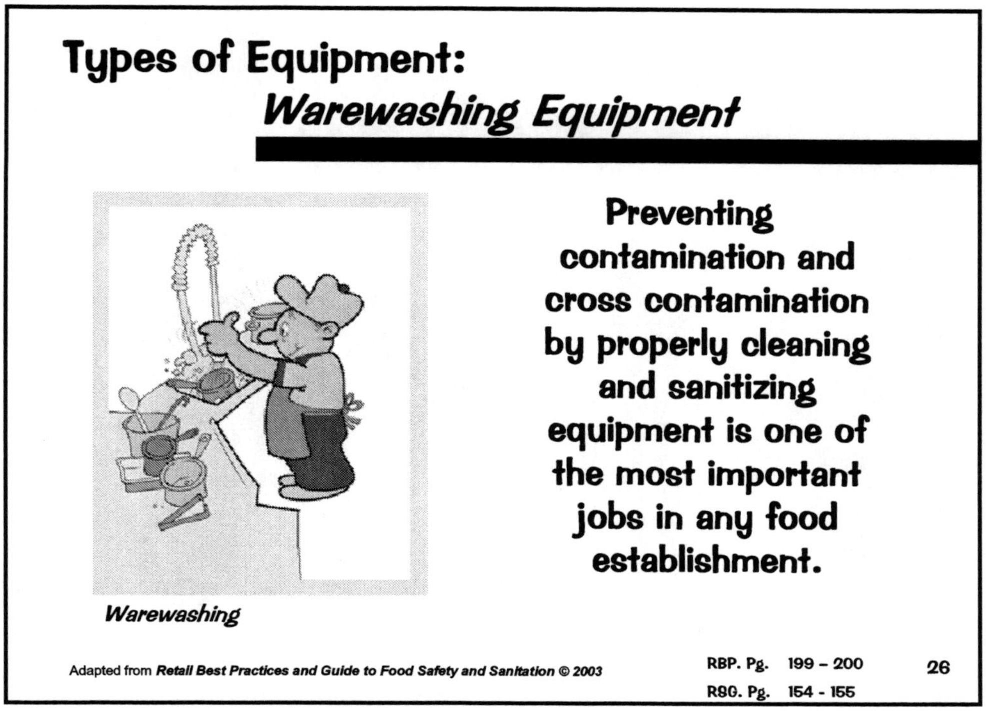

Slide 26

Warewashing:

Cleaning and sanitizing equipment and utensils is one of the most important jobs in any food establishment.

- Warewashing is a two-part operation:
    - Cleaning – removing visible dirt
    - Sanitizing – reducing the number of harmful microbes to safe levels
- Proper warewashing helps to prevent cross contamination.

Warewashing areas should be:

- Designed to handle the anticipated volume of equipment and utensils to be done
- In a separate room or area from food production
- Well lit and ventilated.

The most frequently washed items in food establishments are:

- Equipment parts
- Utensils.

Types of Equipment: *Manual Warewashing*

Use a three-compartment sink, spray units, or the bucket and brush method and properly follow the five-step cleaning process.

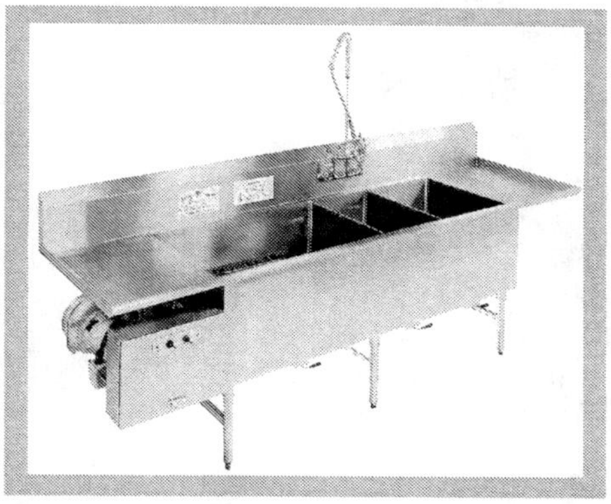

Three–compartment sink

Adapted from ***Retail Best Practices and Guide to Food Safety and Sanitation* © 2003**

RBP. Pg. 200 – 201 27
RSG. Pg. 155

Slide 27

Manual warewashing is used for:

- Cleaning and sanitizing parts of fixed equipment
- Cleaning and sanitizing portable equipment and utensils.

Most manual warewashing is performed in a three-compartment sink:

- The compartments of the sink must be large enough to accommodate the largest pieces of equipment used in the kitchen
- Each compartment must be supplied with hot and cold running potable water
- Drainboards should be provided to allow equipment and utensils to air-dry properly
- The cleaning process includes the following steps:
    1. Pre-scrape or presoak to remove food particles and soil.
    2. Wash with detergent and hot water in the first sink.
    3. Rinse the soap off in the second sink.
    4. Sanitize in the third sink.
    5. Allow the equipment or utensils to air-dry.

Two acceptable methods for sanitizing are:

- Hot-water method [water temperature must be 171°F (77°C) or above]
- Chemical method.

Some other methods of manual warewashing include:

- Bucket and brush
- Wall-mounted hose units
- Spray units.

Cleaning and sanitizing procedures are reviewed in greater depth in Session 6.

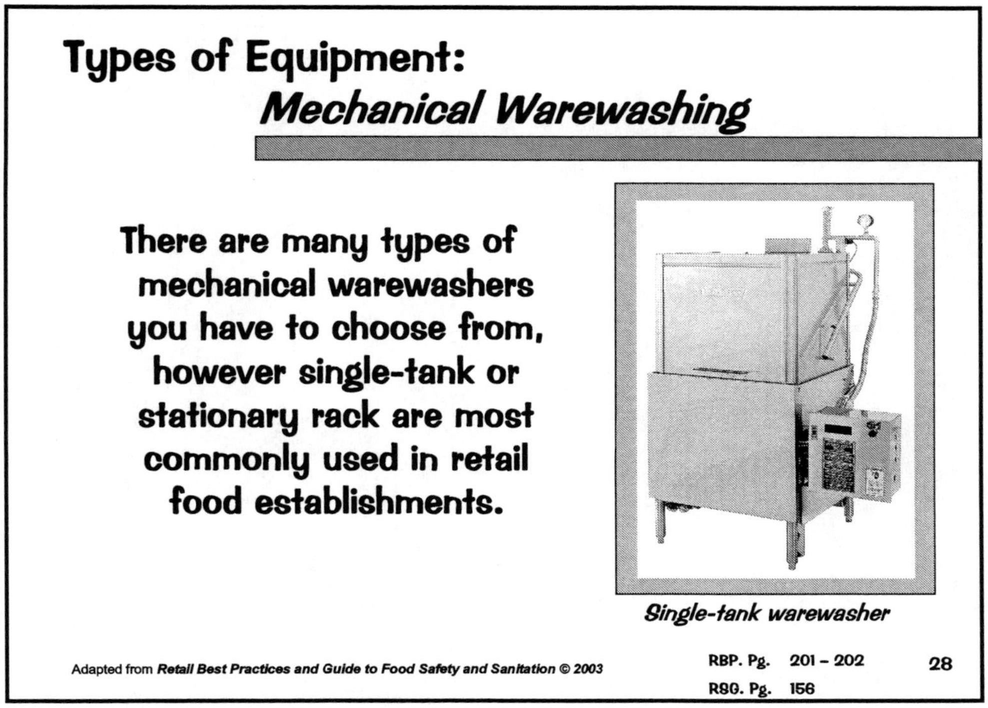

Slide 28

Mechanical warewashing machines are very expensive. Use them to wash equipment and utensils:

- That do not have electrical parts
- That will easily fit into the machine.

All warewashing machines must:

- Automatically dispense detergents and sanitizers
- Have a visual means to verify detergents and sanitizers are delivered, or
- Have visual or audible alarm to signal if the detergent and sanitizer stop being delivered.

There are several types of mechanical dishwashers:

- Immersion
    - Racks are immersed for cleaning
- Single-tank or stationary rack
    - Commonly used in retail food establishments
    - Washes one rack at a time
    - Opening of door controls operation of machine
- Conveyor-rack
    - Conveyor carries racks through machine
- Flight-type
    - Conveyor is one continuous rack and dishes are placed on pegs
- Carousel-type
    - A closed circuit conveyor where dishes are loaded and unloaded
- Low temperature
    - Commonly used in retail food establishments
    - Chemicals are used to sanitize.

When purchasing a mechanical warewashing machine consider:

- Size and design of the machine
- That an adequate supply of very hot water 180°F (82°C) to 195°F (91°C) would be needed
  - A booster heater may need to be installed to reach these temperatures.

Cleaning and sanitizing procedures are reviewed in greater depth in Session 6.

Slide 29

Safety Stop! 5.2

Sample Test Questions

1. The heat in an oven can be distributed by:
   A. Conduction.
   B. Convection.
   C. Radiation.
   D. All of the above.

2 Refrigeration units can be used to:
   A. Reheat foods.
   B. Hold foods below 41°F (5°C).
   C. Hold foods at 135°F (57°C) or above.
   D. None of the above.

3. A potential physical hazard you should be aware of that is associated with cook/chill system packaging is:
   A. The coolant.
   B. The door handle.
   C. The metal clip used to fasten plastic pouches.
   D. The chemical compound used to make the plastic pouches.

4. It is not appropriate to use hot-holding equipment to:
   A. Hold foods at 135°F (57°C) or above.
   B. Keep foods warm using steam, heating elements, or light bulbs.
   C. Reheat foods to 165°F (74°C).
   D. Hold foods for service.

5. Ice must be made from:
   A. Hot water.
   B. Potable water.
   C. Cold water.
   D. Chlorinated water.

6. Misting systems are used to:
   A. Clean produce.
   B. Keep produce fresh.
   C. Keep produce from wilting.
   D. Both B and C.

7. It is important to clean the interior of live seafood display and holding tanks in order to keep them free of algae and:
   A. Prevent slime buildup that may harbor pathogens.
   B. Give employees tasks to perform.
   C. Increase the lifespan of the live seafood.
   D. None of the above.

8. Warewashing equipment should be:
   A. Poorly lit.
   B. Unventilated to allow for steam cleaning.
   C. Large enough to accommodate the facility's needs.
   D. In the same room as food production.

Real-life

Slide 13
Microwave Oven
Pat needed to reheat some chili for the soup and salad bar in the deli. He only needed a small amount so he transferred the bulk chili into a smaller, microwave safe container, covered it with aluminum foil, placed it in the microwave and began to reheat the chili. What was Pat's mistake?

Pat should not have used aluminum foil to cover the chili if he was going to reheat it in the microwave. Electromagnetic waves will not pass through metals such as aluminum foil. Pat should have used plastic wrap or an appropriate lid to cover the chili.

Slide 14
Display Refrigeration
Jess was restocking the cartons of eggs in the dairy section. She had room for about a case and a half. She decided, since it was a busy Saturday afternoon, she would just stock all the eggs in the display unit. Jess went ahead and stocked the eggs on top of the eggs that were stocked the day before and above the safe load line. What did Jess do wrong?

Jess made two mistakes in judgment. First, Jess should have used the FIFO inventory method to stock the eggs, placing the new eggs under the old eggs. Second, foods should never be stocked above maximum load lines in refrigeration display units. Doing this affects the safety of the food product and the operation of the refrigeration unit.

Slide 20
Band Saw
Ralph was in a hurry to cut some porterhouse steaks. He noticed some of the teeth on the blade of the band saw were missing. He cut the porterhouse steaks anyway. What should Ralph have done to keep the food safe, and what could potentially have caused a food safety issue?

Ralph should have taken the time to change the blade on the band saw. If some of the teeth were missing, it could indicate the blade is old and worn. Additional teeth from the blade could break off into the porterhouse steaks and create a physical hazard for the customer. Safety is more important then speed.

Slide 21
Ice
Maria was taking a prescription medication that needed to be refrigerated. She was told where she was supposed to keep any medication needing refrigeration, but she forgot. She did not want to ask her supervisor, because she did not want him to know she was ill; she needed to work. She decided to put her medication in a sealed plastic bag and keep it in an ice machine that was used for display ice in the seafood department. What did Maria do wrong and how could it affect the safety of the seafood?

Maria made a couple of mistakes. First, she should not keep information about any illness from her supervisor. The illness, such as Hepatitis A virus, may affect the safety of the food she is working with. Additionally, she should never allow foreign objects, including plastic bags or medication, to touch ice that will be used for service or display. It could present a physical or chemical hazard that could be transferred onto the ice. (See Session 4 for additional information on proper storage of potentially hazardous foods.)

Slide 24
Live Seafood
Robert is a new employee at a large supermarket. Every day the store puts out a salad bar that includes fresh lobster salad. He was trained on how to make the lobster salad. On his first day without a trainer, Robert went to make the lobster salad. He thought he would be proactive and save the company money by using the lobsters that were dead in the live tank. He cooked them properly and made the rest of the salad to specification. Did Robert do anything wrong?

By using dead or dying lobsters, Robert contaminated the seafood salad. Dead or dying lobsters can cause foodborne illness; only fresh, live lobsters should be cooked and eaten. Dead or dying crustaceans should be removed and discarded.

Making it Real

Single-service and Single-use Articles – Slide 25

- Single-service and single-use articles, because they may not be used by the employee, can have a tendency to be overlooked when it comes to food safety.
    - Supplies
        - Carry-out utensils
        - Bags
        - Straws
        - Plastic wrap
        - Wax Paper
    - Procedure
        - Pass the various single-service/single-use items around the room and allow the learners to handle each of them.
        - Ask them questions regarding:
            - The proper storage of these items
            - The possible ways they could become contaminated
            - How they could become a contaminant
            - How they can be used to protect foods
    - Teaching Tip – Be careful in choosing which of these items to bring to class. Some of the items (i.e., the cutting blade on a roll of plastic wrap) could be hazardous to the class.

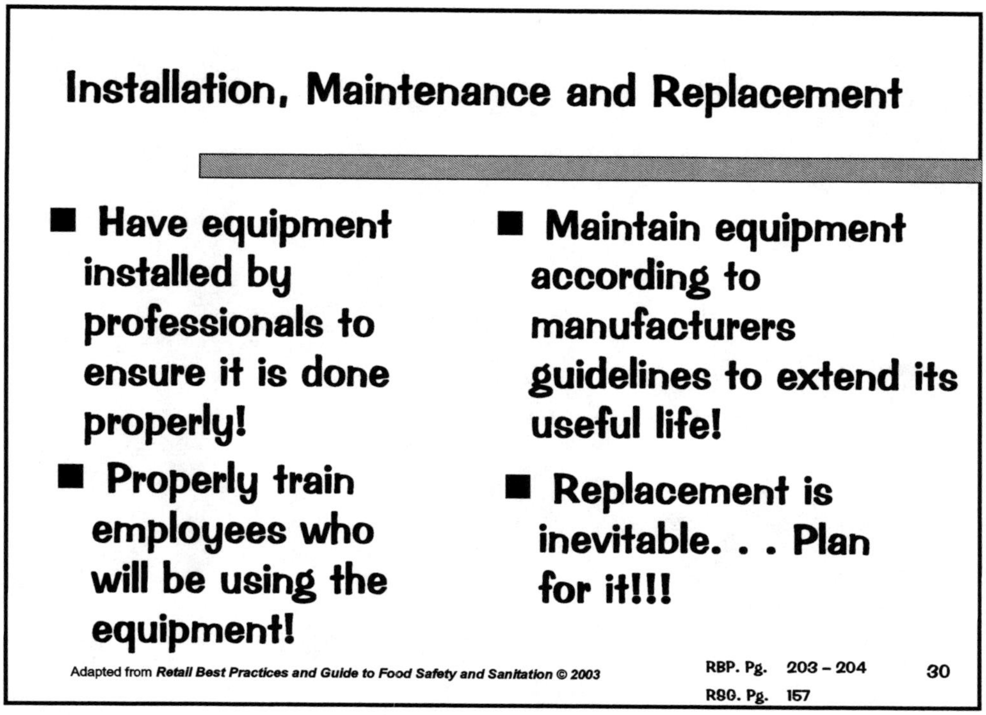

Slide 30

Installation

The best equipment is useless if it is not properly installed.

- Dealers may or may not be responsible for installation (indicated in purchase agreement)
- Architects, engineers, and contractors are responsible for providing and complying with:
    - Adequate plumbing
    - Electrical wiring
    - Adequate ventilation
    - Standards and codes
- Be sure to check the operation of all equipment prior to use
- Be sure to properly train all persons who will be using the equipment prior to use.

Maintenance and replacement

Properly maintain equipment according to manufacturer's instructions to extend its usable life.

Maintenance involves:

- Keeping the equipment clean
- Following the manufacturer's directions for care
- Posting an instruction card on or near equipment
- Careful handling
- Promptly making repairs
- Cleaning daily.

Lubricants used on food equipment may directly or indirectly end up in the food, therefore:

- All lubricants used on food-contact surfaces, bearings and gears, and other components of equipment that are located so the lubricants may leak, drip or be forced into food or onto food-contact surfaces must be approved as food additives or generally recognized as safe
- These lubricants should be used in the smallest amount needed to get the job done.

Replacement is inevitable . . . Plan for it!!!

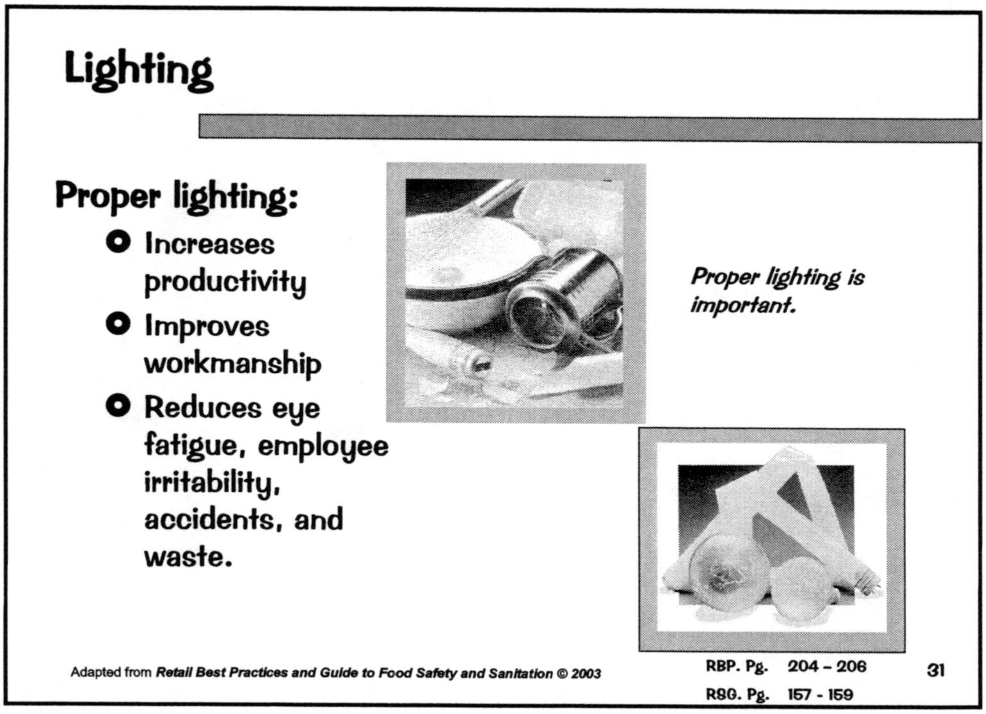

Slide 31

Proper lighting:

- Increases productivity
- Improves workmanship
- Reduces eye fatigue, employee irritability, accidents, and waste.

Proper foot-candles (unit of measure of the intensity of light falling on a surface) are needed for different work centers. FDA lighting intensity guidelines are:

- 10 foot-candles at a distance of 30" above the floor in:
  - Walk-in refrigeration areas
  - Dry food storage areas
  - Other areas during periods of cleaning
- 20 foot-candles at a distance of 30" above the floor in:
  - Areas where fresh produce or packaged foods are sold or offered for consumption
  - Areas used for hand washing
  - Areas used for warewashing
  - Areas used for equipment and utensil storage
  - Toilet rooms
- 50 foot-candles at the work surface
  - When working with unpackaged, potentially hazardous foods
  - When working with food, utensils, and equipment such as knives, slicers, grinders, or saws
  - When employee safety is a factor.

Light bulbs must be shielded, coated, or shatter resistant in areas that:

- Are exposed to unpackaged foods
- Have clean equipment
- Are over self-service buffets and salad bars.

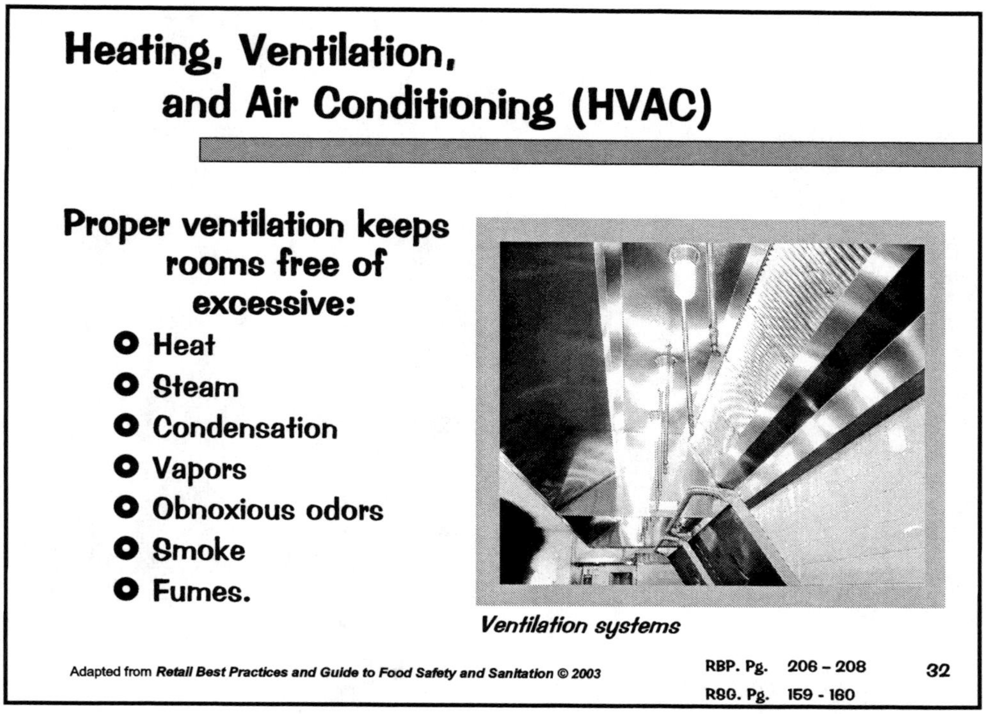

Slide 32

Air conditioning includes:

- Cooling air
- Heating air
- Controlling humidity
- Air circulation
- Air filtering.

Proper ventilation keeps rooms free of excessive:

- Heat
- Steam
- Condensation
- Vapors
- Obnoxious odors
- Smoke
- Fumes.

Hood systems:

- Prevent grease from dripping back onto food, equipment, utensils, linens, or single-service and single-use articles
- Are usually constructed of stainless steel
- Have removable filters for easy cleaning.

Intake and exhaust air ducts must be cleaned so they do not become a source of contamination and a fire hazard.

Slide 33

Safety Stop! 5.3

Sample Test Questions

1. Lubricants used on food equipment must be:
   A. Approved as food additives or generally recognized as safe.
   B. Oil based.
   C. Non-oil based.
   D. Both b and c.

2. Walk-in refrigeration and dry storage areas require:
   A. 10 foot-candles at a distance of 30" above the floor.
   B. 20 foot-candles at a distance of 30" above the floor.
   C. 50 foot-candles.
   D. 10 to 15 foot-candles.

3. Light bulbs must be shielded, coated, or shatter resistant in:
   A. The restrooms.
   B. The break room.
   C. Areas exposed to unpackaged food.
   D. Areas outside the establishment.

4. Single-service and single-use items are designed to be used:
   A. Two times by more than one person and then discarded.
   B. One time by one person and then discarded.
   C. One time by one person and then properly cleaned and sanitized.
   D. One time by multiple people and then discarded.

Real-life

<u>Slide 30</u>
<u>Maintenance</u>
Cam was the manager of the meat department. He had recently purchased a new band saw that had been properly installed by the manufacturer and was warranted for one year on all parts and labor. One afternoon the band saw stopped working. He called the authorized representative to do the repair work, however, they were not available to repair the equipment until the following day. Cam needed the band saw today; frustrated he called a local repairperson he had used on older equipment in the past. Could this create any food safety or additional problems for Cam?

As far as food safety is concerned, so long as the repairperson is qualified to work on the equipment, no problems should result. Having an unauthorized repairperson work on warranted equipment could void a warranty. This is something to be aware of for retail food establishment managers.

<u>Slide 31</u>
<u>Lighting</u>
Elizabeth, a grocery store manager, noticed two light bulbs were not working over the warewashing area. She replaced the bulbs, however, the lights still did not work. She knew she needed to call a repairperson, however, it was close to the end of the month, and she did not want to spend money on it now. Could this create a safety issue and is it an acceptable practice?

The *FDA Food Code* requires there be 20 foot-candles at a distance of 30" above the floor in warewashing areas. By choosing to wait to have the problem repaired, Elizabeth was opening her employees up to injury and her customers up to illness. Sight is a primary sense used by warewashing employees, and not properly lighting the area could present a problem.

Making it Real

Self-audit – Slides *All*

- Now that you have completed the entire session, take another tour of the facility with the class.
  - Supplies
    - Class
    - Facility
  - Procedure
    - You have already taken one tour of the facility with your class; it is time to take another. The title of the session alone indicates a need to get up and review the facilities.
    - This tour should focus on auditing the facility. Point out anything may have been missed during the previous tour. Have the learners address:
      - Various types of equipment
        - There may be some equipment that is not mentioned in this session, so be prepared for those questions
      - Lighting
      - Ventilation

- Have the learners describe the proper maintenance procedures for:
    - Holding tanks
    - Warewashing machines
    - Slicers
    - Grinders
    - Band saws
- Teaching Tip – It is a good idea to plan these tours around a break!

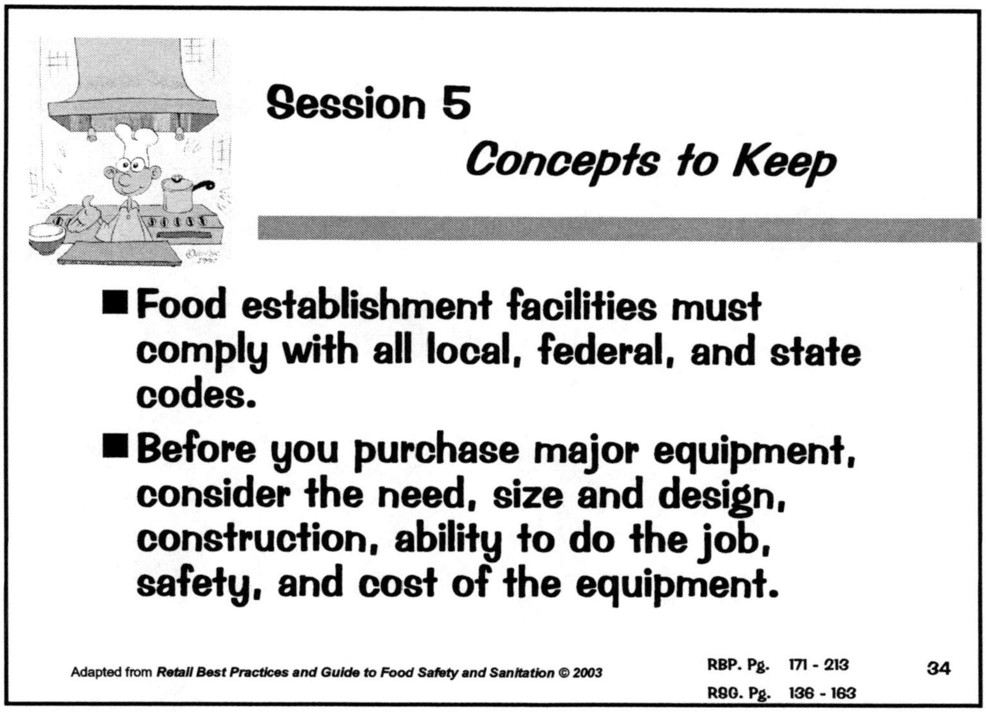

Slide 34

Concepts to Keep

- Food establishment facilities must comply with all local, federal, and state codes.
- Before you purchase major equipment, consider the need, size and design, construction, ability to do the job, safety, and cost of the equipment.

Session 5
More Concepts to Keep

- **Properly maintain equipment according to manufacturer's instructions to extend its usable life.**
- **Food-contact surfaces of equipment must be smooth, seamless, easily cleanable, easy to disassemble and reassemble and have rounded corners and edges.**

Adapted from ***Retail Best Practices and Guide to Food Safety and Sanitation* © 2003**   RBP. Pg. 171 - 213   RSG. Pg. 136 - 163   35

Slide 35

<u>More Concepts to Keep</u>

- Properly maintain equipment according to manufacturer's instructions to extend its usable life.
- Food-contact surfaces of equipment must be smooth, seamless, easily cleanable, easy to disassemble and reassemble and have rounded corners and edges.

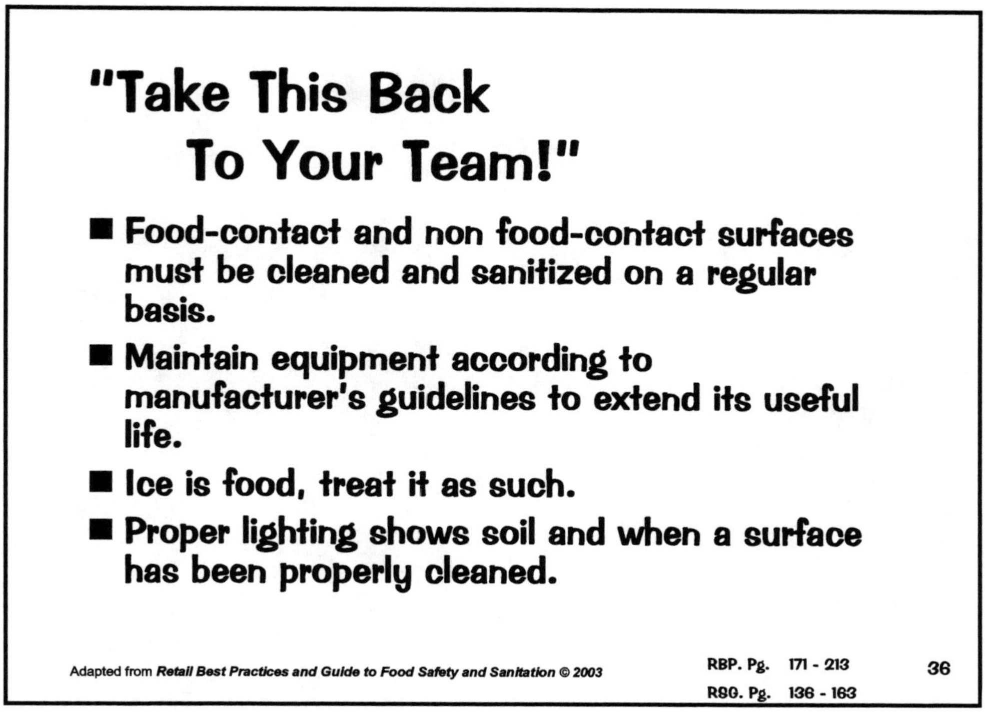

Slide 36

Take This Back to Your Team!

- Food-contact and non food-contact surfaces must be cleaned and sanitized on a regular basis.
- Maintain equipment according to manufacturer's guidelines to extend its useful life.
- Ice is food; treat it as such.
- Proper lighting shows soil and when a surface has been properly cleaned.

Session 6

Cleaning and Sanitizing Operations

6

Adapted from *Retail Best Practices and Guide to Food Safety and Sanitation* © 2003

RBP. Pg. 215 – 258
RSG. Pg. 164 - 195

1

Cleaning and Sanitizing Operations – Session 6 Key Points

Slide 1
Training Tip:
The content of each slide provides the most essential information in the session. The written text below each slide is provided to help you understand and better teach the essential information. The written text may also be used to answer questions that may be posed during the session. Depending on the length of your training session, you may not be able to address all the information contained beneath each slide.

Overview:
Session 6 discusses the need for a good sanitation program in each department of every retail food establishment. It describes:

- The difference between cleaning and sanitizing and why both are equally important
- The manual and mechanical cleaning processes
- The principles involved with sanitizing
- The process of sanitizing using heat and chemicals
- How often various food-contact and non-food-contact surfaces need to be cleaned
- Different types of cleaning chemicals used in retail food establishments and how and when they should be used
- How to keep customers and employees safe when cleaning.

A good sanitation program is a preventive program that anticipates and eliminates potential hazards before they become serious problems. The value of an effective sanitation program will be evident in the form of diminished problems and increased goodwill with customers and regulatory agencies.

Suggested Timeline:
8-hour course: 30 minutes using the PowerPoint© slides only.

- 19 ***Slides.***

16-hour course: 1 hour using the 26 PowerPoint© slides and the 3 ***Safety Stop!*** slides containing:

- 16 ***Sample Test Questions***
- 6 ***Real-life*** scenarios
- 5 ***Making it Real*** interactive exercises.

Video suggestions and supplemental resources:

Videos:
Invisible Challenge, Food Marketing Institute, 1-301-843-3084
Food Safety Success, Food Marketing Institute, 1-301-843-3084

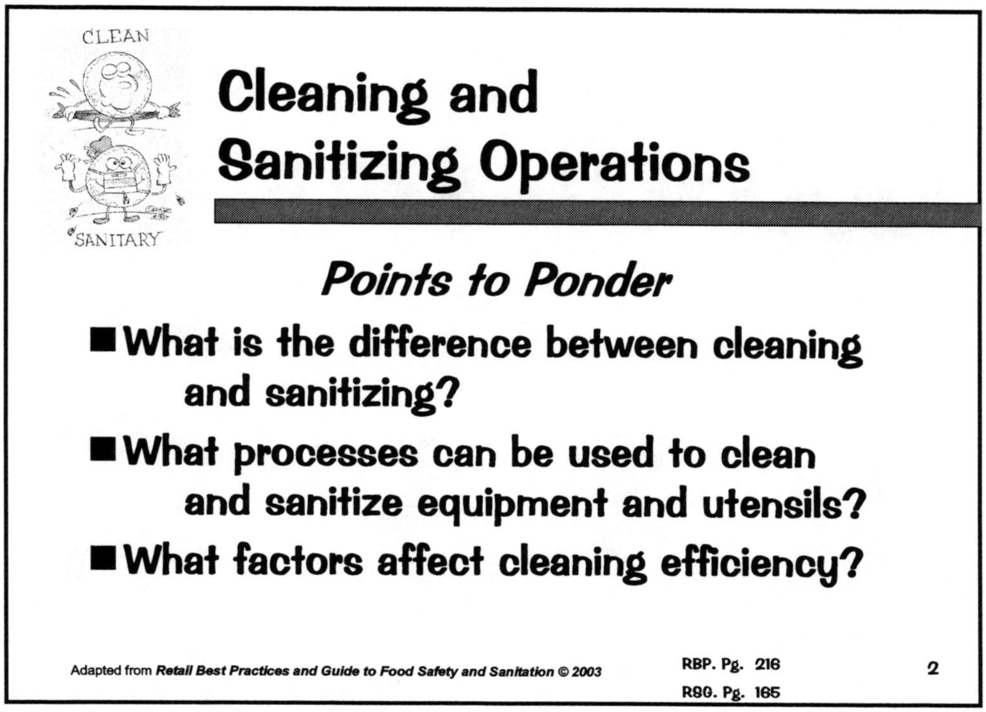

Slide 2

Points to Ponder

- What is the difference between cleaning and sanitizing?

- What processes can be used to clean and sanitize equipment and utensils?

- What factors affect cleaning efficiency?

Principles of Cleaning and Sanitizing

Cleaning is:

- The removal of soil and food residues.

Sanitizing is:

- The treatment of cleaned surfaces to reduce the number of harmful microbes to a safe level.

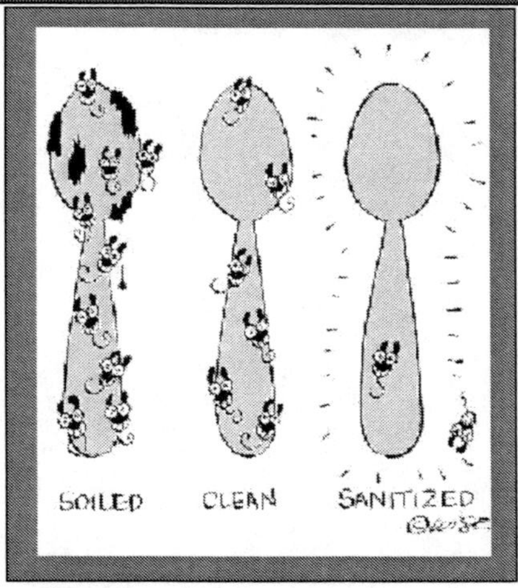

Clean vs. sanitary

Adapted from ***Retail Best Practices and Guide to Food Safety and Sanitation*** © 2003 RBP. Pg. 216 – 217 RSG. Pg. 166 3

Slide 3

Proper cleaning and sanitizing:

- Enhance the safety and quality of food
- Increase the life expectancy of equipment and facilities.

Cleaning is:

- The physical removal of soil and food residues
  - Soil includes:
    - Food wastes
    - Residues
    - Plastic
    - Paper.

Sanitizing is:

- The treatment of cleaned surfaces to reduce the number of harmful microbes to safe levels.

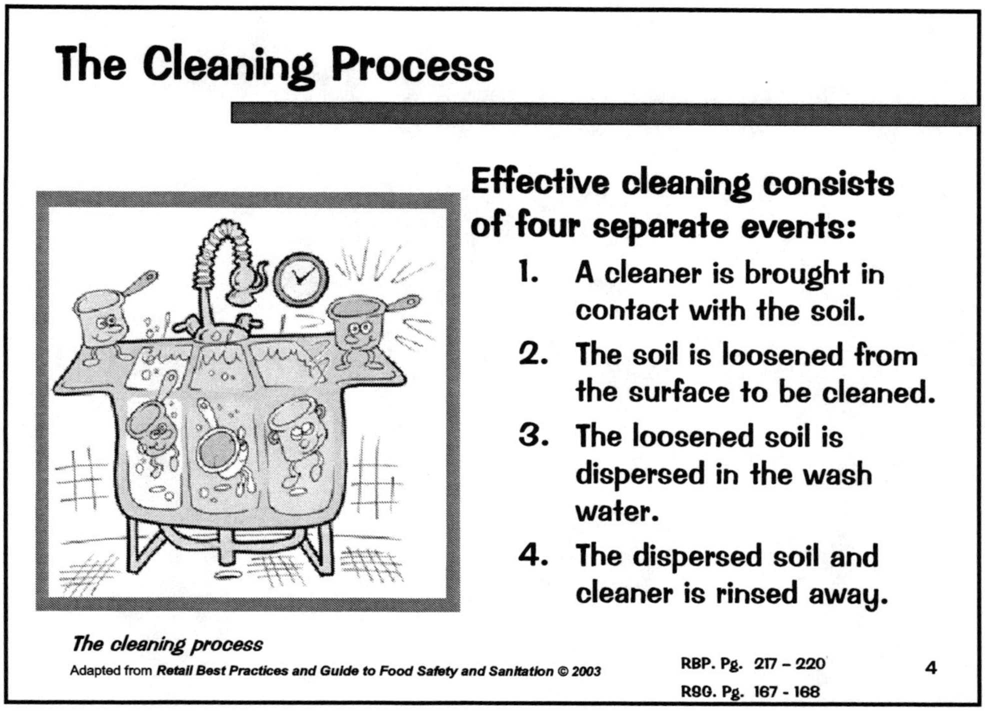

Slide 4

Effective cleaning consists of four separate events:
1. A detergent or cleaner is brought in contact with the soil.
2. The soil is loosened from the surface to be cleaned.
3. The loosened soil is dispersed in the wash water.
4. The dispersed soil and cleaner is rinsed away.

The steps in cleaning are:
1. Scrape large particles from surfaces.
2. Spray with warm water to remove smaller particles.
3. Apply **cleaning agent** (a chemical compound formulated to remove soil)
   - Methods of applying cleaning agents include:
     - Soaking – immerse in solution
     - Spraying – spray with solution
     - **Clean-in-place** – automated system generally used in conjunction with permanent-welded pipeline systems such as milk pasteurization equipment (velocity is primarily responsible for cleaning)
     - Abrasive – powders and pastes for soils that are firmly attached
       - Should not be used on
         - Stainless steel surfaces
         - Food-contact surfaces.
4. Finally, rinse surfaces with hot, potable water to remove soil and the cleaning solution.

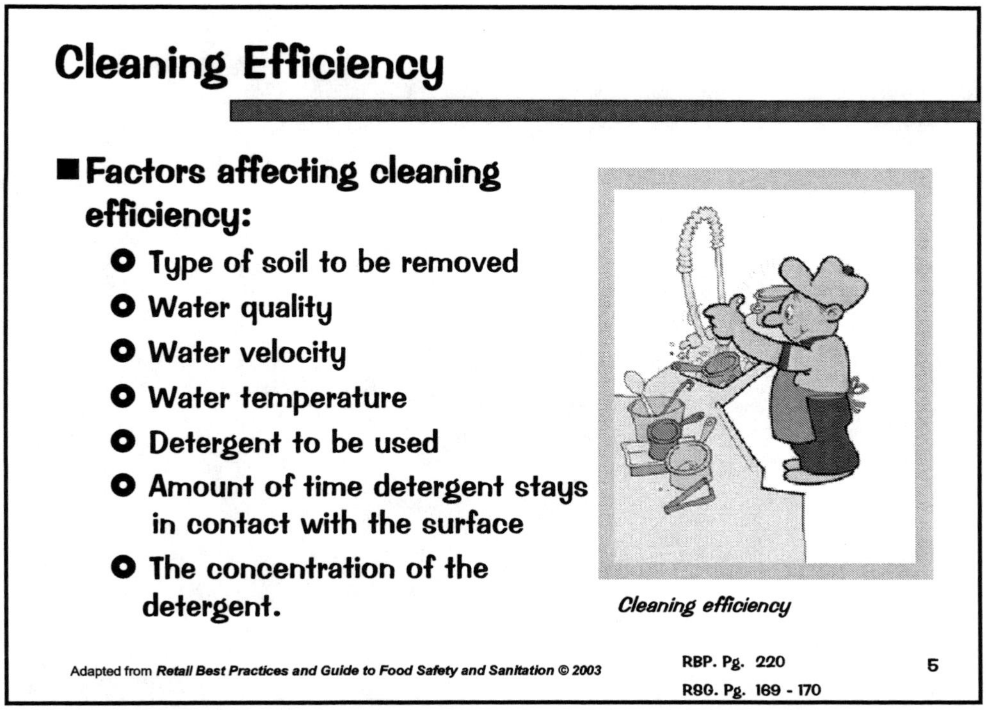

Slide 5

There are several factors that affect how easily soil can be removed:

- Type of soil including:
  - Food deposits
  - Mineral deposits
  - Microorganisms
  - Fats and oils
  - Dirt and debris
- Water quality
- The cleaner to be used
- Water temperature
  - Heat stable detergents work best between 130°F (54°C) and 160°F (71°C) (Specific water temperature guidelines will be discussed on Slides 15 and 16 of this session)
  - Some detergents require cold water
- Water velocity or force
  - The more powerful the detergent, the less velocity ("elbow grease") is needed
- Time cleaner is in contact with surface
  - Soaking items in detergents increases cleaning efficiency
- Concentration of cleaner
  - Stronger detergents increase cleaning efficiency.

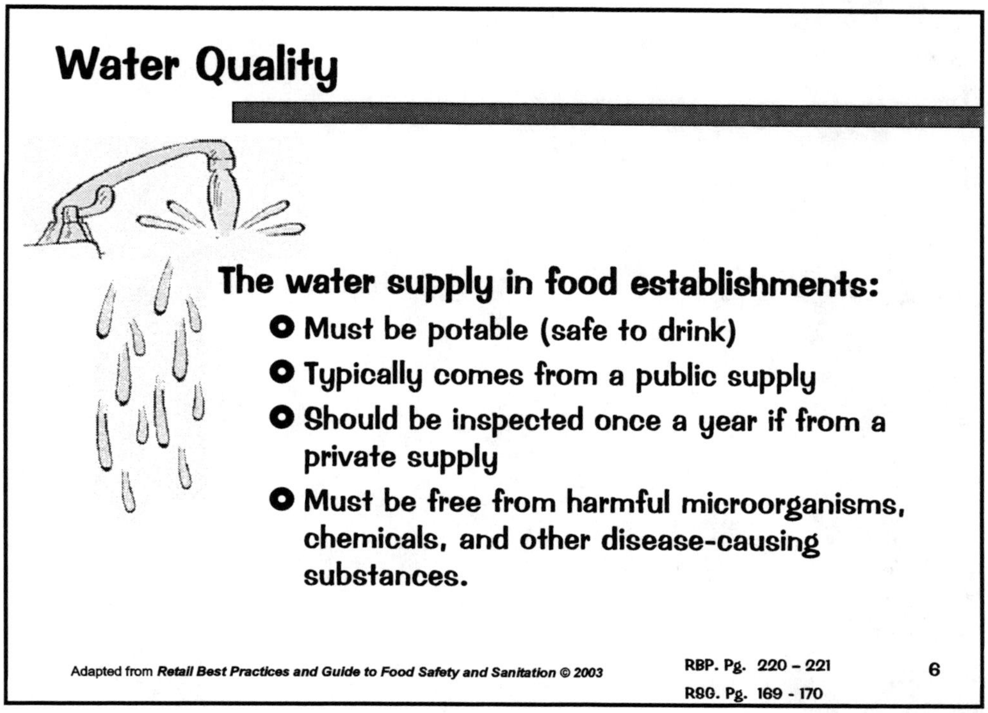

Slide 6

The water supply in food establishments:

- Must be potable (safe to drink)
- Typically comes from a public supply
- Should be inspected once a year if from a private supply
- Must be free from harmful microorganisms, chemicals, and other disease-causing substances.

"Hard" water is caused by dissolved salts of calcium, magnesium, and iron and:

- Reduces the effectiveness of detergents
- Leaves lime scale deposits on surfaces of equipment
- Should be softened for effective cleaning.

If there is an extended interruption in the water supply, the facility should cease operations or sell only prepackaged or pre-prepared food using single-use utensils until water service is restored.

Detergents and Cleaners

Detergents and cleaners improve the cleansing action of water to dissolve dirt and soil.

Adapted from ***Retail Best Practices and Guide to Food Safety and Sanitation*** **© 2003**

RBP. Pg. 221 - 224

RGG. Pg. 171 - 172

7

Slide 7

<u>**Detergents**</u>:

- Are cleaning or purifying agents – solvents
- Improve the cleansing action of water to dissolve dirt and soil
- Must be stored away from food and food-contact surfaces.

Types of detergents include:

- <u>**Soaps**</u>:
    - Are made by chemical reaction of alkali on fats or fatty acids
    - Have limited use in retail food establishments
- Alkaline detergents:
    - Combine with fats to form soaps
    - Combine with proteins to form soluble compounds
    - Corrode aluminum, galvanized metal, and tin
    - Are good general purpose cleaners
- Acid detergents:
    - Dissolve mineral deposits
    - Are frequently used in retail food establishments
    - Are categorized as
        - Inorganic (strong and corrosive) – and –
        - Organic (mild and less corrosive)
- Degreasers:
    - Remove grease or oily soils
    - Are designed for hard surfaces
    - Must be rinsed off after use

- Abrasives:
    - Are natural minerals that are ground into small particles
    - Are useful for jobs that require scouring or scrubbing
    - Can cause scratches on surfaces
- Detergent sanitizers:
    - Contain a detergent and a sanitizer
    - Must be used two times to effectively clean and sanitize surfaces
    - Must be rinsed off after use.

Slide 8

FDA Food Code requires that:

- Food-contact surfaces used with potentially hazardous foods at room temperature be cleaned every four hours including:
  - Before each use with a different type of raw animal food
  - Each time there is a change from working with raw foods to working with ready-to-eat foods
  - Between uses with raw fruits and vegetables and with potentially hazardous foods
  - Before using or storing a food temperature-measuring device
  - At any time during the operation when contamination may have occurred
- Food-contact surfaces used with potentially hazardous foods be cleaned every four hours unless:
  - Containers used to store or hold foods are kept at proper hot- or cold-holding temperatures:
    - Clean every 24 hours
  - In-use utensils intermittently stored in water at or above 135°F (57°C):
    - Clean every 24 hours
  - Equipment is being used in a refrigerated room, requirements are:
    - Room temp is 41°F (5°C) or less clean at least once every 24 hours
    - Room temp is 41°F (5°C) to 45°F (7°C) clean at least once every 20 hours
    - Room temp is 45°F (7°C) to 50°F (10°C) clean at least once every 16 hours
    - Room temp is 50°F (10°C) to 55°F (13°C) clean at least once every 10 hours.

Additional cleaning guidelines include:

- Before each use with different raw animal types
- When changing from raw foods to ready-to-eat foods
- Between uses with potentially hazardous foods and raw fruits and vegetables
- Thermometers should be cleaned before use and storage
- Any time that contamination may have occurred.

FDA Food Code requires that:

- Food-contact surfaces used with non-potentially hazardous foods be cleaned:
    - Any time contamination may have occurred
    - At least every 24 hours for:
        - Iced tea dispensers
        - Customer self-service utensils such as tongs, ladles, scoops, etc.
    - Prior to restocking customer self-service equipment such as condiment dispensers and bulk food display containers
    - Whenever possible following manufacturer's guidelines.

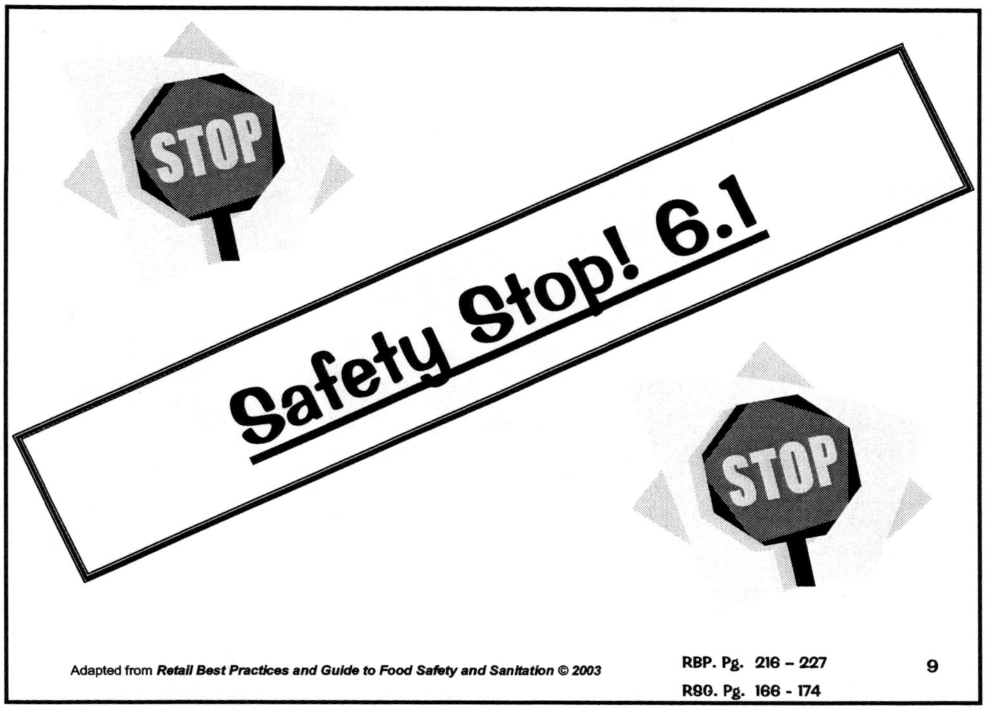

Slide 9

Safety Stop! 6.1

Sample Test Questions

1. Cleaning is
   - **A. The removal of soil and food residues.**
   - B. The treatment of surfaces that reduces the number of harmful microbes to safe levels.
   - C. The process of pre-scraping food particles off equipment and utensils.
   - D. All of the above.

2. Sanitizing is:
   - A. The removal of soil and food residues.
   - **B. The treatment of cleaned surfaces to reduce the number of harmful microbes to safe levels.**
   - C. The process of pre-scraping food particles off equipment and utensils.
   - D. The final step in the cleaning process.

3. All of the following are factors that affect how easily soil can be removed with the exception of:
   - A. Type of soil being removed.
   - B. Water temperature.
   - **C. Amount of water used.**
   - D. Concentration of the cleaner.

4. The water supply in retail food establishments must:
   A. Come from a private well.
   B. Come from a public water source.
   C. Be inspected on a monthly basis.
   D. Be potable.

5. Detergents:
   A. Are cleaning or purifying agents.
   B. Improve the cleansing action of water.
   C. Must be stored away from food and food-contact surfaces.
   D. All of the above.

6. The *FDA Food Code* requires, with a few exceptions, food-contact surfaces used with potentially hazardous foods be cleaned every:
   A. 2 hours.
   B. 4 hours.
   C. 6 hours.
   D. 24 hours.

7. The *FDA Food Code* requires that food-contact surfaces used with nonpotentially hazardous foods be cleaned at least every _________ at room temperature.
   A. 4 hours.
   B. 12 hours.
   C. 24 hours.
   D. Week.

Real-life

Slide 7
Detergents
It was John's third day working in the warewashing area. He was being trained by Maria, who asked him to go to the chemical room and get some more detergent for the manual warewashing sinks and replace the empty one. John went to the chemical closet and grabbed what he thought was the right detergent. He brought it back to the sinks. Maria was not there so he put it in the sanitizer holder. When she returned, he told her he had replaced the old detergent. What was the mistake and how could it have been avoided? What potential problems could stem from the mistake?

Maria, as a trainer, should have followed through on her direction to John. She should have shown him what chemical she needed and ensured it was replaced properly. Using the cleaner in place of a sanitizer may create a chemical hazard on any equipment or utensils that come in contact with it.

Slide 8

Cleaning Equipment in Refrigerated Rooms

Jason was recently transferred from the meat-cutting department, where he worked in a room kept between 45°F (7°C) and 50°F (10°C), to the deli department kept at room temperature (approximately 75°F (24°C). He was now using a meat slicer. On his first day, he was slicing rare roast beef on a meat slicer. He only sliced roast beef on this particular slicer. He only cleaned it at the end of his six-hour shift and not between uses. What was Jason's misconception?

Jason was used to working in a refrigerated room where the cleaning guidelines were different than they are at room temperature. When working in the refrigerated room between 45°F (7°C) and 50°F (10°C), he only had to clean his equipment once every 16 hours. Now that he is working with equipment and potentially hazardous foods at room temperature, he needs to clean it at least every four hours, preferably after each use.

Making it Real

Detergent Recognition – Slide 7

- Identifying detergents and understanding their uses can be difficult. Discussing the types of detergents in your retail food establishment may not be enough to get the information across to the learners. Many chemical companies use color-coding systems to help with this learning process.
    - Supplies
        - A sample, or an empty container of each of the detergents and sanitizers used in the establishment including:
            - Soaps
            - Alkaline detergents
            - Acid detergents
            - Degreasers
            - Abrasives
            - Detergent sanitizers
    - Procedure
        - If you are providing empty containers that have been thoroughly cleaned out or sealed containers, pass the products around the class; if not, show them to the class
        - Have the class identify each item and what it is used for in their establishment
    - Teaching Tip – Photos of the products from the suppliers catalog can also be used for this exercise.

Cleaning Guidelines for Refrigerated Rooms – Slide 8

- Many retail food establishments have food preparation rooms kept below room temperature. These rooms have different cleaning guidelines than rooms kept at room temperature.
    - Supplies
        - Class
        - Description of temperatures at which various rooms are kept
    - Procedure
        - Walk the class through the retail food establishment and have them describe the cleaning frequency according to the FDA of various rooms throughout the establishment
    - Teaching Tip – This exercise can be done in the classroom or as part of a walk around the retail food establishment.

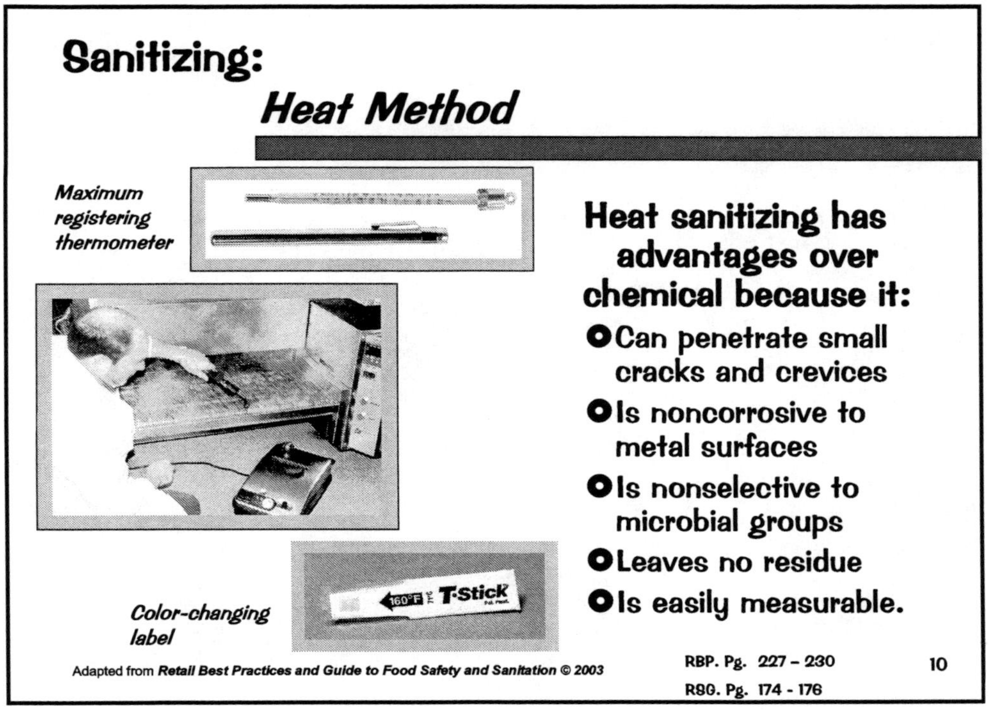

Slide 10

<u>Sanitizers</u> destroy disease-causing organisms that may be present on equipment or utensils after cleaning.

Food-contact surfaces must be thoroughly cleaned prior to sanitizing.

<u>Sanitize</u> using heat or chemical methods.

Heat sanitizing has advantages over chemical because it:

- Can penetrate small cracks and crevices
- Is noncorrosive to metal surfaces
- Kills all types of microorganisms equally effectively
- Leaves no residue
- Is easily measurable.

Maximum registering thermometers are becoming less popular in retail food establishments because they contain mercury and are made of glass, both of which can be contaminants to food and food products.

Heat sanitizing:

- Temperature at the surface of the equipment must reach at least 160°F (71°C)
- Heat sanitizing can be used with both types of warewashing processes:
- Manual
    - Equipment must be immersed in 171°F (77°C) or above for at least 30 seconds
    - Use a dip basket for safety to prevent burns
        - Keep a properly calibrated thermometer on hand to routinely check water temperature

- Color-changing label:
  - Changes color when water reaches required temperature
- Maximum registering thermometer:
  - Holds highest temperature reached until it is shaken down
  - Contains potential hazards:
    - Glass
    - Mercury

- Mechanical – more common method of heat sanitizing
  - Hot water sanitizing rinse must be between 180°F (82°C) and 194°F (90°C) when it leaves the spray nozzles
  - In single-tank, stationary rack, and single-temperature machines the rinse temperature must be 165°F (74°C) as it leaves the spray nozzle.

Heat sanitizing using steam:
Low pressure, high temperature steam cleaning systems:

- Use small amounts of water to clean
- Steam in a flow cabinet should be sufficient enough to achieve a temperature of:
  - 171°F (77°C) for 15 minutes or 200°F (94°C) for 5 minutes.

Remote steam boiler rooms:

- Typically require addition of boiler water additives to prevent buildup of scale and corrosion
  - If the steam from these boilers could potentially come in contact with food, this chemical must be approved as a food additive and labeled for that use.

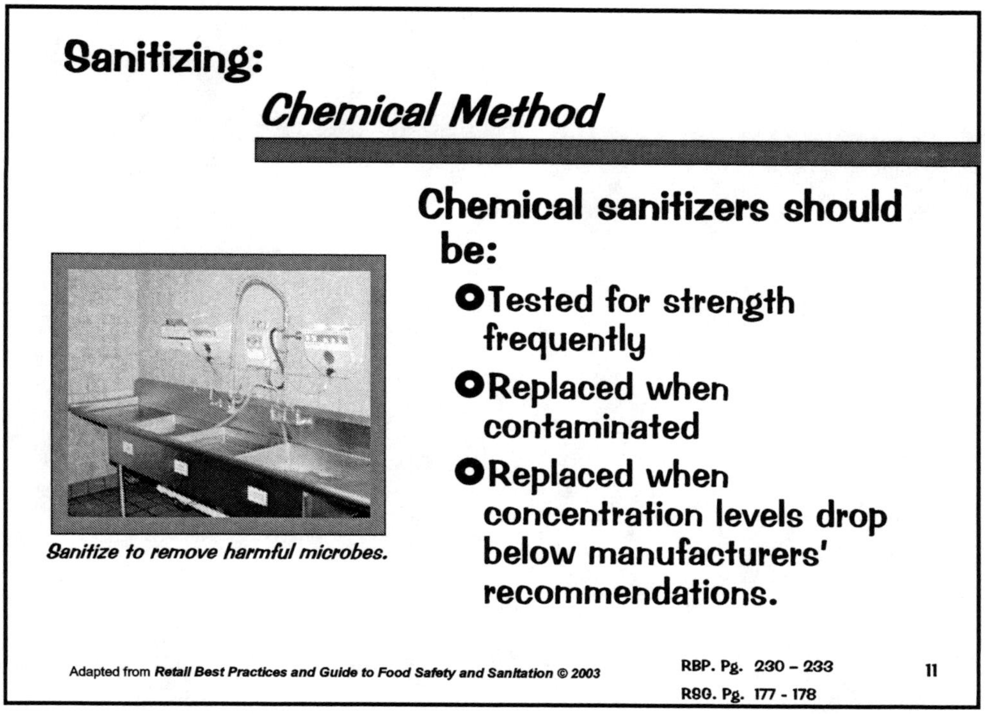

Slide 11

Chemical sanitizers:

- Weaken as microorganisms are destroyed
- Are diluted by water in wash and rinse stages
- Should be tested for strength frequently
- Should be replaced when contaminated
- Should be replaced when concentration levels drop below manufacturers' recommendations.

Chemical sanitizers' effectiveness are affected by:

- Contact of sanitizer
    - Must achieve complete and thorough contact
- **Selectivity** of sanitizer
    - Not all sanitizers destroy all microbes equally
        - Chlorine is relatively nonselective, or will destroy a wide variety of microorganisms
        - Iodophors and quaternary compounds have a selectivity that may limit their use in some retail food establishments
- Concentration of sanitizer
    - Use appropriate levels, more is not always better
- Temperature of solution
    - Recommended 75°F (24°C) to 120°F (49°C)
    - Certain chlorine sanitizers may be used in conjunction with water at a temperature as low as 55°F (13°C)
    - Quats remain active in water as warm as 120°F (49°C) or higher
    - Chlorine and iodophors are not effective in hot water

- pH of solution
    - Sanitizers react differently to pH levels
- Time of exposure
    - Varies based on above factors and amount of microbes and their susceptibility to the sanitizer.

Types of Chemical Sanitizers:
Chlorine, Iodine and Quats

Minimum Concentrations of Sanitizer Solution as expressed in parts per million (ppm) or mg/L	pH 10.0 or less & Minimum Water Temperature	pH 8.0 or less & Minimum Water Temperature	Contact Time
Chlorine 25	120°F (49°C)	120°F (49°C)	Greater than or equal to 10 seconds
Chlorine 50	100°F (38°C)	75°F (24°C)	Greater than or equal to 7 seconds
Chlorine 100	55°F (13°C)	55°F (13°C)	Greater than or equal to 10 seconds
Iodine Greater than or equal to 12.5 and Less than or equal to 25	pH less than or equal to 5.0 or per label and water temperature at 75°F (24°C) or above		Greater than or equal to 30 seconds
Quaternary Ammonium, per label	slightly alkaline pH is preferred, water hardness must be 500 ppm or less or per manufacturer's label, and water temperature at 75°F (24°C) or above		
Hot Water Sanitize, 3-compartment sink with integral heating device.	water temperature must be maintained at 171°F (77°C) or above and equipment and utensils completely immersed in rack or basket		

NOTE: All chemical sanitizers shall be listed in 21 CFR 178.1010 Sanitizing Solutions and used in accordance with the EPA-approved manufacturer's label use instructions.

Adapted from ***Retail Best Practices and Guide to Food Safety and Sanitation* © 2003**

RBP. Pg. 233 – 235
RSG. Pg. 178 - 180

12

Slide 12

Chlorine is a chemical component of chlorine-based hypochlorites.

The *FDA Food Code* requires food-contact surfaces to be exposed to:

- A 50-ppm (parts per million) chlorine sanitizer for at least seven seconds when the chlorine has a pH of 10 or less and a water temperature of 100°F (38°C), or
- At 50 ppm, a pH of 8 or less and a water temperature of at least 75°F (24°C)
    - The equipment and utensils must be exposed to the sanitizer for at least 10 seconds if the strength, pH, and temperature requirements above are not met.

Iodophors are sanitizers containing iodine:

- Function best in acidic solutions at temperatures between 75°F (24°C) and 120°F (49°C)
- Should be applied at 12.5 ppm for immersion
- Should be applied at 25 ppm for swab and spray applications
- Should contact microbes for a minimum of 30 seconds.

Quaternary Ammonium Compounds (quats) are ammonia salts used as chemical sanitizers in food establishments. Quats:

- Should be applied at 200 ppm or below
- Work at temperatures above 75°F (24°C)
- Should contact microbes for a minimum of 30 seconds.

Chemical Sanitizers:
Advantages and Disadvantages

Sanitizer	Advantages	Disadvantages
Chlorine Compounds	• Economical cost • Kills many types of microbes • Good for most sanitizing applications	• Corrosive to equipment • Can irritate human skin and hands
Iodophors	• Less corrosive • Less irritating to skin • Good for killing microbes on hands	• Moderate cost • Can stain equipment
Quats	• Stable at high temperatures • Stable for a longer contact time • Good for in-place sanitizers	• Very expensive • Hard water can reduce effectiveness

Adapted from *Retail Best Practices and Guide to Food Safety and Sanitation* © 2003

RBP. Pg. 233 – 235 13

RSG. Pg. 178 - 180

Slide 13

Advantages of using chlorine as sanitizers are they:

- Destroy many microorganisms
- Deodorize and sanitize
- Are nontoxic to humans when used as recommended
- Are colorless and non-staining
- Are easy to handle
- Are economical.

The effectiveness of chlorine-based hypochlorites is reduced by even small amounts of food soils, so it is important to thoroughly rinse equipment and utensils prior to immersing them in sanitizing solution.

Advantages of using iodophors as sanitizers are they:

- Kill more quickly than chlorine or quats.

Disadvantages of using iodophors as sanitizers are they:

- Are more expensive than hypochlorites and quats
- Can discolor and stain some surfaces
- Are harder to handle than hypochlorites and quats.

Advantages of using quats as sanitizers are they:

- Are effective on certain types of bacteria
- Are noncorrosive and nonirritating to skin
- Have no taste or odor when properly diluted
- Are effective in a wide range of pH levels (most effective in slightly alkaline water).

Slide 14

Safety Stop! 6.2

Sample Test Questions

1. Manual heat sanitizing requires equipment to be immersed in:
   A. 171°F (77°C) or above for at least 30 seconds.
   B. 171°F (77°C) or above for at least 10 seconds.
   C. 180°F (82°C) or above for at least 30 seconds.
   D. 194°F (90°C) or above for at least 30 seconds.

2. Mechanical heat sanitizing rinse water must not exceed:
   A. 212°F (100°C).
   B. 194°F (90°C).
   C. 180°F (82°C).
   D. 171°F (77°C).

3. Chemical sanitizers' effectiveness are affected by:
   A. Selectivity of the sanitizer.
   B. Concentration of the sanitizer.
   C. Time of exposure.
   D. All of the above.

4. It is important to thoroughly rinse off all food soils prior to sanitizing because they:
   A. May be recontaminated in the sanitizing solutions.
   B. May get stuck to another piece of equipment.
   C. May decrease the effectiveness of the sanitizing solution.
   D. May expand in the solution.

5. The only sanitizing agent that remains active above 120°F (49°C) is:
   A. Chlorine.
   B. Iodophors.
   C. Quats.
   D. Hypochlorites.

Real-life

Slide 11
Sanitizer Concentration

Joe was starting his shift in the warewashing area of the meat department. He emptied the three-compartment sinks and refilled the sinks with the proper concentrations of cleaning detergents and sanitizers. As the day went on and Joe washed the dirty equipment and utensils, emptied and refilled his wash and rinse water to keep the water warm and to add more detergent to the wash water. Did Joe do anything wrong? If so, what did he do wrong, and what could he have done differently?

Joe, by changing his wash water and rinse water to keep the water fresh and the concentration of detergent at proper levels, was doing the right things. However, Joe should have used test strips to track the concentration of the sanitizer solution and ensure it was kept at the proper concentration for sanitizing. He should test the level of sanitizer in his third-sink periodically to ensure it is at the proper levels and change it as needed. Since different types of sanitizers require different levels of concentration, Joe should be sure he is using the correct test strips.

Slide 10
Mechanical Sanitizing

Mike was using the single-tank warewashing machine in the bakery. The machine had an external booster heater that was used to bring the sanitizing rinse temperature to 165°F (74°C) to comply with the proper guidelines for the heat sanitizing method. Mike noticed the temperature was only reaching 120°F (49°C) during its final rinse. Mike continued to use the machine as it was. What did Mike do wrong, and what were his other options?

Mike should have informed his manager of the temperature problem with the single-tank dishwashing machine. He should have stopped using the machine and used a three-compartment sink manual warewashing method instead. As you will find out later in the session, if the booster heater on the machine is not working, it cannot be used to wash the utensils since the proper wash and rinse temperature for a single-tank warewashing machine is 165°F (74°C).

Making it Real

Sanitizer Recognition – Slides 12 – 13

- Identifying sanitizers and understanding their uses can be difficult. Discussing the types of sanitizers available in your retail food establishment may not be enough to get the information across to the learners. Many chemical companies use color-coding systems to help with this learning process.
    - Supplies
        - A sample, or an empty container of each of the detergents and sanitizers used in the establishment including:
            - Detergent sanitizers
            - Hypochlorites
            - Iodophors
            - Quats.
    - Procedure
        - If you are providing empty containers that have been thoroughly cleaned out or sealed containers, pass the products around the class, if not, show them to the class
        - Have the class identify each item and what it is used for in their establishment.
    - Teaching Tip – Photos of the products from the suppliers catalog can also be used for this exercise.

Mechanical Warewashing

Mechanical warewashing:

- Pre-rinse
- Rack utensils
- Machine will:
  - Wash
  - Rinse
  - Sanitize
- Allow utensils to air-dry
- Store properly.

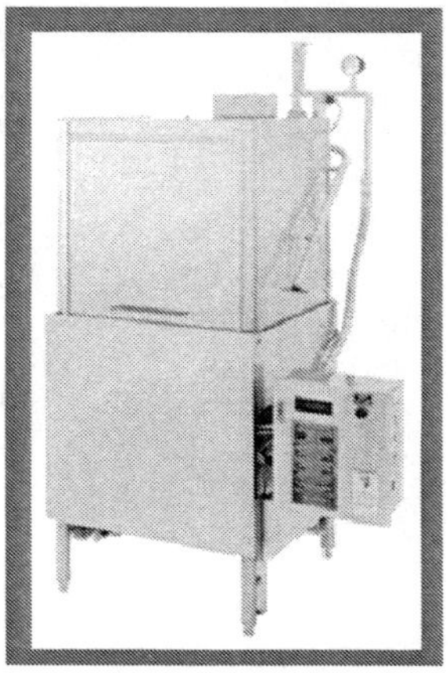

Mechanical warewashing

Adapted from ***Retail Best Practices and Guide to Food Safety and Sanitation* © 2003**

RBP. Pg. 235 – 239
RSG. Pg. 181 - 183

15

Slide 15

During the mechanical warewashing process, you should:

1. Pre-scrape equipment and pre-soak utensils.
2. Rack equipment, utensils so water will spray evenly, and equipment will drain.

The warewashing machine does steps 3 through 5 automatically.

3. Wash in a detergent solution that satisfies temperature requirements
   - Expose to wash water for at least 40 seconds.
4. Rinse in clean water that satisfies temperature requirements.
5. Rinse in sanitizing rinse that satisfies temperature requirements reviewed on Slide 10.
6. Air-dry equipment and utensils
   - *FDA Food Code* prohibits cloth drying.
7. Store in clean, dry area where they are protected from contamination.
8. Clean and maintain the machine to keep it in proper working condition.
9. Ensure the temperatures and water pressure matches the equipment data plate.

Different warewashing machines require different wash and rinse temperatures including:

- Single-tank, stationary-rack, single temperature
  - Wash 165°F (74°C)
  - Rinse 165°F (74°C)
- Single-tank, conveyor, dual temperature
  - Wash 160°F (71°C)
  - Rinse 180°F (82°C)
- Single-tank, stationary-rack, dual temperature
  - Wash 150°F (66°C)
  - Rinse 180°F (82°C)

- Multi-tank, conveyor, multi-temperature
    - Wash 150°F (66°C)
    - Rinse 160°F (71°C) to 180°F (82°C)
- Chemical sanitizing
    - Wash 120°F (49°C)
    - Rinse 120°F (49°C).

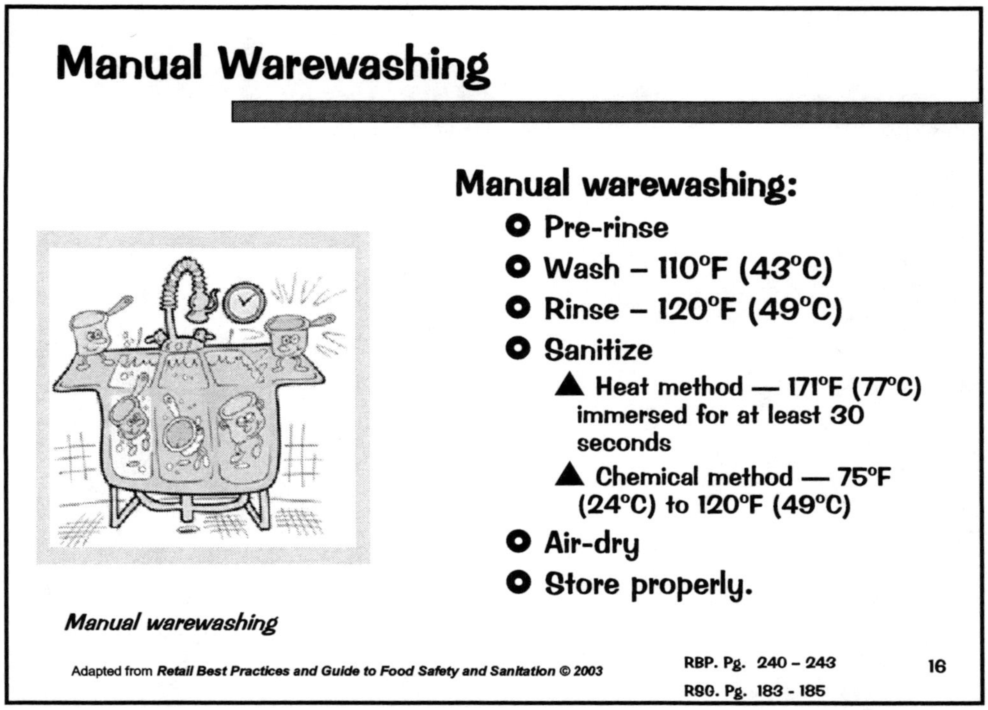

Slide 16

Manual warewashing sinks typically have at least three compartments:

1. First compartment for washing.
2. Second compartment for rinsing.
3. Third compartment for sanitizing.

Manual warewashing sinks:

- Should be large enough to accommodate largest equipment
- Must have sloped drain boards or dish tables
- Must have hot and cold potable water.

Manual warewashing steps with water temperatures include:

1. Pre-rinse and pre-scrape food particles from utensils.
2. Wash – not less than 110°F (43°C) water
   a. Use a cloth, nylon brush, or other approved cleaning device to loosen soil.
3. Rinse – between 110°F (43°C) and 120°F (49°C) water.
4. Sanitize – Sanitizers are ineffective on dirty contact surfaces that have not been through a proper washing procedure.
   a. Heat method – immerse in 171°F (77°C) water for at least 30 seconds
   b. Chemical method – 75°F (24°C) to 120°F (49°C) water depending upon the chemical sanitizer.
5. Air-dry
   a. *FDA Food Code* prohibits cloth drying
   b. Although drying agents are not usually used in the manual warewashing process, if drying agents are used in conjunction with sanitization, they must be Generally Recognized as Safe (GRAS) for the intended purpose, or have been approved for

use as a drying agent under the relevant provisions contained in Title 21 of the Code of Federal Regulations.

6. Store properly.

Wash, rinse, and sanitizing waters should be changed:

- As needed
- When chemical solution concentrations fall below standard
    - Use chemical test strips supplied by manufacturer to test solution concentration.

Regardless of the method used, cleaned and sanitized utensils must be:

- Properly stored in such a way as to protect them from contamination
    - Away from toilet rooms, locker rooms, sewer lines, etc.

In many retail food establishments, manual warewashing sinks are equipped with automated dispensing systems. Be sure these systems are properly cleaned and maintained.

Cleaning Fixed Equipment

Cleaning fixed equipment

For safety reasons, be sure to disconnect the power source prior to cleaning fixed equipment!

Adapted from ***Retail Best Practices and Guide to Food Safety and Sanitation* © 2003**

RBP. Pg. 243 – 247
RSG. Pg. 185 - 188

17

Slide 17

Fixed equipment, such as floor-mounted or table-mounted equipment:

- Should be easy to disassemble so that food contact surfaces can be exposed for proper cleaning
- Manually or mechanically wash all parts that are disassembled prior to reassembly.

The basic steps for cleaning fixed equipment are:

1. Disconnect any power source to the equipment.
2. Disassemble equipment.
3. Scrape accumulated debris off equipment and surroundings.
4. Manually or mechanically wash all disassembled parts.
5. Wash remaining parts with warm soapy water.
6. Rinse thoroughly with fresh, potable water.
7. Swab or spray a chemical sanitizing solution.
8. Allow all parts to air-dry.
9. Reassemble.
10. Re-sanitize parts that may have become contaminated due to handling.

Other cleaning methods include:

- Foam or spray method
    - Used on large pieces of equipment such as band saws
    - The wash, rinse, and sanitizer waters as well as the chemicals flow through the nozzle
    - Keep hoses and nozzles in good condition

- Bucket and brush method
    - Used on equipment that may be damaged by immersion or pressure spraying
    - Use separate buckets for each step (wash, rinse, sanitize)
    - Ensure equipment is properly cleaned and rinsed before it is sanitized
- Clean in Place (CIP)
    - Involves circulating chemical solutions through equipment (i.e., soft serve machines)
    - Guidelines include:
        - Empty food product and waste from equipment
        - Disconnect the power to the equipment
        - Disassemble any removable parts to be cleaned manually or mechanically
        - Reconnect the power to the machine
        - Clean and sanitize the unit by circulating wash, rinse, and sanitizing solutions through the machine
        - Disconnect the power to the machine
        - Reassemble parts and reconnect power to machine
        - Re-sanitize by circulating a manufacturer approved solution through the equipment.

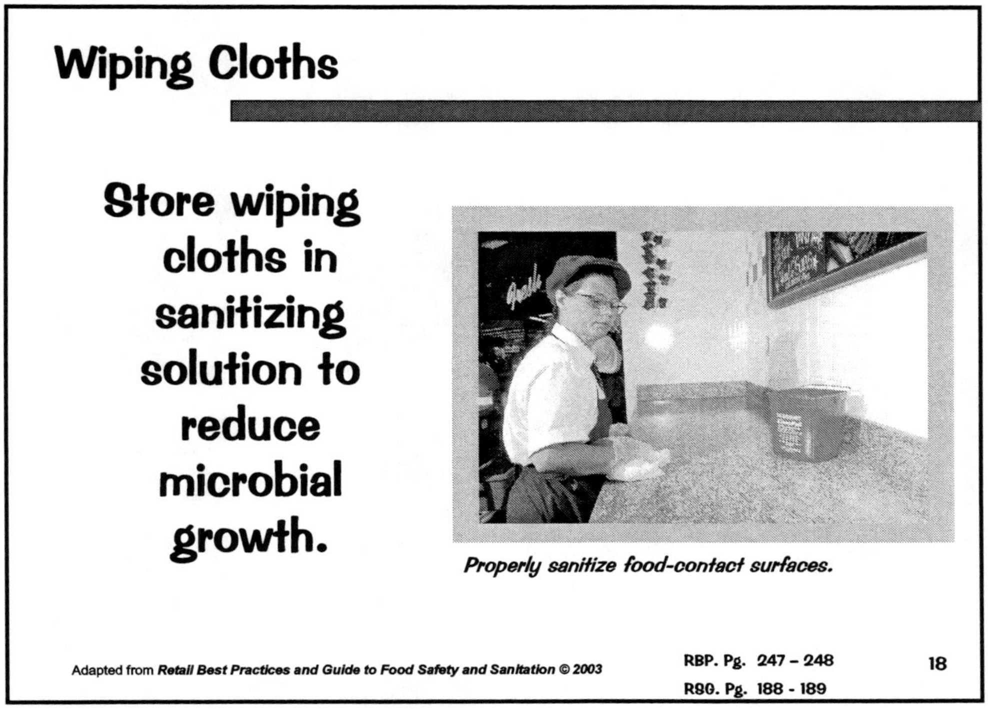

Slide 18

Wiping cloths:

- Should be stored in **in-place sanitizing solution** (sanitizing solution standing in a container) to reduce microbial growth
- Must be changed regularly
- Must be used only for wiping spills
- Must be laundered daily.

Keep wiping cloths separate from cloths that are, or have been, used for cleaning:

- Floors
- Raw food preparation areas
- Equipment.

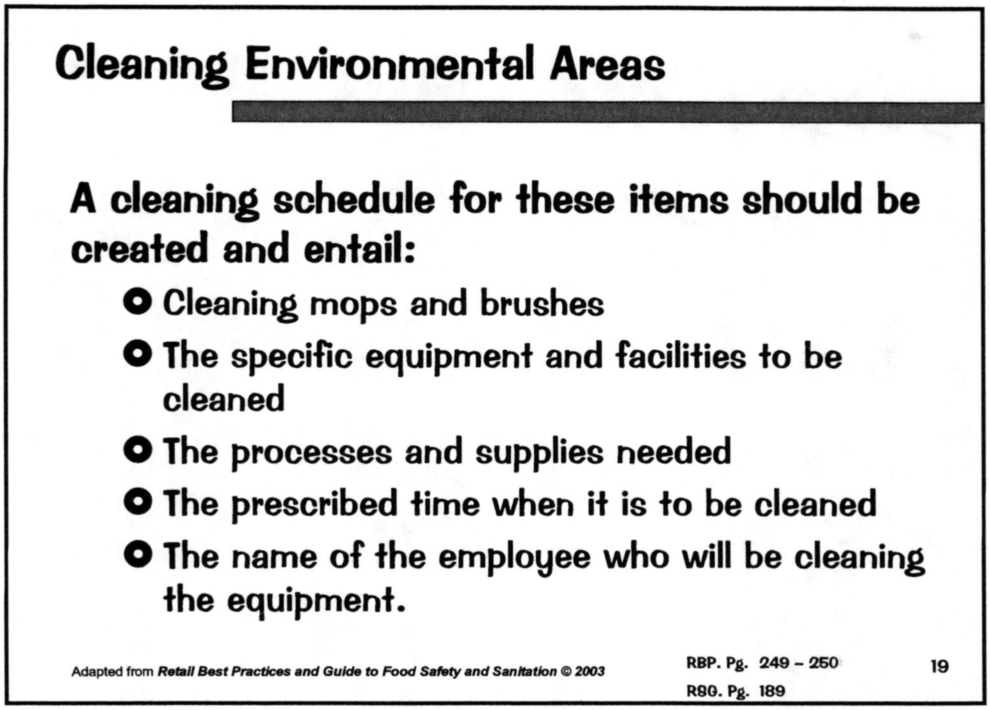

Slide 19

Non-food-contact surfaces must be cleaned regularly. Such areas include:

- Floors and floor drains
- Walls
- Ceilings – lights, fans, fan grates, and light covers.

A cleaning schedule for these items should be created and entail:

- The specific equipment and facilities to be cleaned
- The processes and supplies needed
- The prescribed time when it is to be cleaned
- The name of the employee who will be cleaning the equipment.

Cleaning should be performed:

- At a time when the least amount of food will be exposed
- When service will not be interrupted.

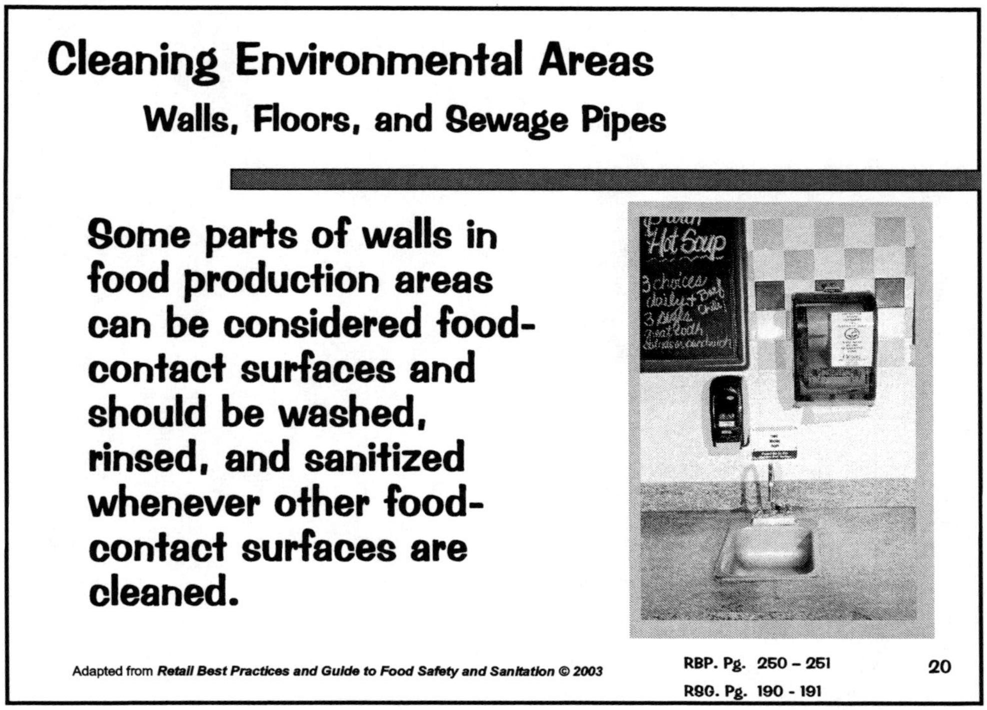

Slide 20

Walls:

- Some parts of walls in food production areas are considered to be food-contact surfaces
    - These should be washed, rinsed and sanitized whenever the other food-contact surfaces are cleaned
- Non-food-contact walls should be cleaned regularly using the bucket and brush or spray methods.
    - Walls and floors should not be cleaned using the spray method if the over spray could contaminate surrounding exposed equipment, supplies, or food in the area.

Floors:

- Floors in customer shopping areas are usually cleaned with floor-cleaning machines
- Floors in food production areas are usually cleaned using a spray system
- Floor drains must be properly maintained. Commonly followed steps include:
    - Remove the grate or cover over the drain
    - Clean out waste and other debris from the drain
    - Use a sprayer or hose to flush the drain and grate or cover
    - Pour in drain cleaner to break up grease and other waste in the drain
    - Wash the drain using a brush or water pressure from the sprayer
    - Rinse the drain with hot water
    - Pour or spray sanitizer into the drain
- Odor problems from floor drains can be reduced by regularly flushing the drain with hot water.

Sewage pipes:

- Properly maintain sewer pipes to prevent health hazards
- If a sewage backup occurs, follow these general guidelines:
    - Immediately:
        - Shut down operation in the affected areas
        - Remove any exposed foods, packaging materials, etc.
    - Remove anything from the floor that might be damaged or contaminated by the water
    - Remove the blockage in the pipe that is causing the backup, and call a professional if needed
    - Wash, rinse, and sanitize all equipment, floors, walls, and other objects that may have been contaminated.

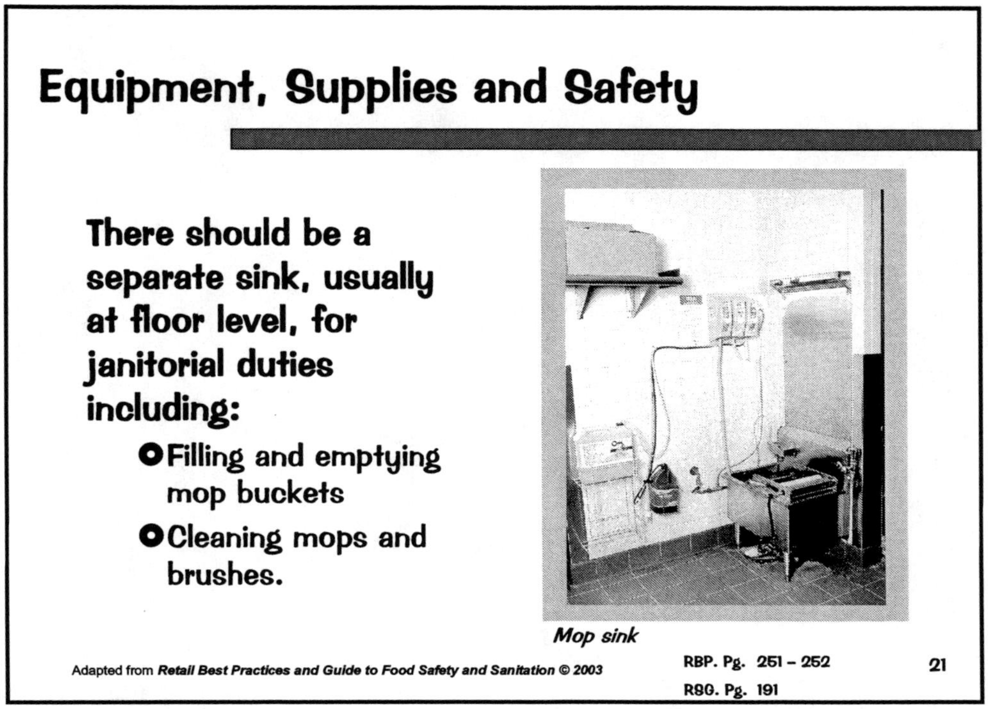

Slide 21

Cleaning equipment and supplies should be stored away from food and utensils.

Examples of commonly used cleaning equipment include:

- Nylon brushes
- Cloths
- Scouring pads
- Squeegees
- Mops and buckets
- Spray bottles
- Hoses and foam guns.

There should be a separate sink, usually at floor level, for janitorial duties including:

- Filling and emptying mop buckets
- Cleaning mops and brushes.

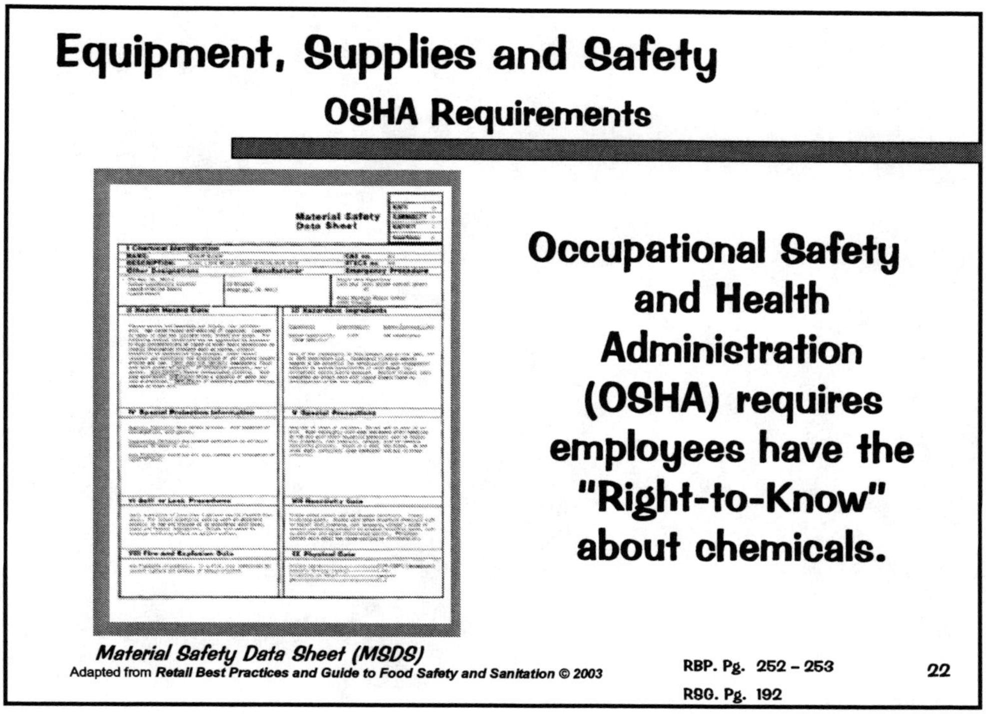

Slide 22

Containers of poisonous or toxic material and personal care items must bear a legible manufacturer's label.

- Working containers used for storing poisonous or toxic materials taken from bulk supplies must be clearly labeled and individually identified with the common name of the material
- A container previously used to store poisonous or toxic materials may not be used to store, transport, or dispense food.

Occupational Safety and Health Administration (OSHA):

- Requires employees have the "Right-to-Know" about chemicals
- Failure to provide Material Safety Data Sheets (MSDS) can result in OSHA fines.

Material Safety Data Sheets (MSDS) provide this information. Information that can be found on MSDS sheets include:

- Ingredients
- Physical and chemical characteristics
- Fire, explosion, reactivity, and health hazard data
- How to handle chemicals safely
- Possible protective equipment that may be needed.

Slide 23

Safety Stop! 6.3

Sample Test Questions

1. To effectively dry equipment and utensils, you should:
   A. Wipe them with a dry towel until all of the moisture is gone.
   B. Allow them to air-dry.
   C. Place them in an approved drying machine.
   D. Wipe them with your apron.

2. When manually washing equipment and utensils, the temperature of the wash water should be:
   A. At least 165°F (74°C).
   B. At least 120°F (49°C).
   C. **At least 110°F (43°C).**
   D. At least 150°F (66°C).

3. The first step in cleaning fixed equipment is to
   A. Disassemble the equipment.
   B. Wash the equipment with a detergent solution.
   C. Disconnect any power source to the equipment.
   D. Scrape off accumulated debris on the equipment.

4. Wiping cloths that are being held between uses should be kept in:
   A. A refrigerator.
   B. An in-place sanitizing solution.
   C. A bucket of warm water.
   A. A laundry bag.

Real-life

<u>Slide 19</u>
<u>Regular Cleaning Schedule</u>
The manager of a 24-hour grocery store that was going to be opening in a couple of days was writing out a daily cleaning schedule. The store was going to be very large and located in a plaza next to an all-night diner and a movie theater. What types of things should the manager consider when writing his cleaning schedule?

There are several things the manager should consider. The first thing the manager will need is flexibility. He may need to adjust his schedule several times before he gets the right fit. Being that the grocery store is located in a plaza with an all-night diner and a movie theater, he may find he encounters some odd times that his business peaks. In general, he should consider:

- Times when the least amount of food will be exposed
- Times when the cleaning and the service to the guests will not be interrupted
- Times when staffing will allow for it.

<u>Slide 21</u>
<u>Discarding Mop Water</u>
Mandy had just finished mopping the floor in the produce section. She took the mop bucket back to the warewashing area so she could empty it and clean the mop. When she got there, she found the janitor's sink was clogged and full of water. Mandy decided to pour the mop water down the food preparation sink and rinse the mop out there. What did Mandy do wrong and what should she have done?

Mandy should have cleared the clog in the janitor's sink and used it. She should never use the food preparation sink to discard dirty mop water or to clean the mop. Pouring dirty water from a mop bucket could potentially contaminate food products that are prepared in the food sink at a later time.

Making it Real

Manual Warewashing Identification – Slide 16

- This exercise will help to ensure all learners understand the key components of a manual warewashing area and their uses.
  - Supplies
    - An illustration of a manual warewashing area (three-compartment sink with pre-wash area)
      - One copy for each learner

- Procedure
    - Have the learners identify for what each station is used
    - Have the learners identify what procedures take place at each location
    - Have the learners identify proper chemical concentrations and exposure time at each location
- Teaching Tip – This can be done on paper or at an actual manual warewashing area. It can also be done for mechanical warewashing.

Making MSDS Sheets User Friendly – Slide 20

- Material Safety Data Sheets are often difficult to read and understand. Unfortunately, when they are needed most, the important information on the sheet may be hard to find. Use this exercise to drive home the importance of the MSDS sheets and to make them more user friendly in a time of need.
    - Supplies
        - A copy of each of the MSDS sheets required to be on-hand in your department or establishment
            - You may choose to bring all the sheets for the entire establishment and break your class into groups
            - Each group would perform this exercise for a different department
    - Procedure
        - Have the class break into groups and read through the MSDS sheets
        - Have them highlight critical information such as:
            - Chemical name
            - Active ingredients
            - Uses of the chemical
            - Storage of the chemical
            - Mixing instructions to reduce concentration
            - Warnings on other chemicals to avoid mixing with
            - First-aid procedures and precautions
    - Teaching Tip – It is a good idea to create a three-ring binder for each department with a MSDS sheet for every chemical used in the department. You may need multiple copies of some MSDS sheets to complete this exercise effectively.

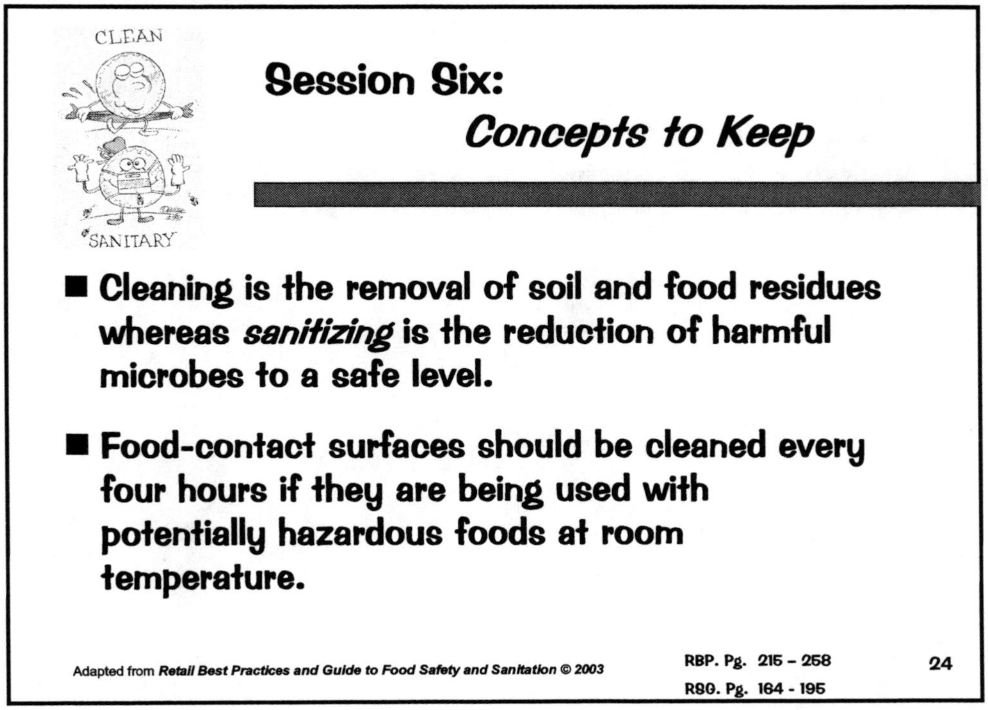

Slide 24

Concepts to Keep

- Cleaning is the removal of soil and food residues whereas sanitizing is the reduction of harmful microbes to a safe level.
- Food-contact surfaces should be cleaned every four hours if being used with potentially hazardous foods at room temperature.

(The real key is to clean well and often.)

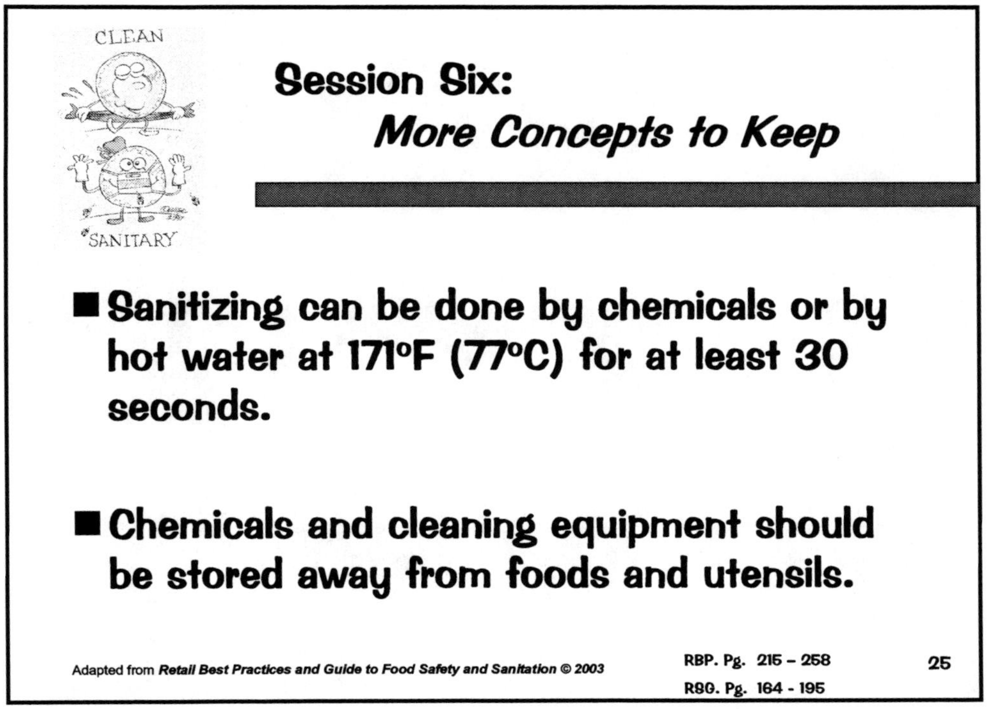

Slide 25

More Concepts to Keep

- Sanitizing can be done by chemicals or by hot water at 171°F (77°C) water for at least 30 seconds.
- Chemicals and cleaning equipment should be stored away from foods and utensils.

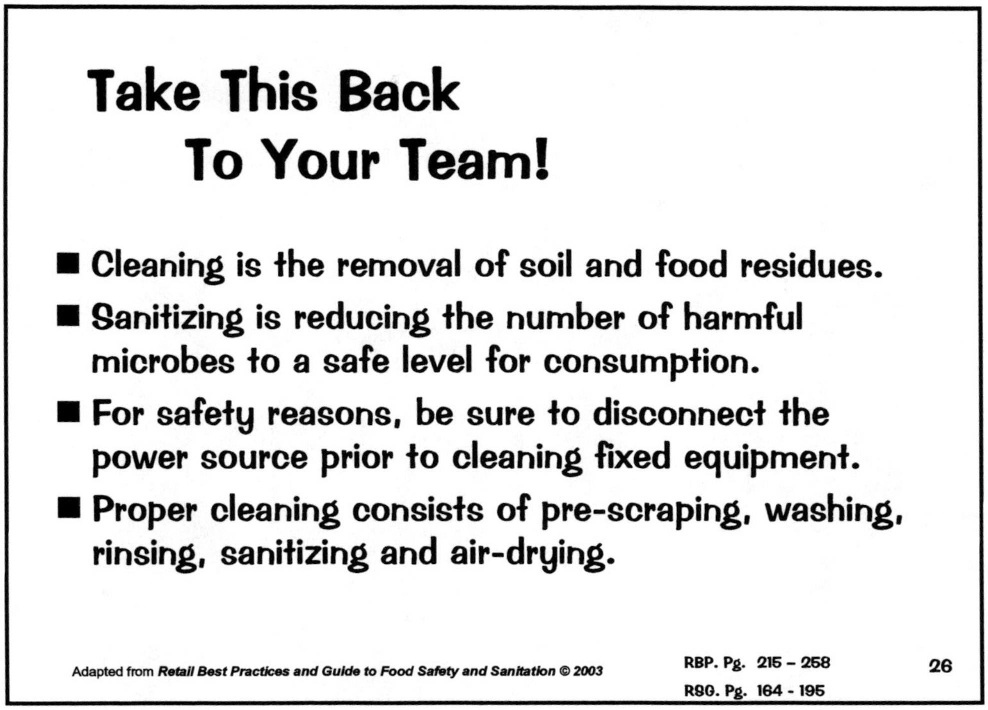

Slide 26

Take This Back to Your Team!

- Cleaning is the removal of soil and food residues.
- Sanitizing is reducing the number of harmful microbes to a safe level for consumption.
- For safety reasons, be sure to disconnect the power source prior to cleaning fixed equipment.
- Proper cleaning consists of pre-scraping, washing, rinsing, sanitizing and air-drying.

Session 7

Environmental Sanitation and Maintenance – Session 7 Key Points

Slide 1

<u>Training Tip:</u>
The content of each slide provides the most essential information in the session. The written text below each slide is provided to help you understand and better teach the essential information. The written text may also be used to answer questions that may be posed during the session. Depending on the length of your training session, you may not be able to address all the information contained beneath each slide.

Overview:
Surveys have shown customers believe cleanliness is a top consideration when choosing a place to eat or shop for food. Customer satisfaction is highest in retail food establishments that are clean and bright, and where quality food products are safely handled and displayed. Session 7 discusses:

- Keeping the building and its surroundings clean and free of debris
- The importance of clean and sanitary restrooms
- Proper water supply
- Proper sewage, garbage and refuse disposal
- The importance of an integrated pest management program.

A clean, sanitary environment is important not only to food safety and sanitation, but to customer relations as well.

Suggested timeline:
8-hour course: 30 minutes using the PowerPoint© slides only.

- 20 ***Slides.***

16-hour course: 45 minutes using the 24 PowerPoint© slides and the 2 ***Safety Stop!*** slides containing:

- 13 ***Sample Test Questions***
- 3 ***Real-life*** scenarios
- 3 ***Making it Real*** interactive exercises

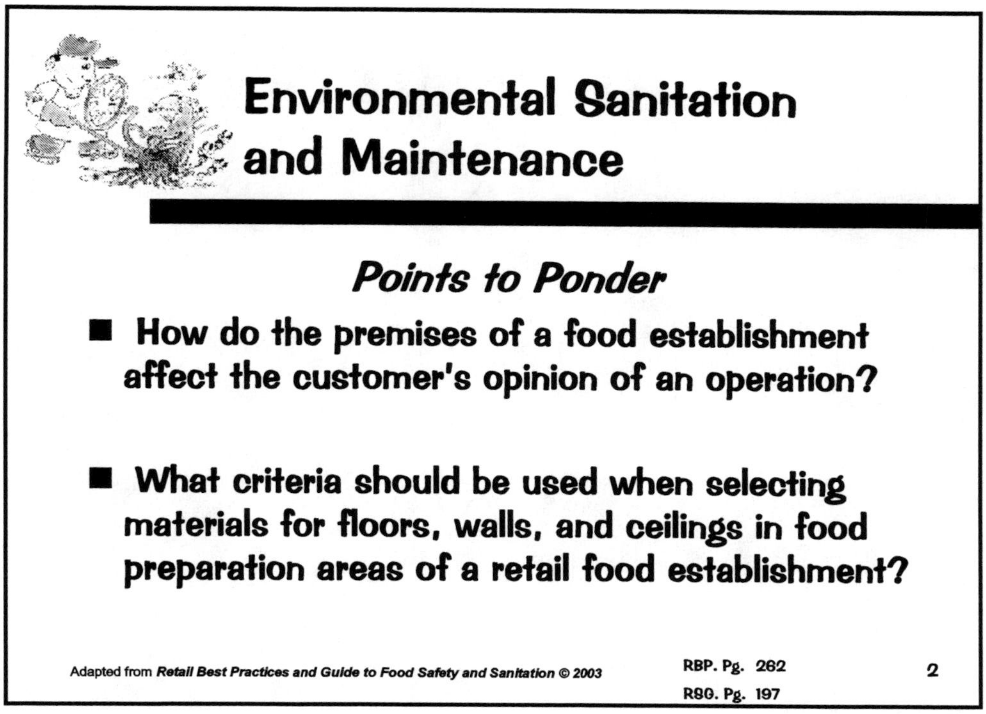

Slide 2

Points to Ponder

- How do the premises of a food establishment affect the customer's opinion of an operation?

- What criteria should be used when selecting materials for floors, walls, and ceilings in food preparation areas of a retail food establishment?

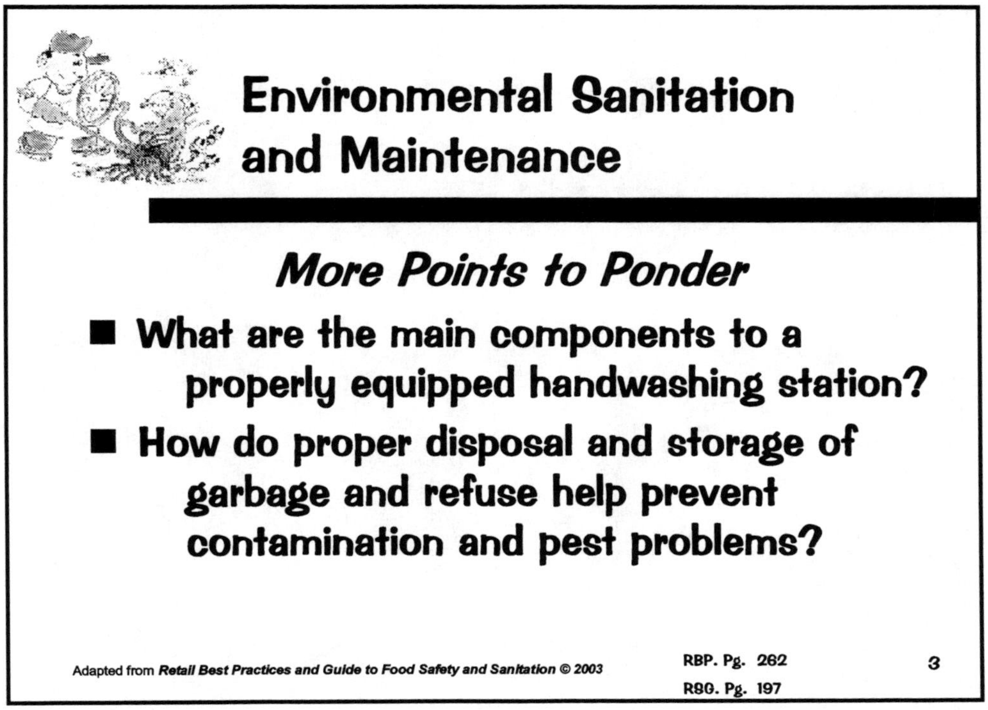

Slide 3

More Points to Ponder

- What are the main components to a properly equipped handwashing station?

- How do proper disposal and storage of garbage and refuse help prevent contamination and pest problems?

Slide 4

The No. 1 factor for customers deciding where to shop is the cleanliness and sanitation of the retail food establishment.

Retail food establishments are made up of:

- The building
- The contents of the building
- The surroundings and property around the building.

The exterior of a food establishment must be kept clean and free of debris and litter. It includes things such as:

- The building structure
- Parking areas
- Landscaping
- Doors
- Windows
- Refuse storage area.

The entrance of the building should be:

- Easy to find
- Easily accessible from streets and parking areas
- Clearly marked
- Well maintained and prevent the entrance of pests.

The interior of the building must be kept clean as well. It consists of such things as:

- Customer service areas
- Aisles
- Checkout and bagging areas
- Work areas
    - Preparation and handling areas
- Restrooms
- Storage areas.

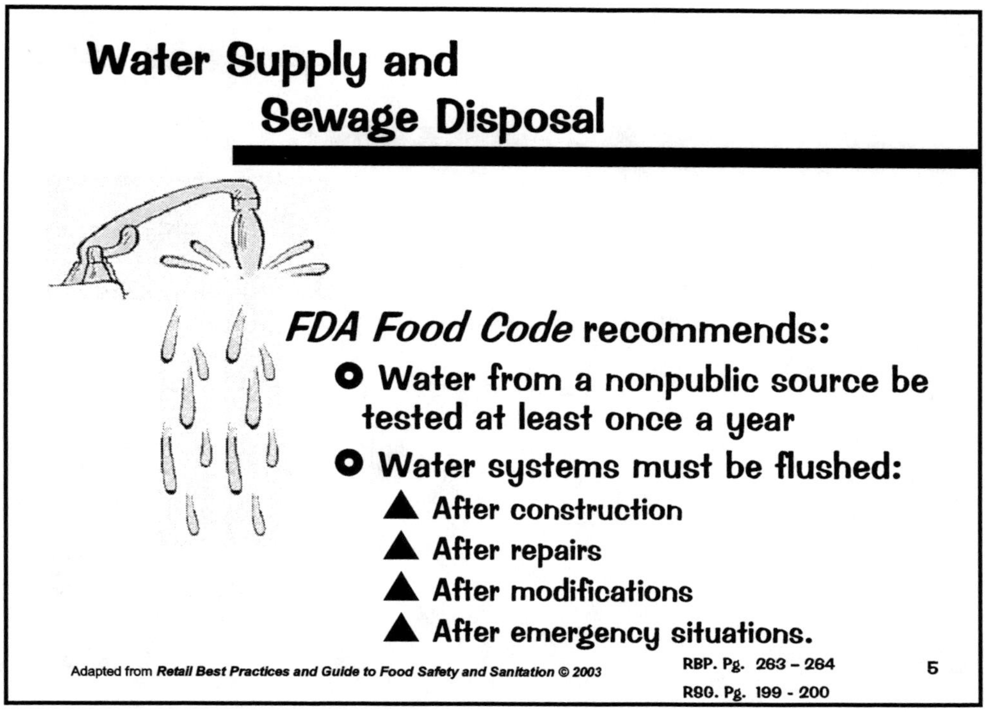

Slide 5

An adequate water supply and proper sewage disposal system are vital to the sanitation of all retail food establishments:

- The water source and the ability to meet hot water generation needs should be sufficient to meet demands of the retail food establishment.

Most establishments are connected to a public water system. Retail food establishments not connected to public water supplies, according to the *FDA Food Code*, should have:

- Wells constructed, maintained, and operated according to the water quality requirements of the jurisdiction
- Water tested at least once a year
- Drinking water systems flushed after:
    - Construction
    - Repairs
    - Modifications
    - Emergency situations.

A reservoir used to supply water to a device such as a produce fogger must:

- Be maintained in accordance with the manufacturer's specifications
- Be cleaned at least once a week in accordance with the manufacturer's specifications or according to the following specifications, whichever is more stringent:
    - Drain and completely disassemble the water and aerosol contact parts
    - Brush-clean the reservoir, aerosol tubing, and discharge nozzles with a suitable detergent solution
    - Flush the complete system with water to remove detergent solution
    - Rinse by immersing, spraying or swabbing the reservoir, aerosol tubing, and discharge nozzles with at least 50 ppm hypochlorite solution.

The use of non-potable water must be approved by the regulatory authority and only used for non-culinary purposes such as:

- Air handling systems
- Cooling systems
- Fire protection systems.

Proper disposal of sewage greatly reduces the risk of food and water contamination.

The *FDA Food Code* recommends sewage be disposed through an approved facility that is:

- A public sewage treatment plant, or
- An individual sewage disposal system that is:
    - Sized
    - Constructed
    - Maintained, and
    - Operated according to rules and regulations of the jurisdiction.

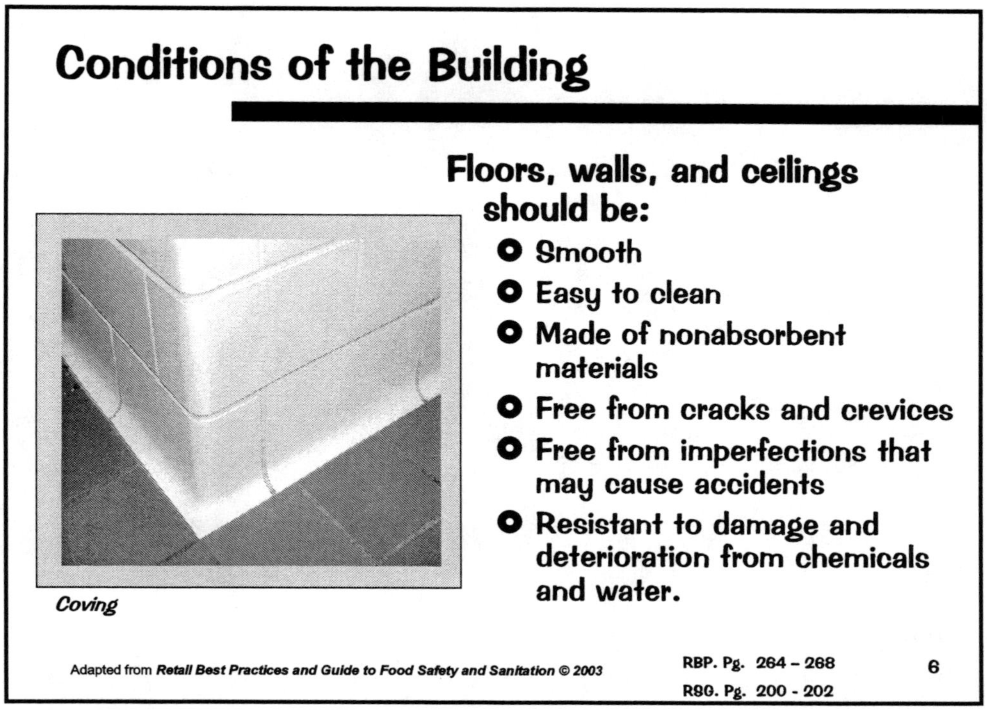

Slide 6

Consider specific needs of individual departments when selecting materials for constructing floors, walls and ceiling. Some general criteria that should be used in all departments includes:

- Sanitation
- Safety
- Durability
- Comfort
- Cost.

Floors, walls, and ceilings should be:

- Smooth
- Easy to clean
- Made of nonabsorbent materials
- Free from cracks and crevices
- Free from imperfections that may cause accidents
- Resistant to damage and deterioration from chemicals and water.

The preferred floor materials in food preparation and warewashing areas include:

- Terrazzo
- Quarry tile
- Asphalt tile
- Ceramic tile
- Concrete sealed with epoxy or similar material to make it durable and nonabsorbent.

Avoid:

- Soft wood
- Vinyl.

The ***FDA Food Code*** prohibits the use of carpeting in:

- Food preparation areas
- Walk-in refrigerators
- Warewashing areas
- Toilet room areas
- Refuse storage rooms
- Other areas subject to moisture.

Floors should be graded toward drains and the floor and walls must be coved.

<u>Coving</u>:

- Is a curved, sealed edge between the floor and wall
- Eliminates sharp corners or gaps.

Use mats and other anti-slip floor coverings to avoid slips and falls!!!

Unless they are being used in small amounts to collect liquid spills or drippage prior to cleaning such as in produce display areas, the *FDA Food Code* prohibits the use of:

- Sawdust
- Diatomaceous earth
- Similar materials.

Walls and ceilings in food preparation and warewashing areas should be:

- Constructed of materials such as ceramic tile, stainless steel or fiberglass that are
    - Smooth
    - Nonabsorbent
    - Easy to clean
- Light in color to enhance artificial lighting (shows dirt and when the surface has been properly cleaned).

Concrete, porous blocks, and bricks should only be used in dry storage areas unless they are finished.

Restrooms

Toilet facilities:

- Are required for all employees
- Must be conveniently located and accessible during all hours of operation
- Must be completely enclosed and provide a tight-fitting and self-closing door.

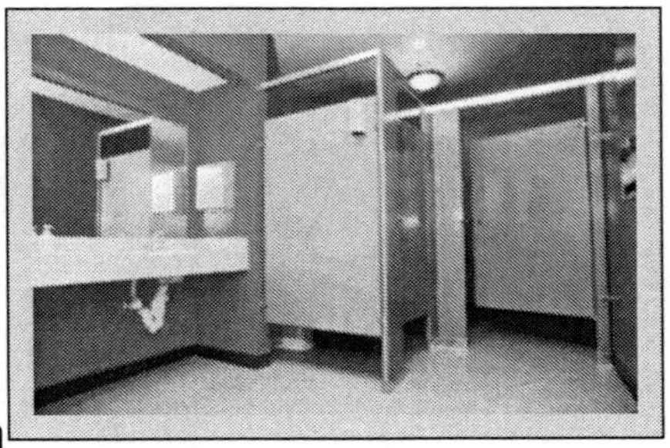

Restrooms should be cleaned frequently!

Adapted from *Retail Best Practices and Guide to Food Safety and Sanitation* © 2003

RBP. Pg. 268 – 269
RSG. Pg. 202 - 203

7

Slide 7

Toilet facilities:

- Are required for employees
- Must be conveniently located and accessible during all hours of operation
- Must be completely enclosed and provide a tight-fitting and self-closing door
- Should be constructed of durable and easily cleanable materials.

Toilet facilities located near work areas can:

- Promote good personal hygiene
- Reduce lost productivity
- Permit closer supervision of employees.

Poor sanitation of toilet areas can spread disease; include them in a regular cleaning schedule.

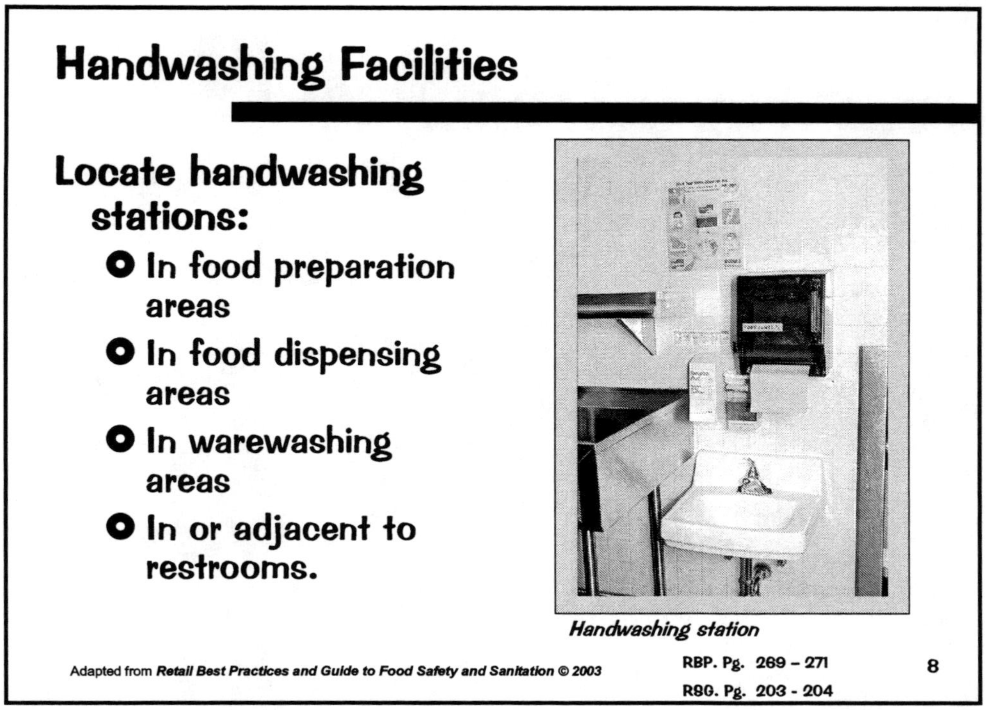

Slide 8

The number of handwashing stations required and their installation are usually set by local codes.

Locate handwashing stations:

- In food preparation areas
- In food dispensing areas
- In warewashing areas
- In or adjacent to restrooms.

Handwashing stations should be equipped with:

- Hot and cold running water under pressure
- A supply of soap
- A means to dry hands (disposable towels or mechanical hot-air are preferred).

Water must be able to reach 100°F (38°C) through a mixing valve or combination faucet.

Self-closing, slow-closing, or metering faucets must provide a minimum of 15 seconds of water flow.

Always keep handwashing stations clean and well maintained.

Never use handwashing stations for purposes other than hand washing.

Plumbing Hazards

Plumbing

A plumbing system includes:

- **Water supply and distribution pipes**
- **Plumbing fixtures and traps**
- **Soil, waste, and vent pipes**
- **Sanitary and storm sewers**
- **Building drains.**

Adapted from ***Retail Best Practices and Guide to Food Safety and Sanitation* © 2003**

RBP. Pg. 271
RSG. Pg. 204 - 205

9

Slide 9

A properly designed/installed plumbing system:

- Prevents contamination of the public water supply
- Is extremely important to food sanitation.

The plumbing system in retail food establishments must be maintained in good repair.

- When repairs are needed they must be completed in accordance with applicable local ordinances and state regulations.

<u>Cross connections</u>:

- Are any physical link through which contaminants can enter a potable water supply
- Direct cross connection:
  - Potable water system is directly connected to a drain, sewer, or other source of contamination
- Indirect cross connection:
  - The source of contamination may be blown across, sucked into, or diverted into a safe water supply.

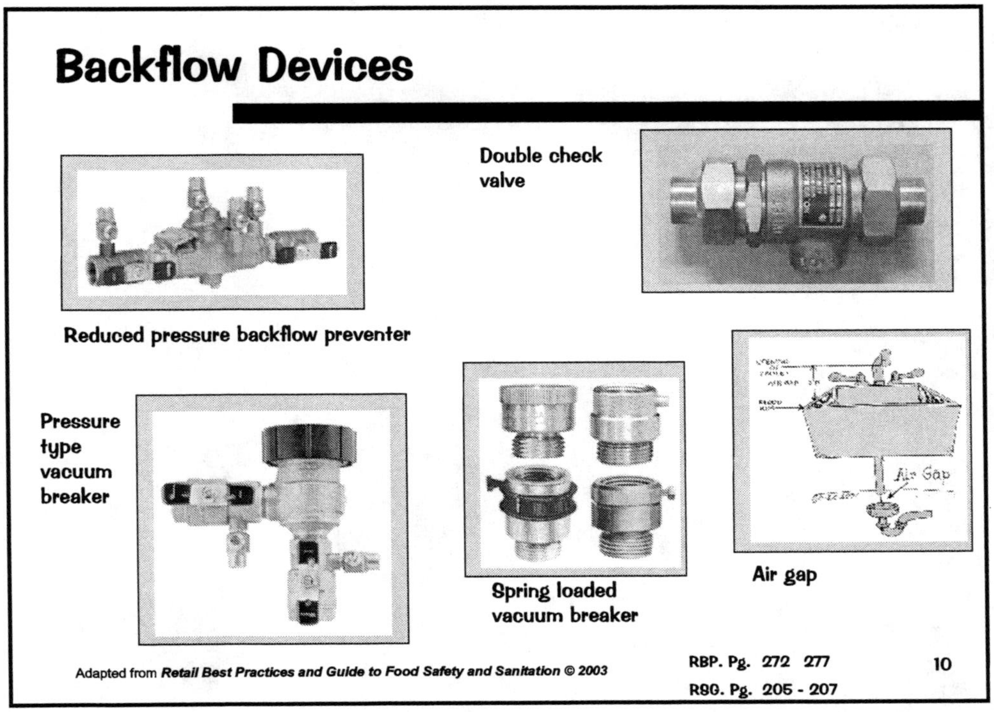

Slide 10

<u>**Backflow**</u>:

- Is the backward flow of contaminated water into a potable water supply.

Backflow can be prevented by using the following devices:

- <u>**Air gap**</u>
  - The physical separation of the potable and non-potable water system by air space
  - Vertical distance between the supply pipe and the flood rim must be two times the diameter of the supply pipe (2 D) of faucet, but never less then 1 inch
- <u>**Vacuum breaker**</u> – a backflow prevention device designed for use under a continuous supply of pressure
  - Atmospheric vacuum breaker
  - Pressure type vacuum breaker
- Double check valve
  - May be used as protection for all direct connections
- Reduced pressure principle backflow preventers
  - May be used on all direct connections that may be subject to backpressure or backsiphonage
  - Used when there is the possibility of contamination by material that is a potential health hazard.

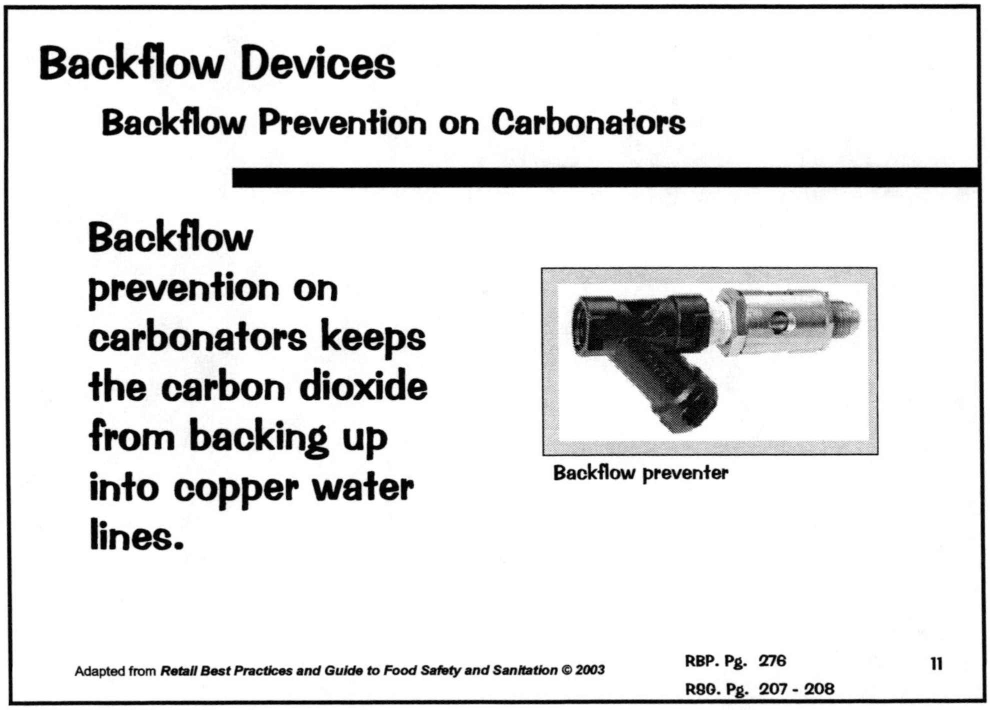

Slide 11

Backflow prevention on carbonators keeps the carbon dioxide from backing up into copper water lines.

- The carbon dioxide could dissolve the copper and mix it into soft drinks
- An air gap or a vented backflow prevention device that meets American Society of Sanitary Engineering (ASSE) Standard No. 12 must be used to prevent this.

Grease traps:

- Remove liquid grease and fats from wastewater
- Must be located for easily accessible cleaning.

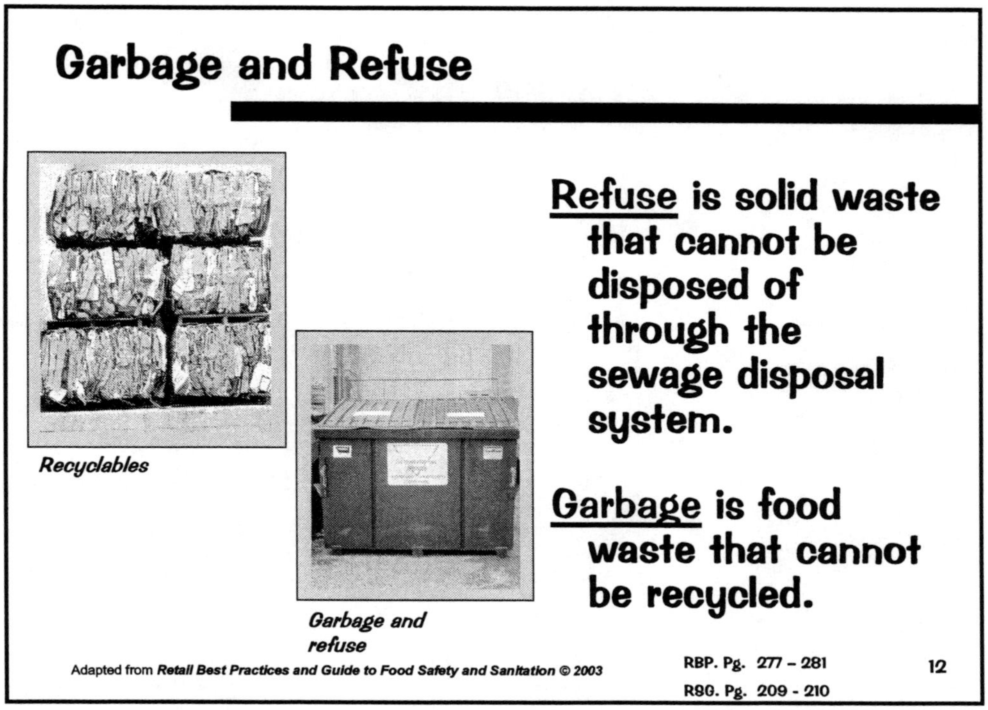

Slide 12

Refuse is solid waste that cannot be disposed of through the sewage disposal system.

Garbage is food waste that cannot be recycled.

Proper waste management protects food and equipment from contamination and decreases the attraction of:

- Insects
- Rodents
- Other pests.

Proper waste management, both inside and outside the establishment, consists of:

- Providing a trash container in each area of the establishment which is:
    - Durable
    - Cleanable
    - Insect and rodent proof
    - Leak proof
    - Nonabsorbent
- Keeping covers on receptacles that contain garbage
- Cleaning waste receptacles on a regular basis
- Providing storage areas for garbage and refuse both inside and outside the establishment
- Keeping storage areas clean and free of litter
- Removing refuse and garbage as often as necessary.

Use plastic bags or wet strength paper bags to line receptacles.

Do not place waste containers in locations where they may create a nuisance or interfere with cleaning.

Install compactors and other equipment for recyclables such as bottles, cans, and cardboard. Recyclables:

- May be stored outside for regular pickup
- A covered receptacle is not required if they are stored so they do not create a rodent-harborage problem
- Redeeming machines may be located in packaged food storage areas or consumer areas.

Slide 13

Safety Stop! 7.1

Sample Test Questions

1. The No. 1 factor that influences customers on where to shop is:
   A. Cost of the food.
   B. Nutrition of the food.
   C. Quality of service.
   D. Cleanliness of food and facilities.

2. The use of non-potable water must be approved by the regulatory authority and only be used for:
   A. Cleaning and sanitizing non-food-contact surfaces.
   B. Cleaning and sanitizing food-contact surfaces.
   C. Non-culinary purposes such as cooling systems.
   D. Drinking water.

3. Each of the following materials are acceptable for constructing floors in food preparation and warewashing areas except:
   A. Terrazzo.
   B. Vinyl.
   C. Ceramic tile.
   D. Quarry tile.

4. The *FDA Food code* prevents the use of carpeting in:
   A. Food preparation areas.
   B. Front entries.
   C. Customer service areas.
   D. Snack bars.

5. Toilet facilities located near work areas can:
   A. Promote good personal hygiene.
   B. Reduce lost productivity.
   C. Permit closer supervision of employees.
   D. All of the above.

6. Trash receptacles, both inside and outside the establishment, that contain garbage should be:
   a Cleaned daily.
   b Kept covered.
   c Discarded after each use.
   d Constructed of absorbent material.

7. Any physical link through which contaminants can enter a potable water supply is known as:
   A. Faucet.
   B. Air gap.
   C. Vacuum breaker.
   D. Cross connection.

8. Backflow can be prevented by which of the following devices:
   A. Coving.
   B. Cross connections.
   C. Air gaps.
   D. Grease traps.

Real-life

Slide 6
Slips and Falls
Robert was having a problem in the food preparation area of the bakery with employees slipping and falling. He decided to lay down some slip-resistant carpeting he bought from a household store. Would this be an acceptable solution to the problem? If not, why?

This solution would not be acceptable. The *FDA Food Code* prohibits the use of carpets in food preparation areas because they present problems with cleaning and with the collection of moisture.

Slide 12
Receptacle Coverings
Fernando worked the morning shift. He was tired of constantly opening and closing the lids on the dumpsters in the garbage area and coming in every morning to find trash piled on top of the closed lids on the dumpster. He decided to remove the lid permanently. Jorge, his manager, noticed what Fernando had done and thanked him for taking the initiative on removing the lids so they would not have the trash sitting on top of the dumpster every morning. Was this acceptable? If not, what should have been done?

It is important to keep lids on trash receptacles both inside and outside the building. Keeping the lids off the receptacle promotes pest infestation. Jorge should have explained this to Fernando and reattached the lids to the receptacle. To solve the problem with the garbage, Jorge should have spoken to the night manager and staff and informed them of the problem with the garbage being left on top of the dumpster rather than in it.

Making it Real

Backflow Preventers – Slide 10

- Take a tour! Many managers take backflow preventers for granted and would not know them if they saw them.
    - Supplies
        - Retail food establishment or any establishment with backflow prevention devices.
    - Procedure
        - Take the class for a walk around the establishment and have them point out the various types of backflow prevention devices
        - Have them describe why the backflow devices are present and what could potentially happen if they were not there
        - Have them list any additional backflow devices that would be appropriate in that setting
    - Teaching Tip – This is a good break to wake the group up! You may choose to hold this activity to the second half of this session and have the class look for any signs of pest infestation as well.
    - Teaching Tip – Many plumbing supply houses will give samples of backflow preventers or sell them at cost. This would be a good activity when training is not at a retail food establishment.

Pest Control

Pests include:

- Insects
  - Flies
  - Cockroaches
  - Moths
  - Beetles
- Rodents
  - Rats
  - House mice.

Pest control

Adapted from *Retail Best Practices and Guide to Food Safety and Sanitation* © 2003 RBP. Pg. 281 RSG. Pg. 211 14

Slide 14

All retail food establishments should have a pest control program in place.

Pests include:

- Insects
  - Flies
  - Cockroaches
  - Moths
  - Beetles
- Rodents
  - Rats
  - Mice.

To keep pests away from your establishment, you must:

- Prevent entry of insects and rodents into the establishment
- Eliminate food, water, and hiding places for rodents and building areas
- Implement an integrated pest management program
- Keep the interior and exterior of the establishment clean and sanitary.

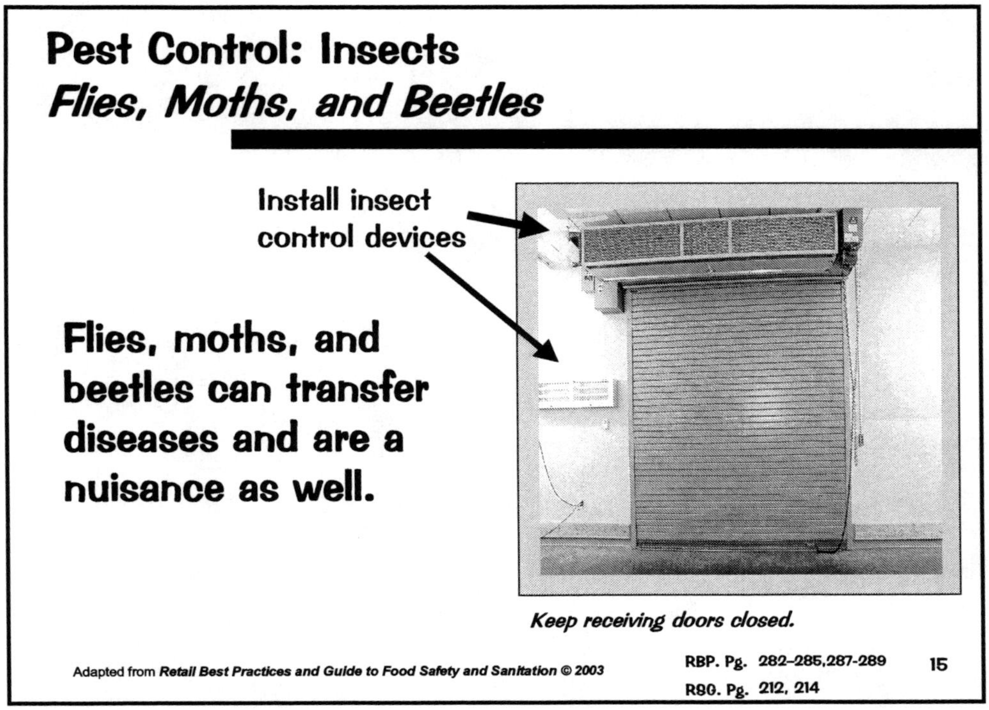

Slide 15

Pest control

- Insects:
  - Flies:
    - Common types – Houseflies, blowflies, and fruit flies
    - Concerns
      1. Flies regurgitate as a means of feeding.
      2. Most carry disease-causing bacteria on their body and in their intestinal tract.
    - Control
      1. Eliminate food and water supply by proper food storage and proper waste management.
      2. Use screens and air curtains on doors and windows.
      3. Install insect electrocution traps in non-food prep areas.
      4. Non-electrocuting systems may be used in areas where electrocuting devices are not.
      5. A professional pest control operator may apply chemical insecticides.

- Moths and beetles:
  - Common types – Indian meal moth, saw-toothed grain beetle, flour weevil, rice weevil
  - Concern – Loss to damaged food and nuisance is more of a concern than disease
  - Controls
    1. Proper stock rotation (FIFO).
    2. Proper storage of product.
    3. Clean spills completely and quickly.

4. Check incoming supplies for signs of insect infestation or damage.
5. Keep storage areas cool and dry.
6. A professional pest control operator may apply residual chemical insecticides and provide pheromone traps.
7. Utilize an active integrated pest management program.

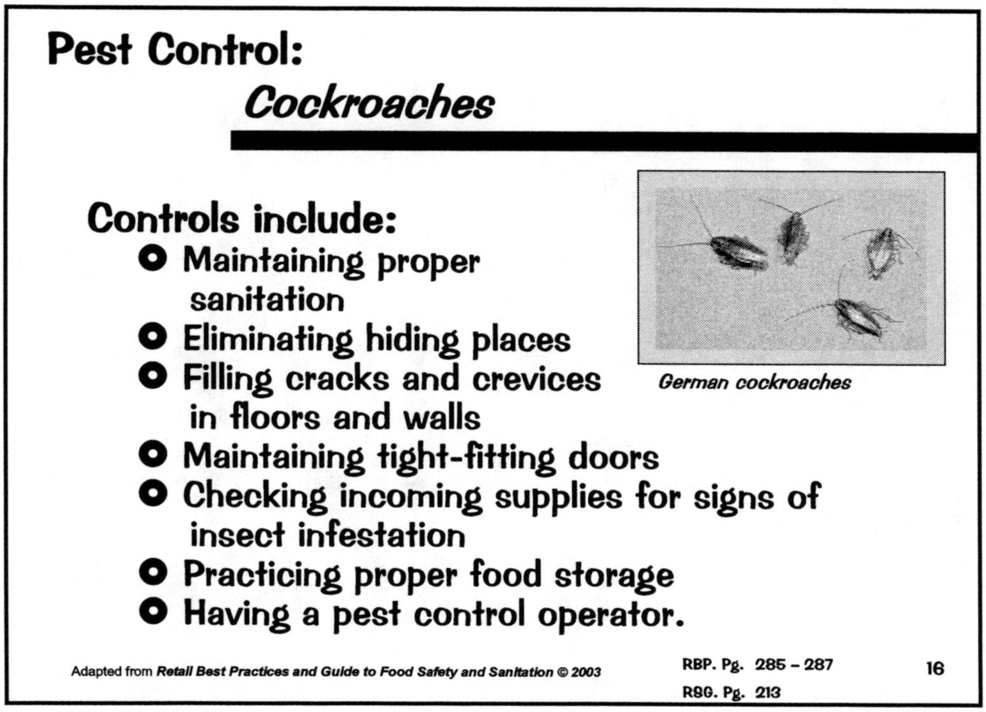

Slide 16

Pest control:

- Cockroaches:
  - Common types – American, Oriental, German, and brown banded
  - Concern – Most carry bacteria on their body and in their intestinal tract such as *Salmonella*
  - Controls –
    - Maintain proper sanitation inside and outside facilities
    - Eliminate hiding places such as cardboard boxes and rags
    - Fill cracks and crevices in floors and walls
    - Have tight-fitting doors and windows
    - Check incoming supplies for signs of insect infestation and eggs
    - Practice proper storage of foods
    - As a last alternative, a professional pest control operator may apply chemical insecticides.

Pest Control:
Rodents

Rodents:

- **Carry germs that can cause many diseases**
- **Can consume large quantities of foods.**

Rodents
Adapted from ***Retail Best Practices and Guide to Food Safety and Sanitation © 2003***

RBP. Pg. 289 – 290
RGG. Pg. 214 - 215

17

Slide 17

Rodents:

- Carry germs that can cause many diseases including salmonellosis, plague, and murine typhus
- Can consume and/or damage large quantities of foods.

Types of "domestic" rodents include:

- Norway rats (most common in United States) – also known as brown rats, sewer rats, and wharf rats
    - They are the largest of the domestic rodents
    - Primarily burrow and hide in the ground and around buildings
    - Will eat almost any food but prefer garbage, meat, fish, and cereal
    - Stay within 100 – 150 feet of food and water
- Roof rats –
    - Smaller than the Norway rat
    - Agile climbers and generally live on upper floors of buildings
    - Will eat vegetables, fruits, cereal, and grain
    - Stay within 100 – 150 feet of food and water
- House mice
    - Smallest of the domestic rodents
    - Found around buildings, nesting in walls, cabinets, and stored goods
    - Will eat grains and cereals
    - Stay within 10 – 30 feet of food and water.

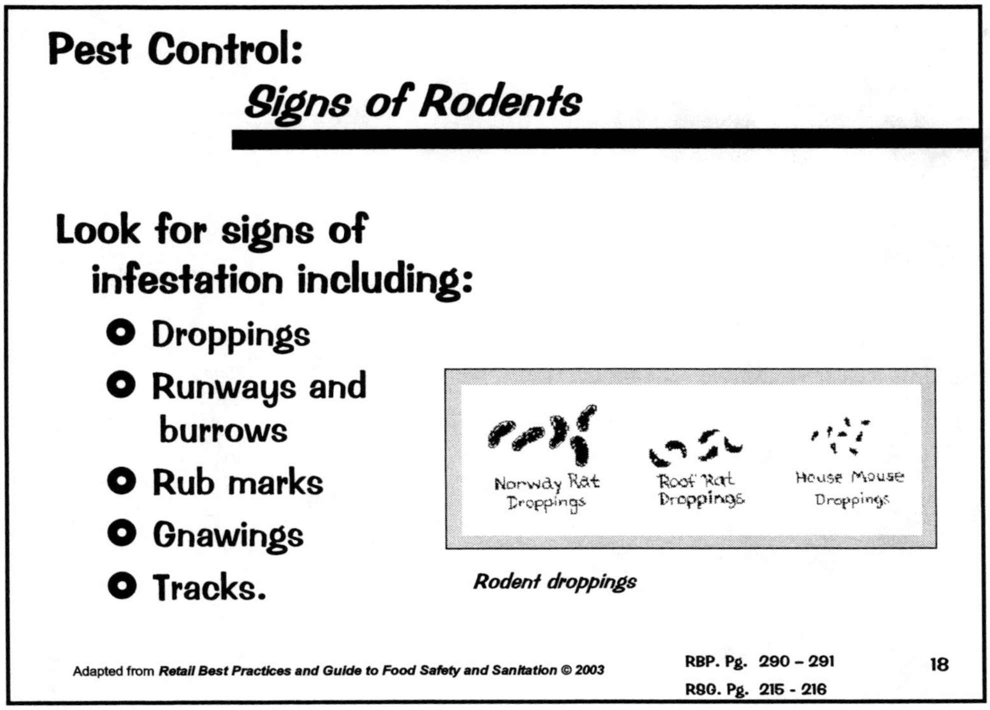

Slide 18

Rodents are nocturnal and are rarely seen in the daytime, therefore, you will need to look for the following signs of infestation including:

- Droppings – surest sign of an infestation
  - Fresh ones will be moist, soft, and shiny
  - Old ones will be dry and hard
- Runways and burrows
  - Look for paths that are 2 to 3 inches wide in grass/weeds
  - Look for holes that are 3 inches in diameter in earth banks, walls, and under rubbish
  - Active burrows will be free of cobwebs and dust and will have freshly dug earth at the entrance
- Rub marks
  - Rats stay close to walls and the oil and filth from the body will create rub marks
- Gnawings
  - Rats' incisor teeth grow 4 to 6 inches each year, gnawing keeps them short enough to use
  - Gnawings are fresh if they are light colored and have well defined teeth marks
- Tracks
  - Look for:
    - Tracks in dust in little used rooms
    - 1½" long mud streaks around puddles.

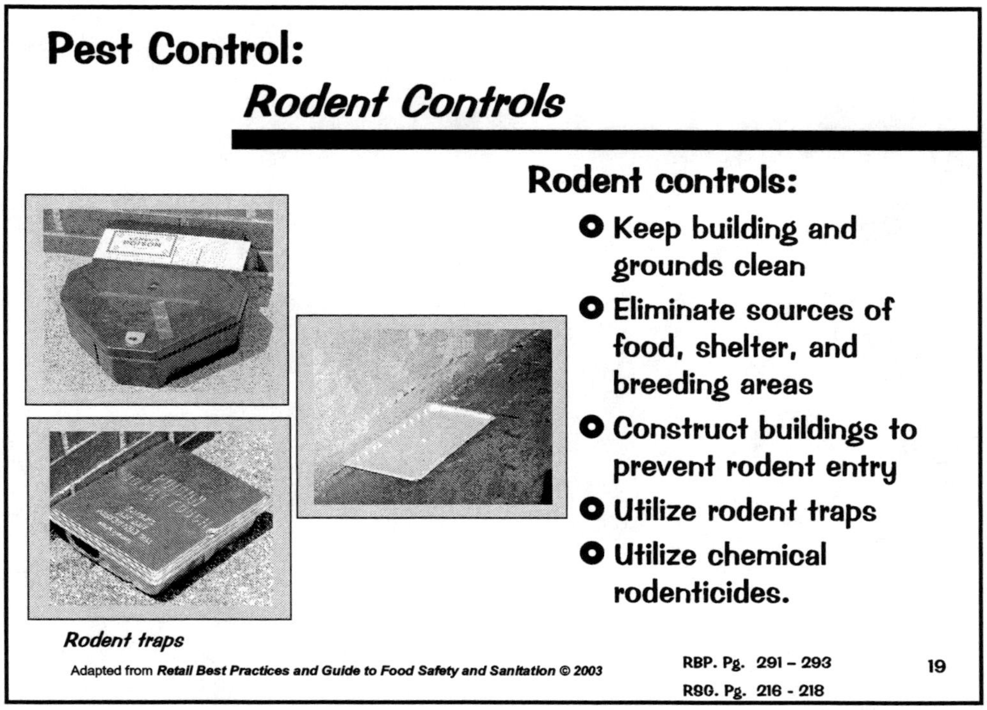

Slide 19

Rodent controls:

- Keep building and grounds clean and free of litter, waste, refuse, and uncut weeds/grass
- Eliminate sources of food, shelter, and breeding areas
- Construct buildings to prevent rodent entry
- Utilize rodent traps such as bait boxes and glue boards
- A professional pest control operator may apply chemical rodenticides.

The use of tracking powder pesticides is prohibited in retail food establishments.

- They can directly or indirectly contaminate
  - Food
  - Equipment
  - Utensils
  - Linens
  - Single-service/single-use items
- This contamination could adversely affect both the safety of the food and the general environment.

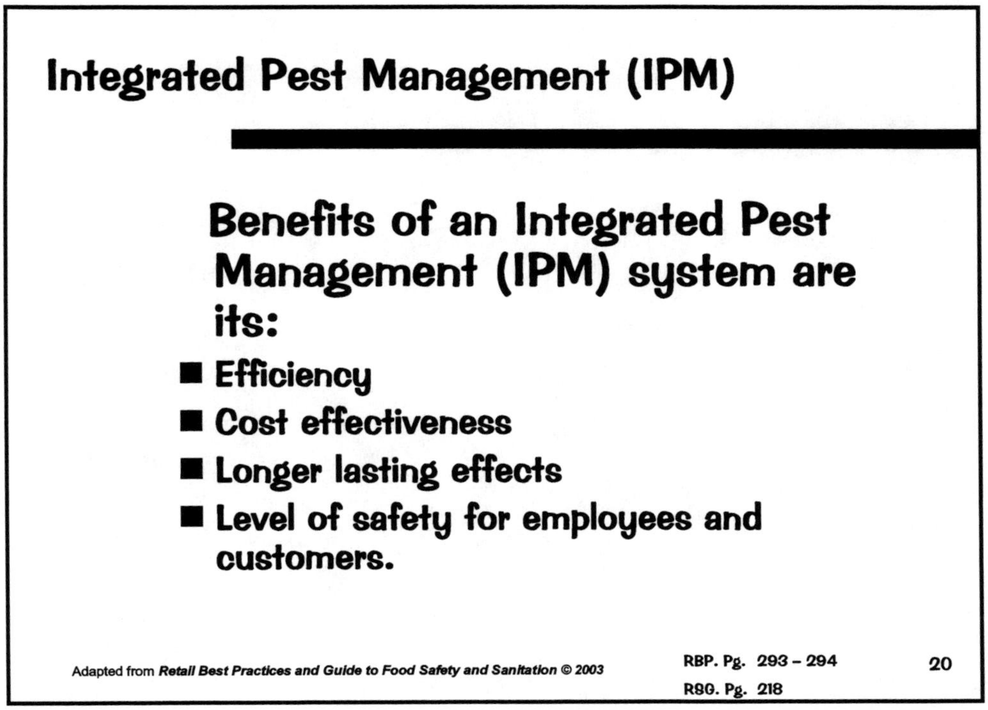

Slide 20

<u>Integrated Pest Management (IPM):</u>

- Uses a combination of sanitation, mechanical, and chemical procedures to control pests
- Chemical pesticides are used only as a last resort
    - If chemical pesticides are used, they must be approved for use as specified in the Code of Federal Regulations (40 CFR 152).

The National Pest Management Association recommends a five-step program:

1. Inspection.
2. Identification.
3. Sanitation.
4. Application of two or more pest management procedures.
5. Evaluation of effectiveness through follow-up inspections.

Benefits of IPM:

- More efficient
- Cost effective
- Longer lasting
- Safer for employees and customers.

Slide 21

Safety Stop! 7.2

Sample Test Questions

1. The primary responsibility of retail food establishment managers in pest control is to ensure:
   A. Good sanitation that will eliminate food, water, and harborage areas.
   B. Pesticides are applied safely.
   C. The pest control operator they use employs proper procedures.
   D. The exterior of the building is kept free of litter.

2. Common controls for eliminating moths and beetles include all of the following except:
   A. Proper stock rotation (FIFO).
   B. Keeping storage areas warm.
   C. Cleaning spills completely and quickly.
   D. Checking incoming supplies for signs of insect infestation or damage.

3. Rodents create food safety issues by:
   A. Being a nuisance.
   B. Consuming large quantities of food.
   C. Carrying and transferring germs that can cause diseases.
   D. All of the above.

4. Common signs of rodent infestation include:
   A. Gnawings.
   B. Rub marks.
   C. Droppings.
   D. All of the above.

5. A program that uses a combination of sanitation, mechanical, and chemical procedures to control pests is called:
   A. Rodent Control System.
   B. Integrated Pest Management.
   C. Animal Awareness Program.
   D. Pest Management Control System.

Real-life

<u>Slide 20</u>
<u>IPM</u>
Jeff was having difficulty meeting his bottom line and needed to find a way to increase sales or cut costs. Jeff went through his budget and decided on some areas to cut costs. He decided to eliminate the services of his pest control company. He thought he could do the same thing they were doing for half the price. Was this a good decision? Why or why not?

This was a bad decision on Jeff's part. By eliminating the professional services of a pest control company, Jeff opened the store up to potential pest infestation. Jeff does not have the experience or expertise to do this alone and should keep the services of the professionals.

Making it Real

Pest Identification – Slides 14 - 18

- Where do all the pests come from?
    - Supplies
        - Class
    - Procedure
        - Break the class into groups and have them discuss and identify the following:
            - Create a list of things that attract pests to retail food establishments
            - Create a list of common signs of insect and rodent infestation
            - Create a list of things a retail food establishment can do to control the pests commonly found in and around their operation
    - Teaching Tip – Have the learners present their findings to one another through a group leader. Allow them to determine what they retained and what information they lost!

Pest Control Professional – Slide 17

- Invite a pest control professional into your class and give them time to discuss the various types of pest control management.
    - Supplies
        - Pest control professional
            - You can usually get the pest control company you are currently using to participate in this activity at no charge.
    - Procedure
        - Encourage the pest control professional to discuss:
            - Pests common to the area
            - Steps that can be done by food employees that will support their services
    - Teaching Tip – This can be done with your local health inspector as well. Showing an interest in training and education to your local health representatives is always a good idea!

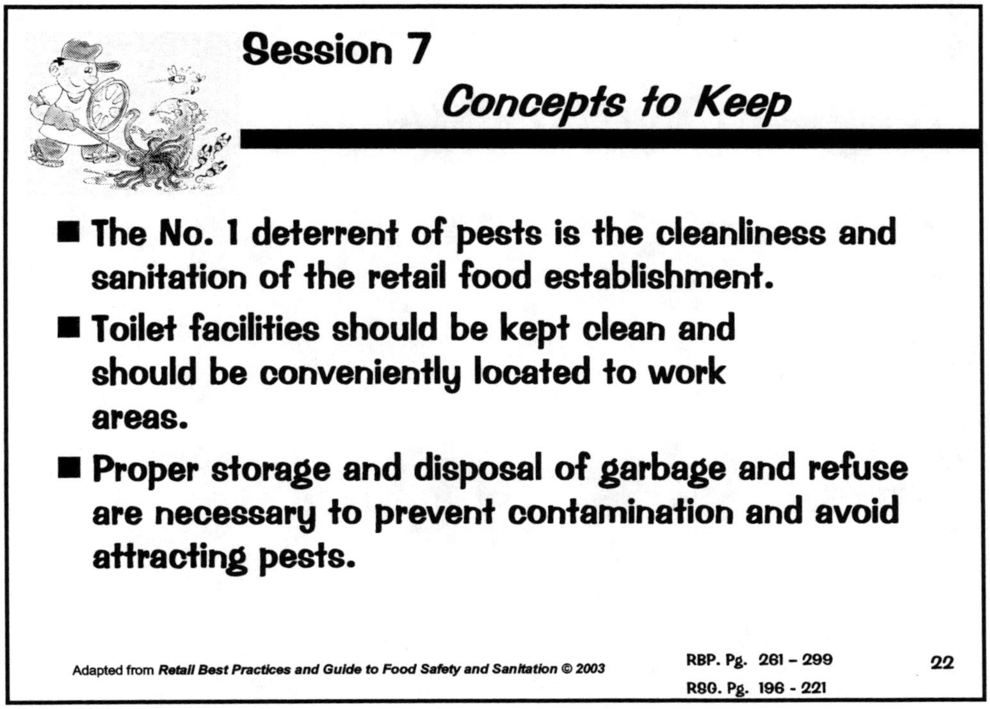

Slide 22

<u>**Concepts to Keep**</u>

- The No. 1 deterrent of pests is the cleanliness and sanitation of the retail food establishment.
- Toilet facilities should be kept clean and should be conveniently located to work areas.
- Proper storage and disposal of garbage and refuse are necessary to prevent contamination and avoid attracting pests.

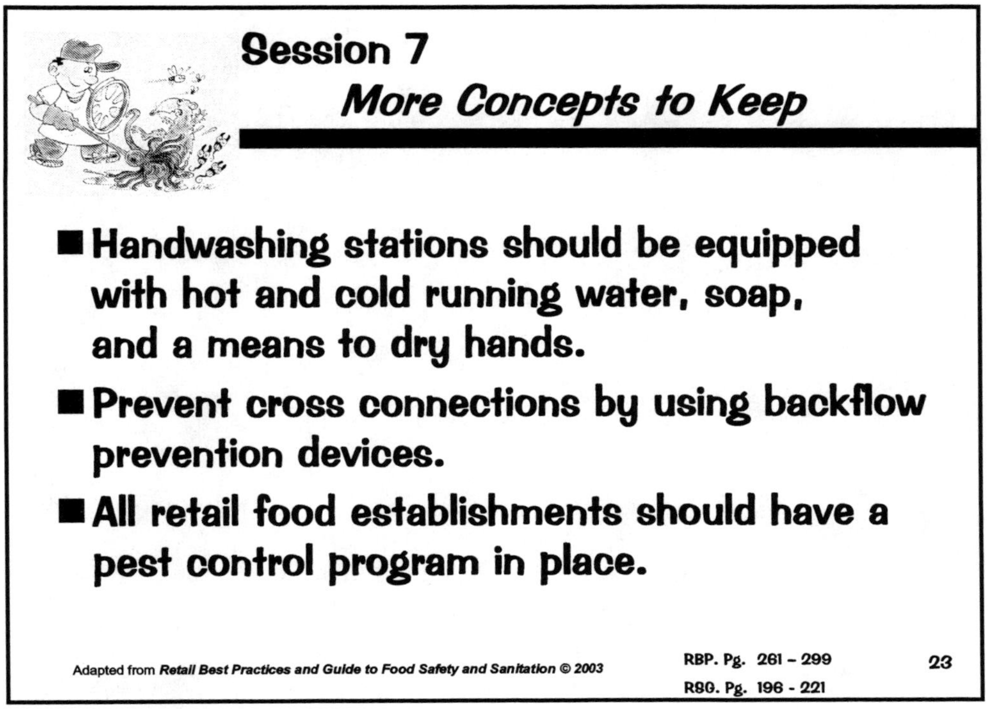

Slide 23

More Concepts to Keep

- Handwashing stations should be equipped with hot and cold running water, soap, and a means to dry hands.
- Prevent cross connections by using backflow prevention devices.
- All retail food establishments should have a pest control program in place.

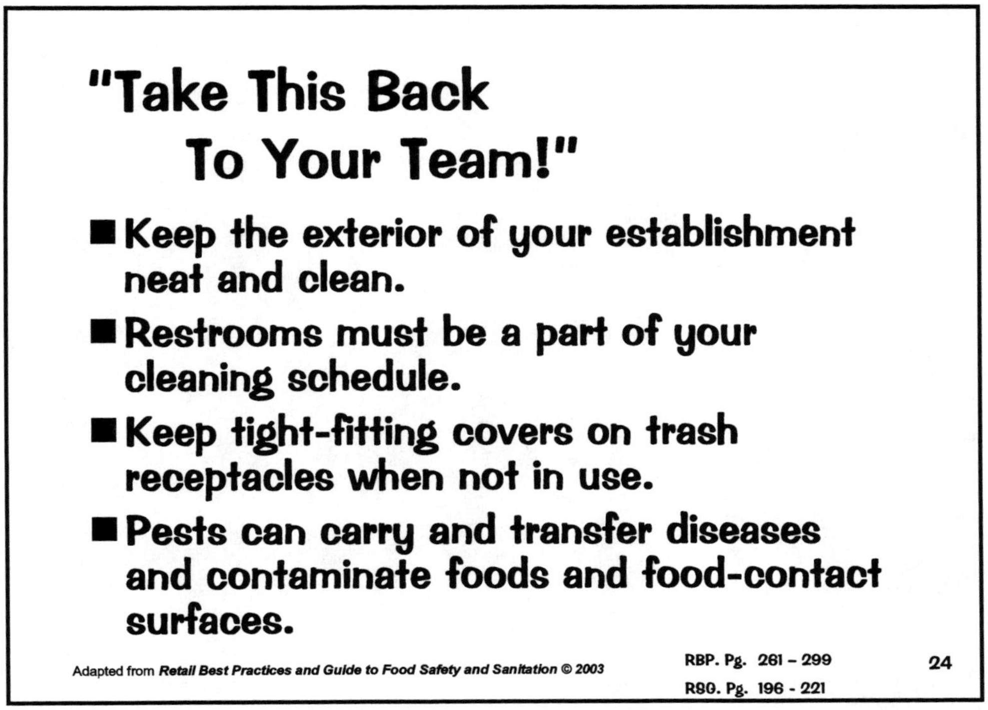

Slide 24

Take This Back to Your Team!

- Keep the exterior of your establishment neat and clean.
- Restrooms must be a part of your cleaning schedule.
- Keep tight-fitting covers on trash receptacles when not in use.
- Pests can carry and transfer diseases and contaminate foods and food-contact surfaces.

Session 8

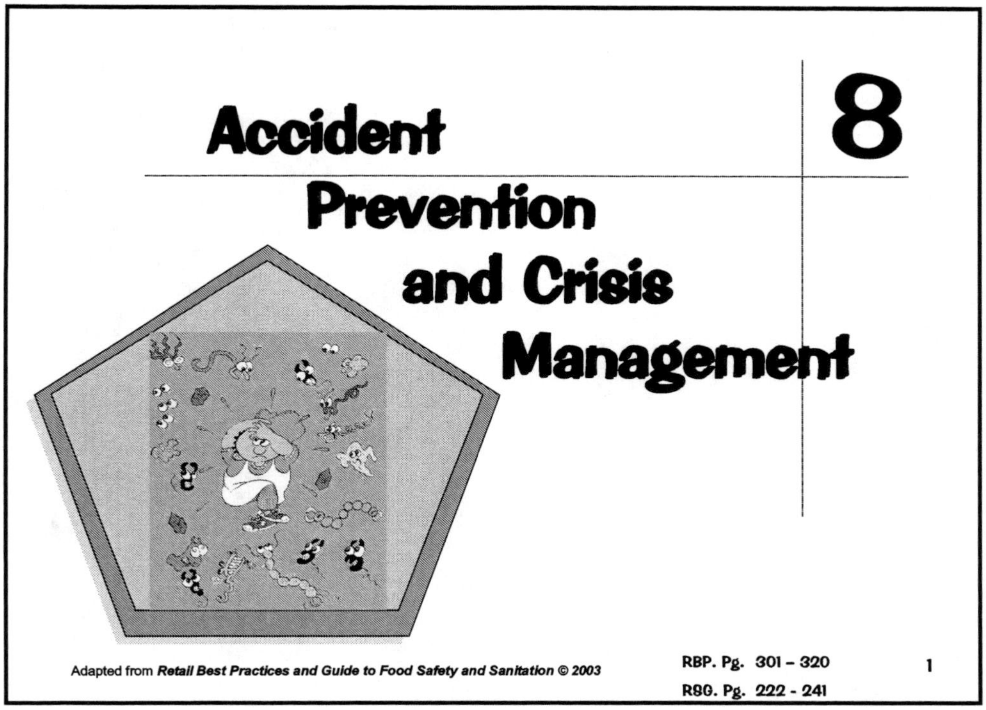

Accident Prevention & Crisis Management – Session 8 Key Points

Slide 1

<u>Training Tip:</u>

The content of each slide provides the most essential information in the session. The written text below each slide is provided to help you understand and better teach the essential information. The written text may also be used to answer questions that may be posed during the session. Depending on the length of your training session, you may not be able to address all the information contained beneath each slide.

Overview:

Session 8 reviews the concepts of accident prevention and crisis management. The learner should leave this session with an understanding of ways to prevent accidents, common types of accidents, and what to do when an accident happens. The learner will also understand what to do when a crisis occurs that may adversely affect the management, staff, and operations of the retail establishment.

Suggested Timeline:

8-hour course: 15 minutes using the PowerPoint© slides only.

- 11 ***Slides.***

16-hour course: 45 minutes using the 21 PowerPoint© slides and the 1 ***Safety Stop!*** slide containing:

- 8 ***Sample Test Questions***
- 4 ***Real-life*** scenarios
- 3 ***Making it Real*** interactive exercises.

Video suggestions and supplemental resources:

Videos:
Chemical Safety, Food Marketing Institute, 1-301-843-3084
An Ounce of Prevention: Protecting Yourself from Bloodborne Pathogen, Food Marketing Institute, 1-301-843-3084
Preventing Customer Slips and Falls, Food Marketing Institute, 1-301-843-3084
Preventing Cuts, Food Marketing Institute, 1-301-843-3084

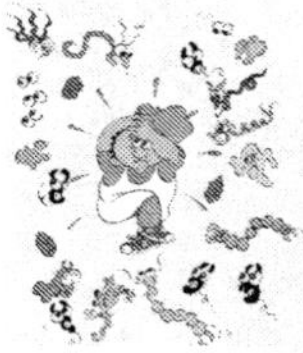

Accident Prevention and Crisis Management

Points to Ponder

- How can you prevent accidents?
- What should you do when a food recall or food crisis occurs?
- What do you do when an accident occurs?

Adapted from ***Retail Best Practices and Guide to Food Safety and Sanitation*** © 2003

RBP. Pg. 301 – 302
RSG. Pg. 223

2

Slide 2

Points to Ponder

- How can you prevent accidents?

- What should you do when a food recall or food crisis occurs?

- What do you do when an accident occurs?

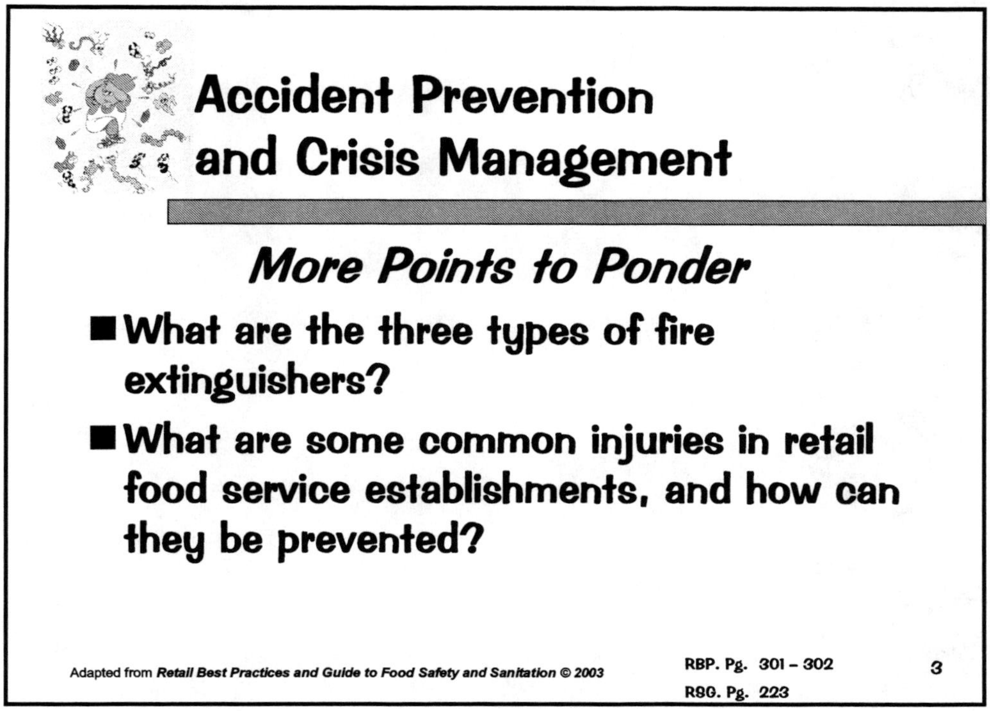

Slide 3

More points to Ponder

- What are three types of fire extinguishers?

- What are some common injuries in retail food service establishments, and how can they be prevented?

Accidents Can Be Prevented!

Accident prevention programs typically involve:

1. **Eliminating hazards in the environment.**
2. **Training employees about hazards and how to prevent or avoid them.**
3. **Providing personal protective equipment for employees.**

Accident prevention

Adapted from ***Retail Best Practices and Guide to Food Safety and Sanitation* © 2003**

RBP. Pg. 302 – 303
RGS. Pg. 224 - 225

4

Slide 4

Accidents, like foodborne illnesses, can be prevented.

Accident prevention programs typically involve three types of activities:

- Eliminating hazards in the environment
  - Use built-in safety guards on equipment
  - Make sure equipment is unplugged before cleaning or servicing
  - Buy equipment that is right for the job (proper design and construction)
- Training employees about potential hazards and how to prevent or avoid them
  - Good employee training on policies and procedures – teach food employees proper use of equipment and procedures:
    - Using ladders
    - Lifting things properly
    - How to use large heavy machinery
    - Using safety devices on equipment
    - Safe clothing choices
- Providing personal protective equipment for employees
  - Wear steel and nylon slash resistant fiber gloves when slicing and using cutting equipment
  - Using hot pads to protect hands against burns
  - Wearing work shoes with closed toes and rubber soles to protect against falls and damage to feet.

- Post a list of important telephone numbers beside the telephone in a readily accessible area. Include:
    - Fire department
    - Police
    - Emergency medical system
    - Nearest hospital emergency room.

When an Accident Happens...

- **What to do when an accident occurs:**
  1. Stay calm!!!
  2. Check to see if the victim is responsive - "Are you OK?"
  3. Examine the injury for seriousness - Do you need outside help?
  4. Administer first aid as needed.
  5. Keep unnecessary personnel away from the victim.
  6. Record necessary information.

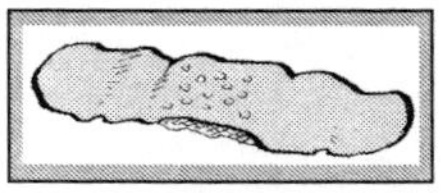

Be prepared!

Adapted from ***Retail Best Practices and Guide to Food Safety and Sanitation* © 2003**

RBP. Pg. 304
RSG. Pg. 226

5

Slide 5

What to do if an accident occurs:

1. **Stay calm.** Information is needed to determine the seriousness of the injury. Begin by checking to see if the victim is responsive. Kneel and ask, "Are you OK? Would you like me to call for emergency medical services?"
2. **Examine the injury** and decide whether outside help is needed. If in doubt, call for help. Victims who are not breathing and do not have a heartbeat have a much greater chance of survival if they receive prompt medical care in a hospital or by trained paramedics.
3. **Administer first aid** according to the type of injury and your level of first-aid training.
4. **Keep unnecessary personnel away** from the victim.
5. **Record the victim's name, the date and time of accident,** type of injury or illness, any treatment given, and amount of time it takes for emergency assistance to arrive.

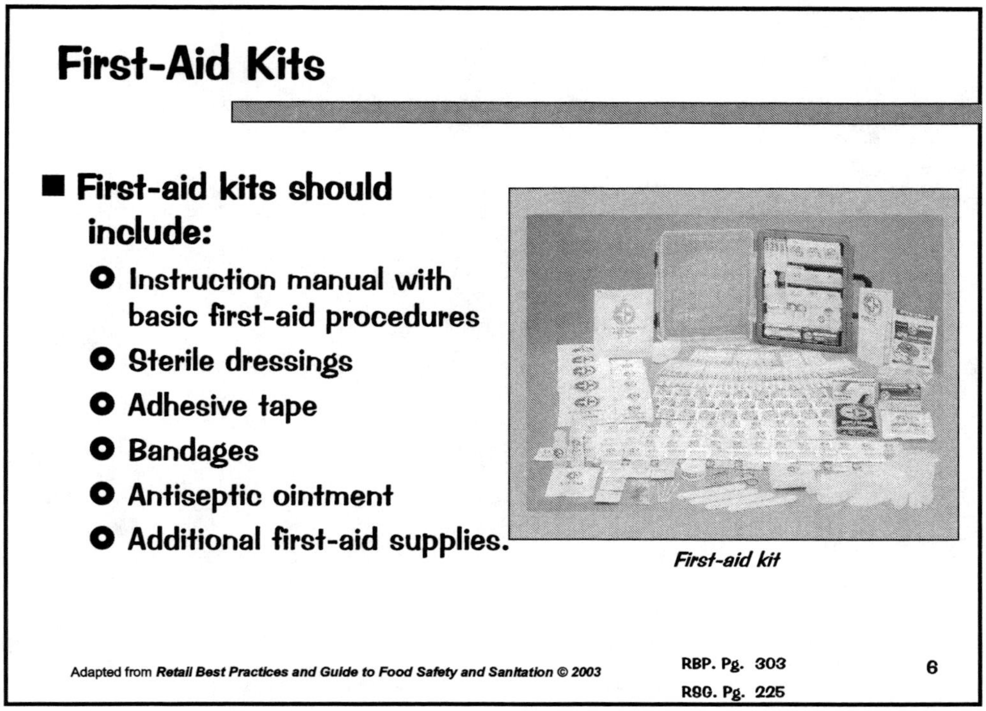

Slide 6

First-aid kit

Each department should have a first-aid kit. Store the kit away from food, equipment, utensils, linens, and single-use articles.

A first-aid kit should include:

- Instruction manual describing basic first-aid treatments
- Sterile dressings
- Adhesive tape and bandages
- Antiseptic ointment
- Other basic first-aid supplies.

Check expiration dates on supplies monthly and replace them as needed.

Common Injuries: *Falls*

Prevention:

- Keep floors clean
- Use signs to alert others of hazards
- Use anti-slip mats.

Adapted from ***Retail Best Practices and Guide to Food Safety and Sanitation* © 2003**

RBP. Pg. 304
RSG. Pg. 227

7

Slide 7

Common Injuries – Falls

Prevention

- Keep floors clean and free from debris
- Use signs to alert others of hazards
- Use anti-slip mats in high traffic areas.

First aid

- Keep victim still
- Seek medical assistance if the victim has difficulty moving.

Common Injuries: *Cuts*

Prevention:

- **Keep knives sharp and stored properly**
- **Wear slash-resistant gloves**
- **Don't leave knives on counters or in sinks.**

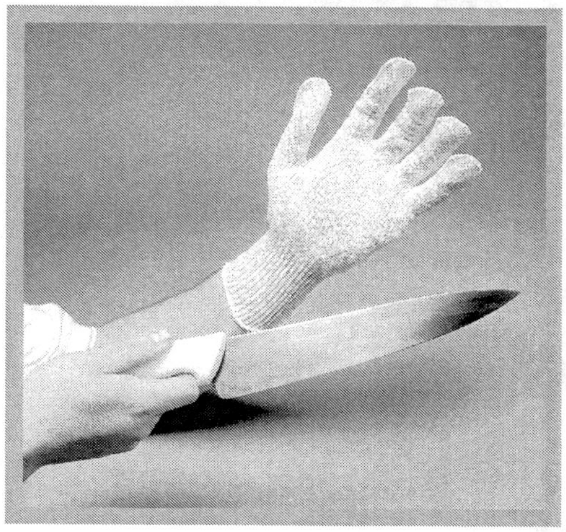

Adapted from ***Retail Best Practices and Guide to Food Safety and Sanitation*** © 2003

RBP. Pg. 305 – 307
RSG. Pg. 226

8

Slide 8

Common Injuries – Cuts

Prevention

- Keep knives and box cutters sharp and stored properly
- Wear approved slash-resistant gloves
- Don't leave knives on counters or submerged in dish water
- Use proper tools for the task.

First aid

- Protect yourself from contact with body fluid
- Apply pressure to cut with clean cloth or towel
- Elevate the injury
- Clean wound and apply antibiotic ointment and water-resistant bandage
- A medical professional should treat severe cuts.

Common Injuries: *Burns*

Prevention:

- **Provide hot pads**
- **Don't use wet hot pads**
- **Turn pot handles away from traffic**
- **Vent steam from covered containers.**

Adapted from ***Retail Best Practices and Guide to Food Safety and Sanitation*** © 2003

RBP. Pg. 307
RSG. Pg. 227 - 228

9

Slide 9

Common Injuries – Burns

Prevention

- Provide hot pads for use
- Don't use wet hot pads
- Appropriately turn pot handles away from traffic areas
- Vent steam from covered containers.

First aid

- Remove the source of the burn (including shutting off electricity if needed).
- Soak in cool water.
- Cover the burn with a sterile bandage. Don't puncture blisters.
- Use a pain reliever to reduce swelling.

A medical professional must treat third-degree burns.

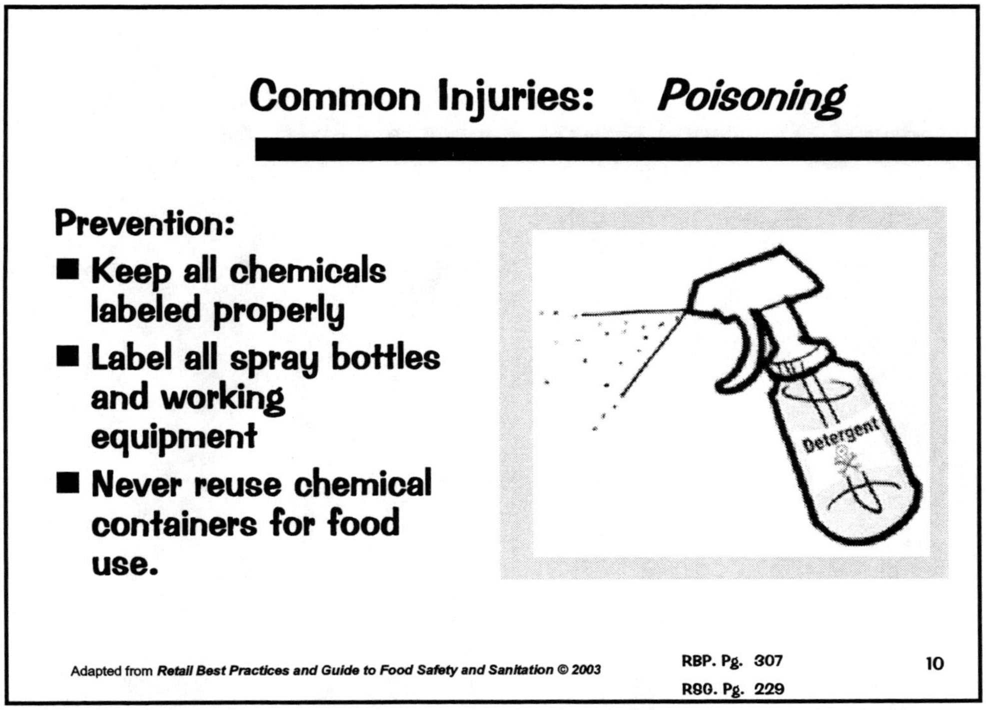

Slide 10

Common Injuries – Poisoning

Prevention

- Make sure all chemicals and toxic materials are clearly labeled
- Working containers such as spray bottles and buckets must be clearly labeled
- Never reuse a container that stored chemicals or toxic materials to store, transport or dispense food.

First aid

- Call for medical assistance immediately
- Begin internal recall crisis management on food products
- Contact the local health department if the product has left the premises.

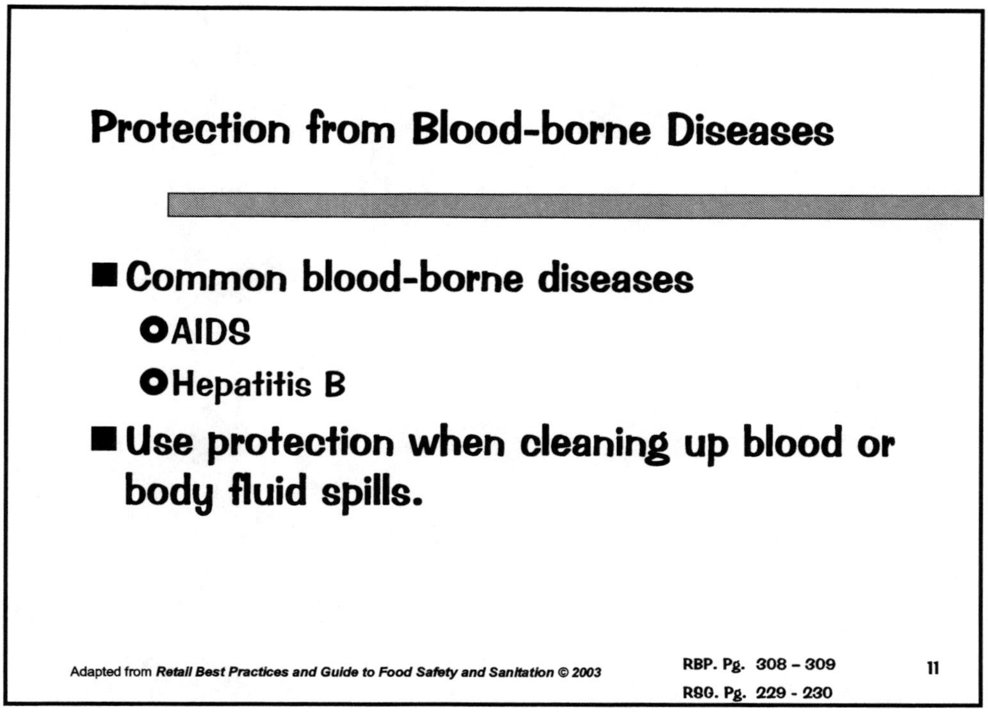

Slide 11

Blood and body fluid-borne diseases

Common blood-borne diseases:

- AIDS
- Hepatitis B.

Safety procedures:

- Employees must wear protective gloves and outer garments
- Use cleaning and sanitizing procedures to remove blood or body fluids
- Use a solution of 1 part chlorine bleach to 10 parts water to disinfect areas contaminated by blood
- Always use gloves and protective gear before cleaning up any accidents involving blood and body fluids
- Remove contaminated gloves by turning them inside out and then wash hands
- Remove contaminated clothing for washing.

Employee Medications

Employee medication must be stored away from:

- Food
- Equipment
- Utensils
- Linens
- Single-use items (straws, butchers paper eating utensils, etc.).

Do not store medication in food refrigerators!

Adapted from *Retail Best Practices and Guide to Food Safety and Sanitation* © 2003

RBP. Pg. 309
RSG. Pg. 232

12

Slide 12

Employee medication must be stored away from:

- Food
- Equipment
- Utensils
- Linens
- Single-use items (straws, butchers paper, eating utensils, etc.)
- Medications stored in a refrigerator where food is stored must be kept in a clearly labeled and covered container located on the lowest shelf of the refrigerator
- Medications cannot be stored in display or storage cases or walk-in units.

Safety Training Programs

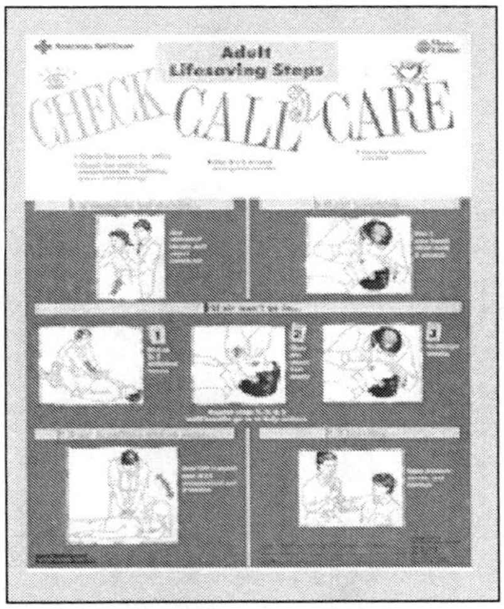

Cardiopulmonary resuscitation poster

It is desirable for someone on your staff to know how to administer:

- CPR
- First aid for choking

Safety charts may be posted near work areas.

Adapted from ***Retail Best Practices and Guide to Food Safety and Sanitation* © 2003**

RBP. Pg. 309 - 310
RSG. Pg. 230 - 231

13

Slide 13

Safety training programs

- It is desirable for someone on your staff to know how to administer:
    - CPR
    - First aid for choking.
- Safety charts are available from the American Red Cross, National Safety Council, and American Heart Association.

Fire Prevention

Fire extinguishing equipment

Top 2 causes of fire:

- Grease
- Electrical problems.

Adapted from ***Retail Best Practices and Guide to Food Safety and Sanitation* © 2003**

RBP. Pg. 311
RSG. Pg. 232

14

Slide 14

Causes of fire:

- Grease
- Electrical problems.

Prevention:

- Grease – keep hoods, ventilating fans, filters, ducts, and walls free of grease buildup
- Electric – Regularly inspect equipment, repair faulty wiring, avoid overloading circuits and use proper grounding
- Preparation –
    - Have the proper fire extinguishers available
    - Make sure all exits are properly marked
    - Conduct periodic fire drills
    - Periodically test fire extinguishers
    - Have a safety blanket easily accessible.

In case of a fire:

- Get customers and employees out of the building
- Call the fire department.

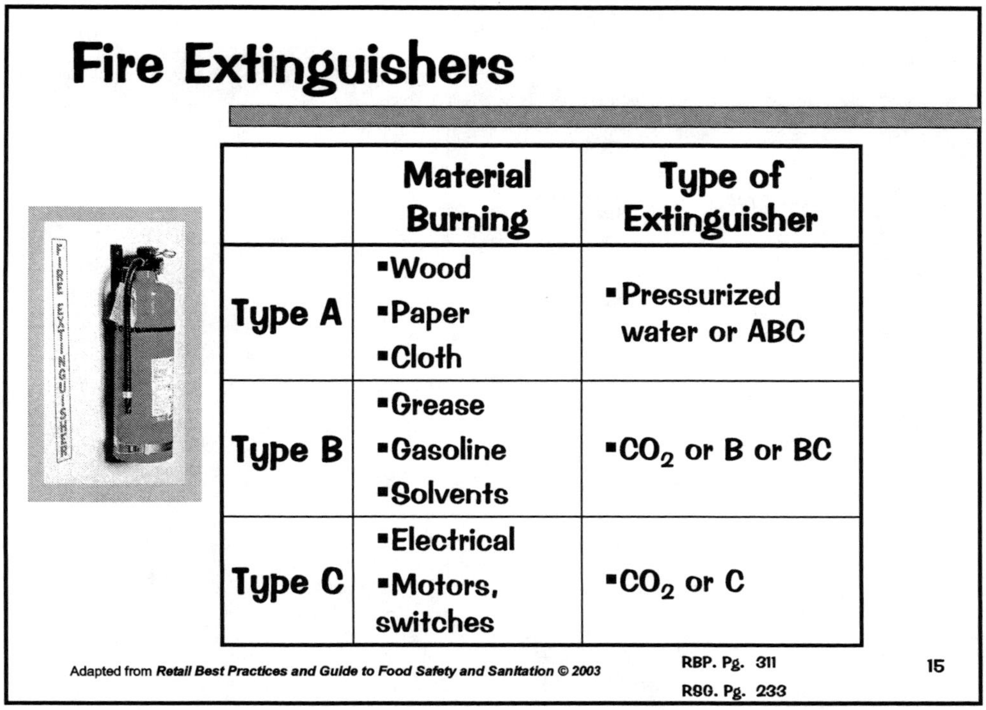

Slide 15

Fire Extinguishers

Type	Material Burning	Most Common Type of Extinguisher Used
Type A	Wood, paper, cloth	Pressurized water or ABC
Type B	Grease, gasoline, solvents	Carbon Dioxide (CO_2) or B or BC
Type C	Motors, switches, electrical	CO_2 or C

Never use a type A extinguisher on a grease or electrical fire!

- Extinguishers weighing less than 40 pounds can be mounted 5 feet from the ground.
- Extinguishers weighing more than 40 pounds must be stored no more than 3 feet off the ground.

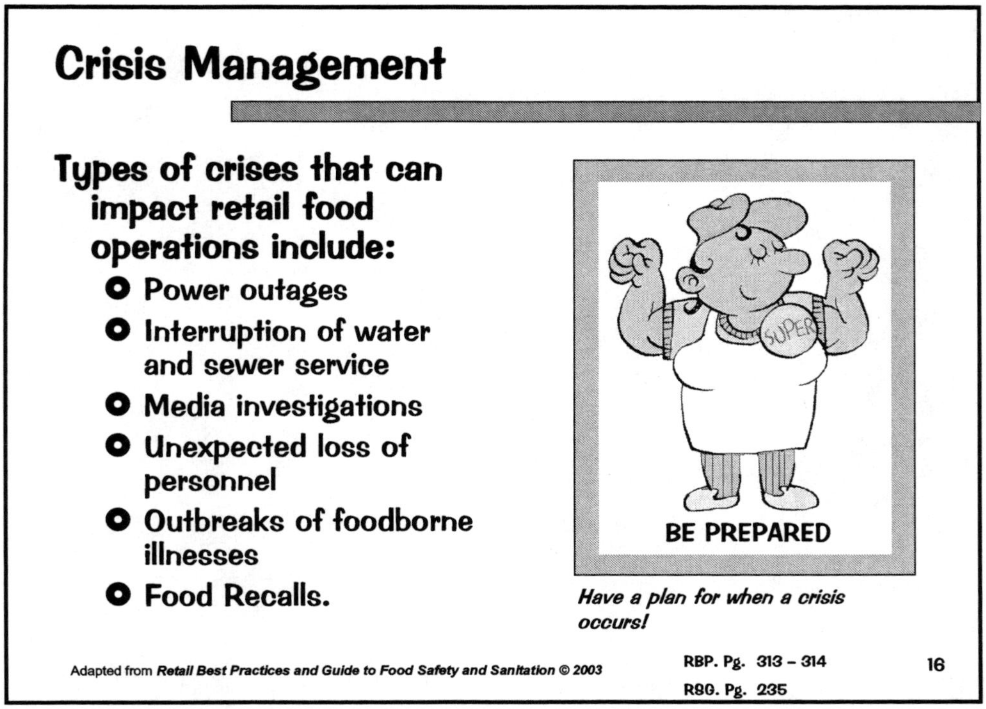

Slide 16

Types of crises that can impact retail food operations include:

- Power outages
- Interruption of water and sewer service
- Media investigations
- Unexpected loss of personnel
- Outbreaks of foodborne illnesses
- Food recalls

Have <u>one</u> spokesperson for the establishment to deal with public and media

- Be honest and say "I don't know" if that is the case.

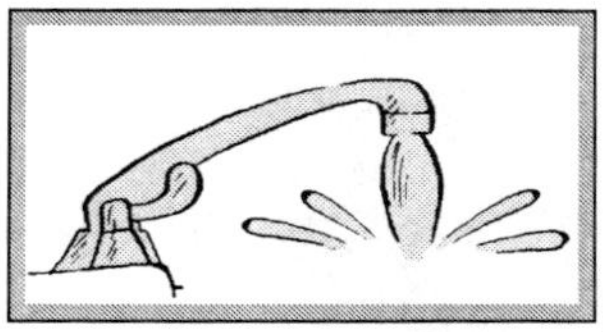

Water Supply Emergencies

- **When you have an interruption in water service:**
  - Use safe drinking water from another source
  - Use bottled water or haul potable water to the site
  - Disinfect water by chlorination or boiling
  - Use bagged ice instead of ice machines
  - Use disposable single-use dishes and utensils
  - Use prepared foods from other sources.

Adapted from *Retail Best Practices and Guide to Food Safety and Sanitation* © 2003

RBP. Pg. 314 – 315
RSG. Pg. 235 - 236

17

Slide 17

Boil order – water must be boiled before drinking due to bacteria in the water.

To continue operating under "boil water advisories/notices" or "interrupted water service" from all water supplies:

- Secure and use safe drinking water from an approved source
- Use bottled water obtained from approved sources or water hauled to the site in tank trucks
- Disinfect the water by either chlorination or boiling.

To chlorinate water:

- Add six drops of liquid chlorine bleach to one gallon of water and mix. Check the label on the container to make sure the active ingredient, sodium hypochlorite, is available at a concentration of 5.25 ppm. The important things to remember when chlorinating water are:
  - Wait 30 minutes after adding the chlorine before using the water for drinking or cooking purposes
  - If the six drops of bleach does not give the water a slight taste of chlorine, double the amount of chlorine until a slight taste is present. The taste of chlorine will be evidence the water is safe to drink.

Purifying water by boiling:

- Bring the water to a full boil for at least five minutes
- Cool and aerate the boiled water by pouring it from one clean container to another.

Alternative Procedures to Minimize Water Usage During an Outage:
• Commercially packaged ice may be substituted for ice made on-site • Single-service items or disposable utensils may be substituted for reusable dishes and utensils • Prepared foods from approved sources in place of complex preparation on-site • Restrict menu choices or hours of operation • Portable toilets may be utilized for sanitary purposes.

After the water emergency is officially lifted or water service resumes, these precautionary measures must be followed:

- Flush the building water lines and clean faucet screens and water line strainers on mechanical dishwashing machines and similar equipment
- Flush and sanitize all water-using fixtures and appliances such as ice machines, beverage dispensers, hot water heaters, etc.
- Clean and sanitize all fixtures, sinks, and equipment connected to waterlines.

There must be water pressure before resuming operations in a retail food establishment, and the water should be sampled for bacteriological quality.

The safety of water cannot be judged by color, odor, or taste.

Don't hesitate to contact your local regulatory agency if you have any questions regarding appropriate operations at your establishment during water emergency orders.

Foodborne Illness Incidents or Outbreaks

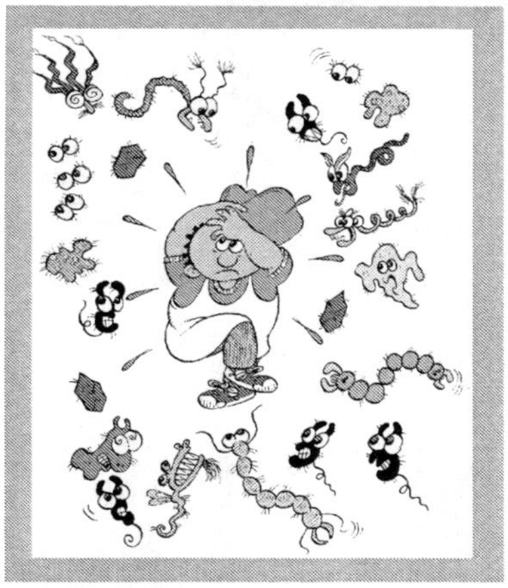

If you suspect you may have a foodborne illness problem:

- **DO NOT admit fault until the facts have been checked and confirmed.**

Follow food safety procedures to eliminate hazards and to avoid foodborne illness incidents!

Adapted from ***Retail Best Practices and Guide to Food Safety and Sanitation*** © 2003

RBP. Pg. 317 – 318
RSG. Pg. 237 - 239

18

Slide 18

Foodborne Illness Incident or Outbreak

Action Plan When Foodborne Illness Is Suspected:

- Remove all suspected food from the sales areas
- Isolate it in a separate area and preserve it for analysis
- Label it so it does not become mixed with other foods
- Do not sell any more of the suspect product
- Listen to everything the customer says and look for clues that might explain the problem
- Do not try to diagnose the problem or belittle the customer's complaint
- Keep a record of the interview
- Assure the customer you will get back to him or her after investigating the problem
- Make sure you do contact the customer again after facts are gathered
- Do not admit your establishment is at fault–you have no proof of that until facts are checked
- Initiate the crisis management team.

Crisis management team includes members from :

- Media affairs
- Food safety
- Loss prevention
- Legal counsel.

Slide 19

Safety Stop! 8.1

Sample Test Questions

1. Which of the following activities is most likely to be a part of an accident prevention program?
   A. Greeting customers as they enter the retail food establishment.
   B. Mopping the floor without using a wet floor sign.
   C Smoking in designated areas only.
   D. Training employees about potential hazards and how to prevent or avoid the hazards.

2. Which of the following is the first action you should take if an accident occurs?
   A. Administer first aid.
   B. Stay calm and examine the injury.
   C. Keep unnecessary personnel away from the victim.
   D. Run and hide.

3. First-aid kits should be kept in every department and stored away from:
   A. Food.
   B. Equipment.
   C. Single-use articles.
   D. All of the above.

4. All of the following are acceptable first-aid procedures for a cut except:
   A. Protect yourself from contact with body fluid.
   B. Apply pressure to the cut with a clean cloth or towel.
   C. Keep the wound below the head.
   D. Clean the wound and apply antibiotic ointment and a water-resistant bandage.

5. All of the following are acceptable methods to prevent burns except:
   A. Providing hot pads for food employees to use.
   B. Using damp or wet hot pads to handle a hot pot or pan.
   C. Turning pot handles away from traffic areas.
   D. Venting the steam from covered containers.

6. Which of the following types of fire extinguishers should be used to put out a grease fire?
   A. Type A.
   B. Type B.
   C. Type C.
   D. Type A or C.

7. Which of the following is an acceptable method to make water safe to drink?
   A. Freezing and then thawing the water.
   B. Chlorinating the water by adding six drops of liquid chlorine bleach to one gallon of water and mixing it well.
   C. Boiling water at a full boil for a full five minutes.
   D. Both B and C are acceptable.

8. A crisis management team can include members from:
   A. Food safety professionals.
   B. Loss prevention.
   C. Legal council and media affairs.
   D. All of the above.

Real-life

Slide 8
Employee injury
Laura is halving the day's fresh cantaloupe. She cuts her hand deep enough that it bleeds, but not so severely that she needs medical attention. What should she do now?

Laura should start by calling for help if she feels she needs it, and sitting down if she feels lightheaded. Anyone that comes to assist Laura should put on impermeable gloves to protect themselves from her body

fluids. Direct pressure should then be applied to the wound with a clean cloth or towel. Her hand should be raised above her head (elevated) to keep the blood from rushing to the injury. When the bleeding stops, apply antibiotic ointment, cover the wound with a water-resistant bandage and an impermeable glove.

The fruit, and any additional food that may have been contaminated by Laura's blood should be discarded and her knife, cutting board, workstation, and any food-contact and non-food-contact surfaces that may have been contaminated by her blood should be properly cleaned and sanitized.

Ensure the person in charge (manager or supervisor) is made aware of the situation. And fill out any incident reports that may be needed. Follow any additional in-house regulations that may be in place for this incident.

This exercise can be repeated with any type of injury on slides 7 – 11

Slide 15
Fire Extinguishers
Marcus was frying chicken in the deli for the hotline. He reached into his freezer to grab some more chicken. When he turned around; the grease in the fryer was on fire. Marcus grabbed the nearest fire extinguisher, a Type A extinguisher, and sprayed the fire. This only seemed to make the fire spread. What was Marcus' mistake and how could this problem have been prevented?

Marcus should have checked the type of fire extinguisher he grabbed. He should have used a Type B fire extinguisher suitable for putting out grease fires. The Type A extinguisher is only suitable for putting out wood, paper or cloth fires. The Type A extinguisher should not have been in the kitchen in the first place. Marcus could have checked the oil in the fryer prior to starting his shift: chances are the oil was old and filled with debris and that is probably what caused the fire.

Slide 16
Food Recall
Review and discuss a food recall crisis from the different perspectives of managers, owners, employees, customers, media, and government. What is the reaction of each to the crisis? How does the crisis affect each group of people? Have class participants take on the roles of each participant and discuss.

Slide 17
Water Supply Emergency
Shawn, the manager of a convenience store, received a call from the local health department informing him the area was under a "Boil Water Order" until further notice. What should Shawn do?

Shawn should start by securing and using safe drinking water (the bottled water he sells in his establishment would work for this) for any of his fresh water needs. He should proceed to turn off the soda fountain and any drinking fountains that may be in his establishment. Shawn may also choose to post signs regarding the "Boil Water Order" so his customers know what is happening. Shawn should contact his direct superior and inform them of the situation and follow any pre-planned crisis management plan in place for the water supply emergency.

Making it Real

First-aid Kit Review – Slide 6

Bring in a properly filled first-aid kit. Discuss what each item is and describe briefly why the item is in the kit.

- Supplies
    - Standard first-aid kit
    - Various amounts of water
- A good variety to this exercise is to allow the learners to describe what each item is and in what scenarios it may be used. It can also be helpful to allow them to practice with the more unusual items in the kit.

Simulate a Crisis – Slides 16 – 17

Use a simulated crisis involving a retail food establishment. Have the class learners develop a crisis-management strategy for handling the situation. Have the members of the class discuss what they would do in order to maintain control of the situation. In what order would they perform these activities?

- Supplies
    - Sample crisis management plan
- Discuss and review the steps involved in the crisis management plan.

Make Water Safe – Slide 17

Have the learners practice making water safe to drink by adding the appropriate amount of liquid chlorine bleach to various amounts of water.

- Supplies
    - Household bleach
    - Various amounts of water
        - 1 gallon
        - 1 quart
        - 1 pint
        - 5 gallons, etc.
- Surprise the students with this activity and give them a time limit to perform the task, such as 15 seconds. This will help to simulate the pressure they will feel in just such an occurrence.

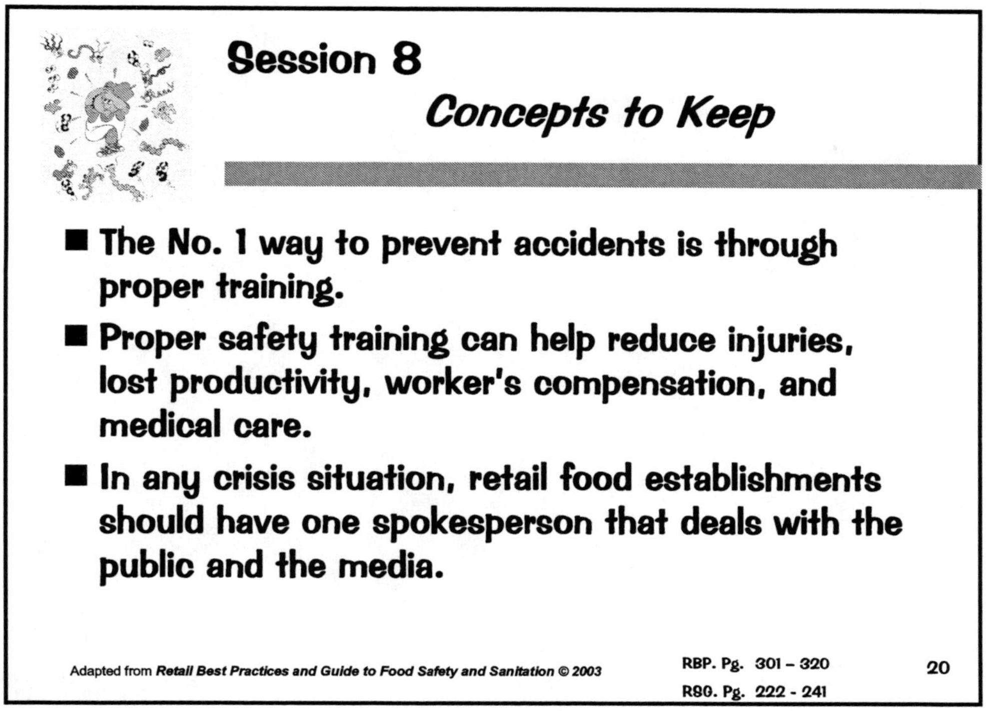

Slide 20

Concepts to Keep

- The No. 1 way to prevent accidents is through proper training.
- Proper safety training can help reduce injuries, lost productivity, worker's compensation, and medical care.
- In any crisis situation, retail food establishments should have one spokesperson that deals with the public and the media.

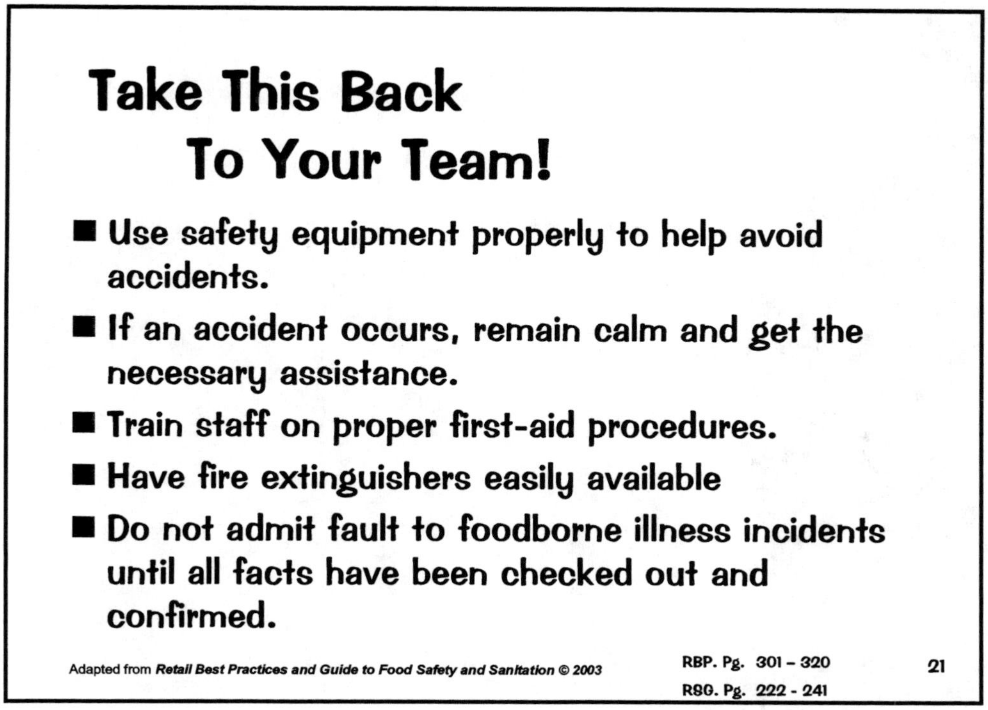

Slide 21

Take This Back to Your Team

- Use safety equipment properly to help avoid accidents.
- If an accident occurs, remain calm and get the necessary assistance.
- Train staff on proper first-aid procedures.
- Store medication properly.
- Do not admit fault to foodborne illness incidents until all facts have been checked out and confirmed.

Session 9

Education and Training – Session 9 Key Points

Slide 1

Training Tip:

The content of each slide provides the most essential information in the session. The written text below each slide is provided to help you understand and better teach the essential information. The written text may also be used to answer questions that may be posed during the session. Depending on the length of your training session, you may not be able to address all the information contained beneath each slide.

Overview:

Session 9 looks at the importance of employee training in food safety. The learner should leave this session with an understanding of the importance of training in preventing foodborne illness and accidents. This session gives an overview of the process of good training, from organization and planning to delivering the training and evaluating its effectiveness. Training and education in retail food safety is provided here to give basic training techniques and ideas to people who may not have a background in training, but will be asked to do training in the process of their jobs.

Suggested Timeline:

8-hour course: 15 minutes using the PowerPoint© slides only.

- 12 ***Slides.***

16-hour course: 30 minutes using the 15 PowerPoint© slides and the 1 ***Safety Stop!*** slide containing:

- 5 ***Sample Test Questions***
- 2 ***Real-life*** scenarios
- 2 ***Making it Real*** interactive exercises

Video suggestions and supplemental resources:

Books:

Training the Trainer: Performance Based Training for Today's Workplace, Prentice Hall, 1-800-282-0693

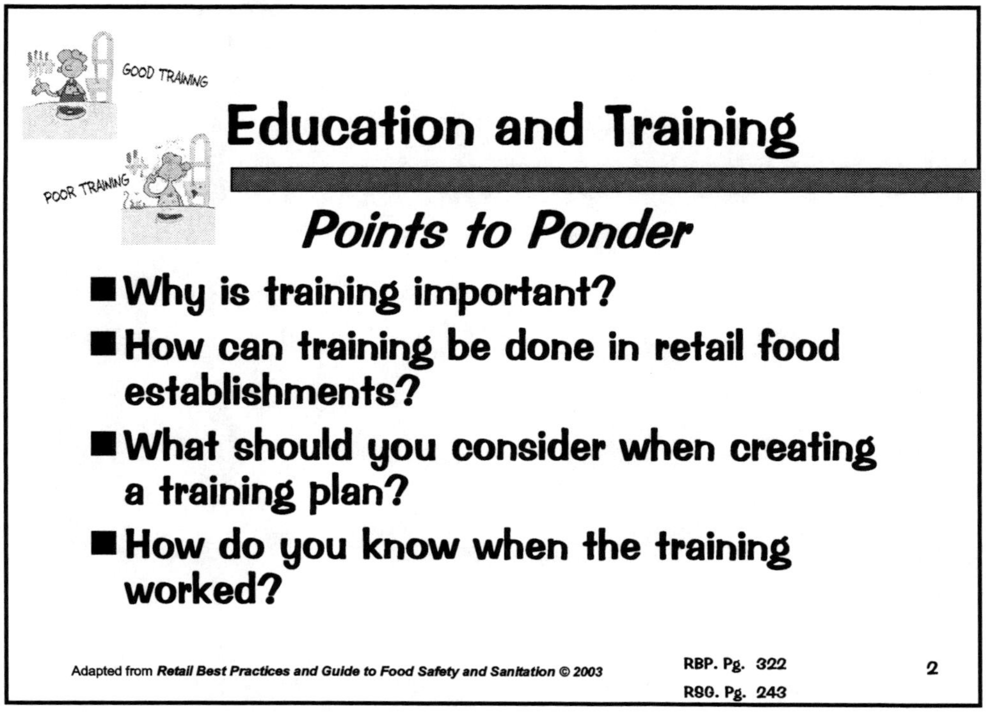

Slide 2

Points to Ponder

- Why is training important in retail food establishments?

- How can training be done in retail food establishments?

- What should you consider when creating a training plan?

- How do you know if the training worked?

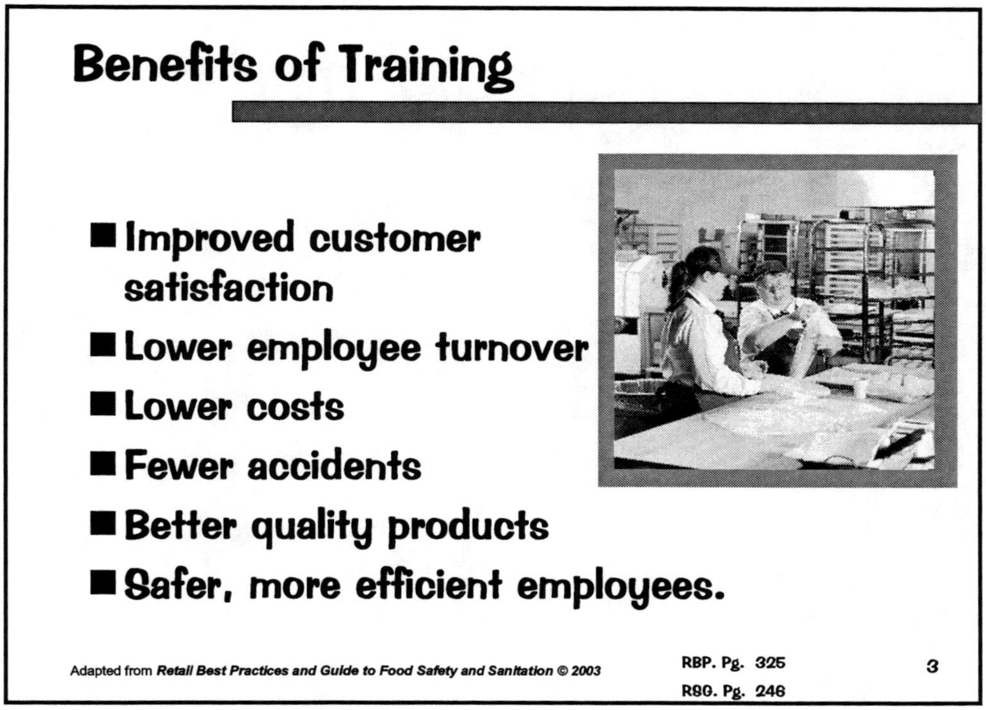

Slide 3

Benefits of training

- Improved customer satisfaction
- Lower employee turnover
- Lower costs
- Fewer accidents
- Better quality
- Safer, more efficient employees.

It's a proactive approach to active managerial control of the entire system with great payback.

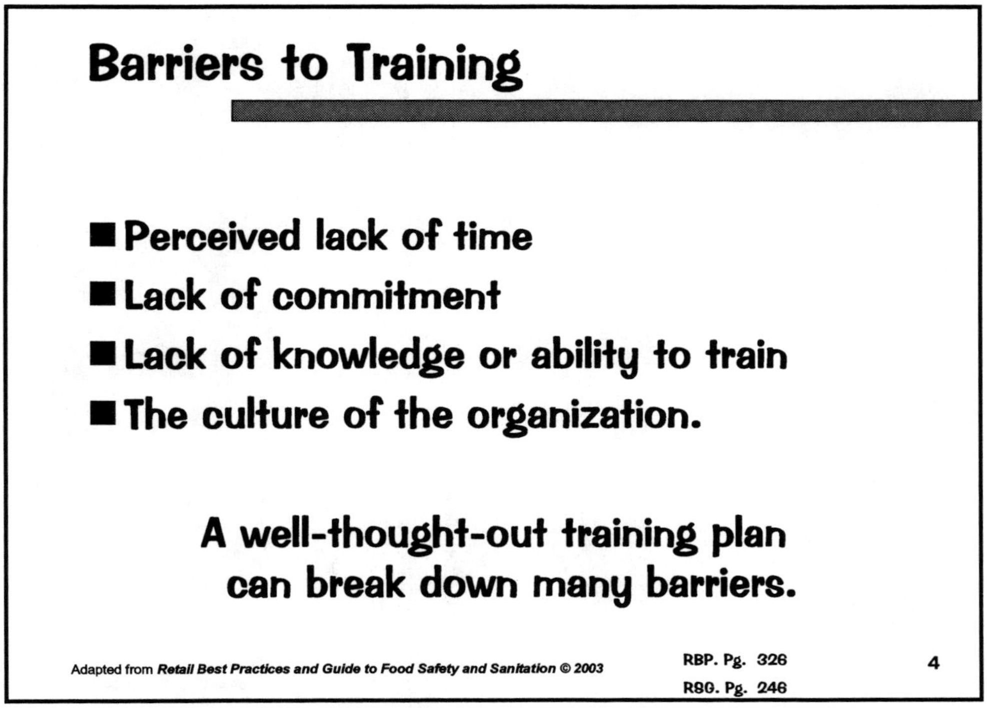

Slide 4

Barriers to training

- Perceived lack of time
- Lack of commitment
- Lack of knowledge or ability to train
- The culture of the organization
- Lack of a well-thought-out training plan.

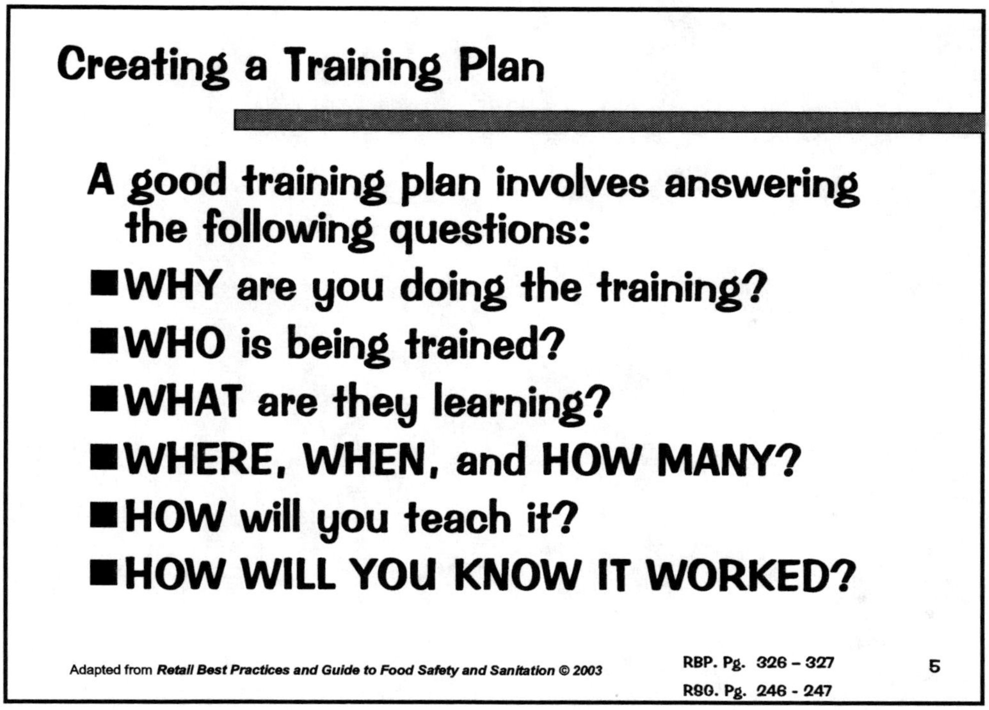

Slide 5

A good training plan involves answering the following questions:

- **WHY are you doing the training?**
- **WHO is being trained?**
- **WHAT are they learning?**
  - What are you going to be teaching – what do learners need to know?
  - What teaching and learning resources are available?
- **WHERE, WHEN, AND HOW MANY?**
  - Where will the training take place?
  - What amount of time is allotted for the program?
  - How many learners will be in each group?
- **HOW will you teach it?**
- **How will you know it worked?**
  - Do evaluation tools exist for this site or will you need to provide them?

WHY are you doing the training?

- Are there any current issues that need immediate attention?
- What outcomes does upper-level management expect?

Adapted from *Retail Best Practices and Guide to Food Safety and Sanitation* © 2003 — RBP. Pg. 327 – 328 / RSG. Pg. 247 — 6

Slide 6

WHY are you doing the training?

- Are there any current issues that need immediate attention such as:
    - New standards of operation
    - Failed sections on an inspection
    - Updated food code elements
- What outcomes does upper-level management expect
    - Are they seeking a higher return on investment
    - Are they seeking to reduce turnover
    - Are they seeking to reduce labor cost?

Determine if the training is:

- Department specific or cross functional, cross-departments
- General store training.

WHO is being trained?

Things to consider about the trainee:

- What department are they from?
- Are they managers or employees?
- What is their average level of knowledge, skills, and abilities?
- Are there any language barriers?
- How old are they?
- Do they have varying levels of education?

Adapted from ***Retail Best Practices and Guide to Food Safety and Sanitation* © 2003**

RBP. Pg. 323
RSG. Pg. 247

7

Slide 7

Things to consider about the trainee:

- What department are they from?
- Are they managers or employees?
- What is their average level of knowledge, skills, and abilities?
- Are there any language barriers?
- How old are they?
- Do they have varying levels of education?

Age can make a difference

Born:	Veterans 1922-1942, 52 million	Baby Boomers 1943-1960, 73.2 million	GenXers 1960-1980, 70.1 million	Nexters 1980-2000, 6.7 million
Work ethic	Dedicated	Driven	Balanced	Determined
View of authority	Respectful	Love/Hate	Unimpressed	Polite
Training type	Traditional Classroom	Lifelong learning Read selectively	Computer-based Interactive Read less	Interactive Technology-based Read more

Adapted from *Retail Best Practices and Guide to Food Safety and Sanitation* © 2003

RBP. Pg. 323 – 324
RSG. Pg. 244 - 245

8

Slide 8

Learners today influence training

- Today's work force is made up of a diverse group of people:
  - Culture
  - Diversity
  - Age
  - Nationality
  - Ability
  - Knowledge
  - Motivation.

- For the first time, we have four generations of workers in the workplace at the same time.
- Each has its own way of looking at the world and its own preferred methods of learning.
  - Veterans
  - Baby boomers
  - Gen Xers
  - Nexters.

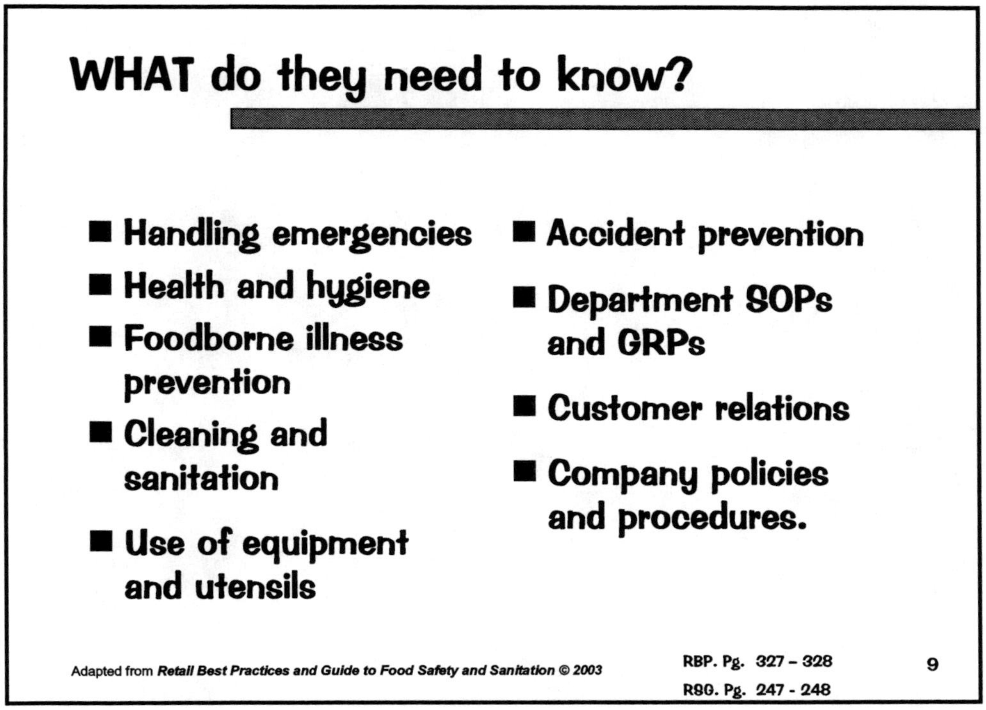

Slide 9

WHAT are they learning?
What do they need to know?

Training can be done to solve problems or prevent them from occurring.

Situations where training is important:

- Emergency situations and response systems
- Issues related to health and hygiene
- Common causes of foodborne illness
- Food safety management procedures including time and temperature controls
- Cleaning and sanitation procedures
- How to safely use and clean equipment and utensils
- Accident prevention measures
- Customer relations and workplace etiquette
- Company policies and procedures
- Department specific procedures and protocols
    - SOPs & GRPs
    - HACCP procedures
    - Reportable illnesses and awareness.
- What teaching and learning resources are available?

WHERE, WHEN, and HOW Many?

- Where will the training be held?
- How much time should be spent on training segments?
  - Short sessions – 30 to 45 minutes at a time
  - Break up longer classes with activities and interaction
  - Give people a chance to try out their new skills
- How many people will be trained at a time?

Adapted from *Retail Best Practices and Guide to Food Safety and Sanitation* © 2003 RBP. Pg. 330 – 332 RSG. Pg. 248 10

Slide 10

Logistics
WHERE, WHEN, AND HOW MANY?

- Schedule training far enough in advance to be sure everyone has time to plan for it.
- Try not to cancel any planned training because it will make the training seem less important.
- Where will the training take place?
- What amount of time is allotted for the program?
- How many learners will be in each group?

Recommended times for training:

- Short sessions–30 to 45 minutes at a time are easier to handle
- Most adults prefer to learn a task or fact and then immediately try it out
- When longer classes are needed, break up the time with activities and interaction that allows participants to get up and move around
- Keep your training within the scheduled time period; do not start early and do not run long.

How Will You Teach It?

- Keep it simple
  - Don't overload the session with information
- Have a clear goal for the training
- Make it fun
  - Use activities
  - Involve the participants
  - Use visual aids, props, computers when appropriate
- Treat everyone with respect
- Take advantage of training moments.

Adapted from *Retail Best Practices and Guide to Food Safety and Sanitation* © 2003

RBP. Pg. 330 – 332

RGG. Pg. 249

11

Slide 11

How will you teach it?
An effective training session includes the following points:

- Keep it simple
  - Trying to teach too much at a time can be counterproductive
- Stay focused
  - Set clear objectives for the training session
- Make it fun
  - Use color
  - Use activities
  - Involve the audience
  - Use audio/visual aids as an enhancement, not a replacement for instruction
- Treat everyone with respect. Ridicule and sarcasm do not belong in training sessions.

Training moments - learning can occur continuously in retail food establishments.

- Be a positive role model for your employees.
- Use normal happenings as a training moment to improve performance.
- When employees are not practicing proper procedures, be a coach. Teach them the right way to do it.
- Reinforce good habits. Reward employees who perform their jobs correctly.

How Will You Know it Worked?

- **How will you know that your training was affective?**
  - Provide follow-up and reminders
  - Observe people doing the job
  - Retrain employees on skills as needed
  - Use written tests with caution
  - Give praise when the job is well done.

Adapted from *Retail Best Practices and Guide to Food Safety and Sanitation* © 2003 RBP. Pg. 331 - 332 RSG. Pg. 249 - 250 12

Slide 12

Evaluation – How will you know the training was effective?

To make sure learning has taken place, consider the following actions:

- Provide follow-up discussions and reminders when the employee is back on the job. This helps to reinforce the learning that took place in the training session.
- Observe whether or not people perform the job as taught.
  - Watch your employees perform their jobs after the training is completed
  - Look for the new skills they have learned
  - When they use their new skill, praise them
  - When they do not, coach them on the right way.
- If an employee has trouble with a new skill, retrain until he or she does it right.
- Use written tests with caution.
  - Employees who are not literate need special consideration
  - Verbal questions work better than written tests.
- Recognize outstanding performance.
  - Follow up verbal recognition with a hard copy in the employee's record.
- Communicate positive and consistent feedback.
  - Provide verbal and written feedback to the employee.
- Ask the employee for suggestions that would help in learning the task.

Slide 13

Safety Stop! 9.1

Sample Test Questions

1. All of the following are outcomes of good training with the exception of:
   A. Lower employee turnover.
   B. Safer, more efficient employees.
   C. **Lack of customers.**
   D. Lower costs.

2. The most effective training takes place in sessions consisting of:
   A. 10 to 12 minutes.
   B. 30 to 45 minutes.
   C. 1 hour.
   D. 2 hours.

3. People in the "Generation X" learn best
   A. With computer-based, interactive instruction.
   B. In a traditional classroom.
   C. With materials that require a lot of reading.
   D. With hands-on training.

4. An example of a training moment is:
   A. A scheduled staff training on a new piece of equipment.
   B. Praising an employee for handling a disgruntled customer in a positive way.
   C. Reminding a deli employee about proper glove usage when seeing her handle food after handling money.
   D. All of the above.

5. The best way to evaluate the effectiveness of a training session is to
   A. Give a written test at the end of the training.
   B. Observe job performance after the training.
   C. Ask for co-worker feedback on an employee's performance.
   D. Have the employee do a self-evaluation.

Real-life

Slide 10-12
On-the-Job Training
It is a busy day in the bakery and Mary, the bakery supervisor, is training two new people on muffin prep. Mary is an excellent production person and teaches by having people watch her work. Mary washed her hands and then started prepping the batter for the muffins. As she prepped the batter she told them how to fill the pans and put them in the oven. Then she gave the two trainees the recipe and told them to finish the muffins while she started working on the cookies. The trainees washed their hands in the prep sink and then began to scoop batter into the pans. They looked in several drawers for a #8 scoop, but could only find a # 12 scoop, so they used that instead. Everyone was upset when the muffins came out small and overcooked. What could Mary have done to prevent this?

Mary does a lot of things automatically and forgets that new employees need more step-by–step instruction. As she washed her hands, she should have explained the proper place and method for washing hands. She should then have the trainees wash their hands and give them feedback on their process. Mary should have worked with the trainees through the entire process of making the muffins, from mixing the batter, to baking and finishing. If she had done this she could have shown them where to find the right size scoop, explained how the scoop size affects the finished product, and helped them have a successful training experience.

Slide 10
Training Meeting
John was conducting the weekly Wednesday morning staff meeting with his deli employees. At the end of the hour meeting people were getting restless because they had a large box lunch order to assemble in a short time frame. John decided to take an extra 10 minutes to demonstrate the new safety features of the new automated meat slicer. Later in the week one of the deli employees got cut while cleaning the new machine. How could this have been prevented?

John should have scheduled a short training session on the safety features of the slicer at a different time when people were not preoccupied with other things. Training should also be scheduled when people are alert and fresh. After sitting in a meeting for an hour people will have a harder time remembering the information. If John wanted to include the slicer training in the meeting time, he should have done it at the beginning of the hour and made sure adequate time was available for participants to ask questions and work with the equipment.

Making it Real

Design a Mini-Lesson – Slide 5

Many of the people in your class are trainers or are aspiring to be trainers. Be sure they understand how to train properly using this exercise.

- Supplies
  - Group
  - Common training topics
- Procedure:
  - Write five common topics of training required in retail food establishments on slips of paper and drop them into a paper bag.
  - Divide the class into small groups of 3 to 4, and have each group draw out a topic.
  - Have each group spend 5 to 10 minutes developing a brief training plan for their topic.
  - Each group should answer the basic questions needed for a good training plan including:
    - WHY are you doing the training?
    - WHO is being trained?
    - WHAT are they learning?
    - WHERE, WHEN, AND HOW MANY?
    - HOW will you teach it?
    - HOW WILL YOU KNOW IT WORKED?
  - Have each group spend 2 to 3 minutes presenting their training plan to the rest of the class. If time permits, evaluate and discuss the training plans presented.

One-on-One Training – Slide 8 & 10

- Supplies
  - Previous Make it Real materials
- Procedure:
  - Think back to all the other Make it Real exercises you have done to this point such as calibrating a thermometer, making water safe to drink, measuring the temperature of unpackaged foods, etc.
  - Break the class into groups of two and have them practice training each other on the procedures.
  - Pay close attention to details – did they wash their hands, did they put on gloves, did they clean the thermometer, etc.

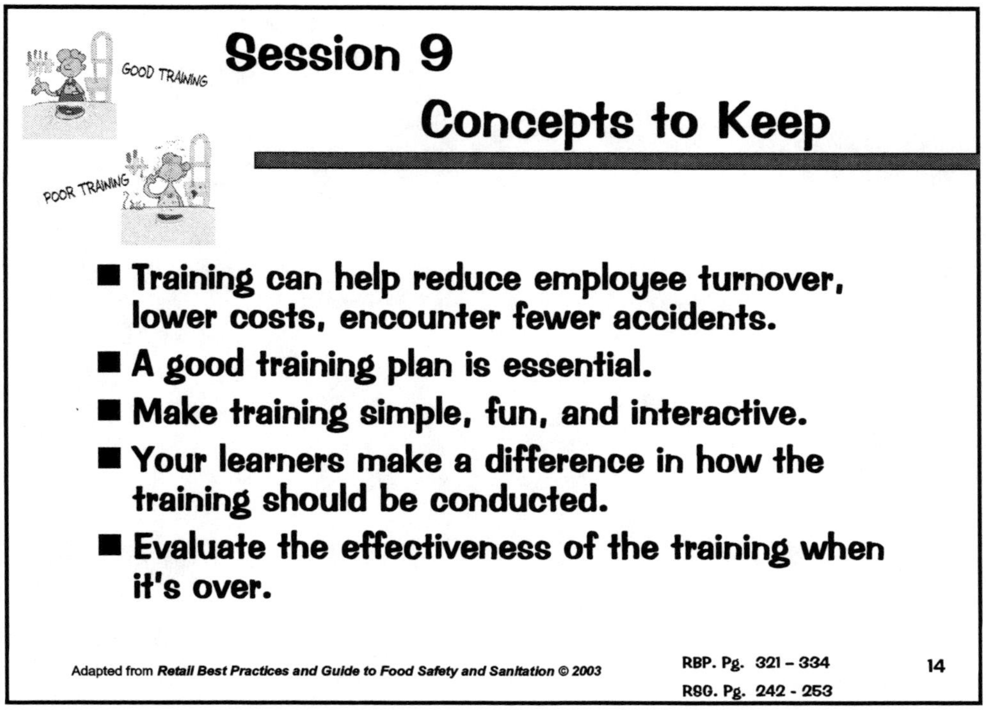

Slide 14

Concepts to Keep

- Training can help reduce employee turnover, lower costs, encounter fewer accidents.
- A good training plan is necessary for effective training.
- Make training simple, fun, and interactive.
- Your learners influence the type of training you need.
- Remember to evaluate the effectiveness of the training when it's over.

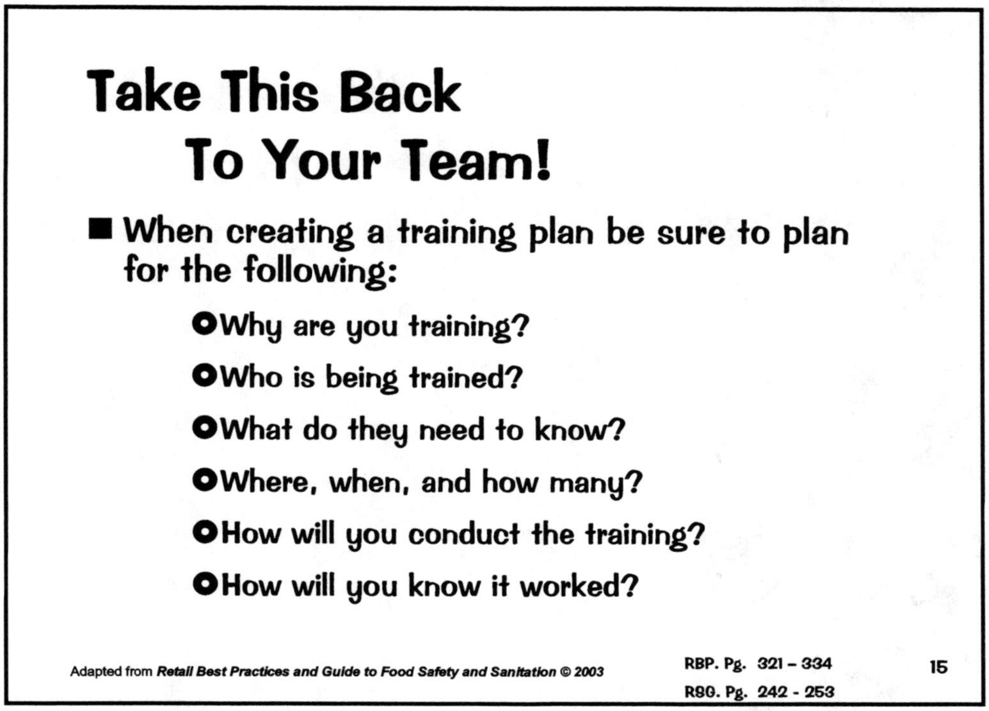

Slide 15

Take This Back to Your Team!

- When creating a training plan be sure to plan for the following:
  - Why are you training?
  - Who is being trained?
  - What do they need to know?
  - Where, when, and how many?
  - How will you conduct the training?
  - How will you know it worked?

Session 10

Food Safety Management Programs

10

Adapted from ***Retail Best Practices and Guide to Food Safety and Sanitation* © 2003**

RBP. Pg. 337 - 362
RSG. Pg. 254 - 277

1

Food Safety Management Programs – Session 10 Key Points

Slide 1

Training Tip:

The content of each slide provides the most essential information in the session. The written text below each slide is provided to help you understand and better teach the essential information. The written text may also be used to answer questions that may be posed during the session. Depending on the length of your training session, you may not be able to address all the information contained beneath each slide.

Overview:

This session discusses various food safety management programs available to help prevent foodborne illness. Much of the chapter focuses on the HACCP system, examining the different steps in creating a HACCP program. Best Practices, SOPs, GMPs, GAPs and GRPs are also emphasized, along with the need for a food recall system and the importance of product coding.

Suggested Timeline:

8-hour course: 45 minutes using the PowerPoint© slides only

- 23 ***Slides***.

16-hour course: 1 hour 15 minutes using the 30 PowerPoint© slides and the 1 ***Safety Stop!*** slide containing:

- 10 ***Sample Test Questions***
- 3 ***Real-life*** scenarios
- 2 ***Making it Real*** interactive exercises.

Video suggestions and supplemental resources:

Videos:
Food Safety Zone: Cross Contamination, 1-877-564-2333
Food Safety Zone: Personal Hygiene, 1-877-564-2333
Food Safety Zone: Sanitation, 1-877-564-2333
Invisible Challenge, Food Marketing Institute, 1-301-843-3084

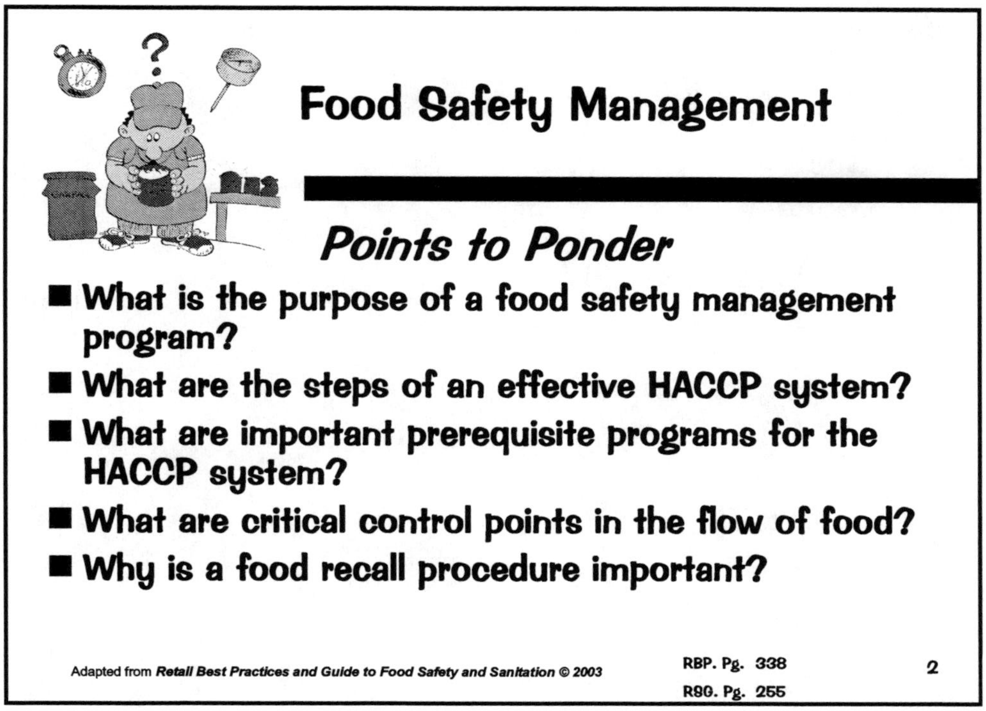

Slide 2

Points to Ponder

- What is the purpose of a food safety management program?
- What are the steps of an effective HACCP system?
- What are important prerequisite programs for the HACCP system?
- What are critical control points in the flow of food?
- Why is a food recall procedure important?

The Need

Food safety programs help individual food establishments effectively prevent foodborne illness from occurring.

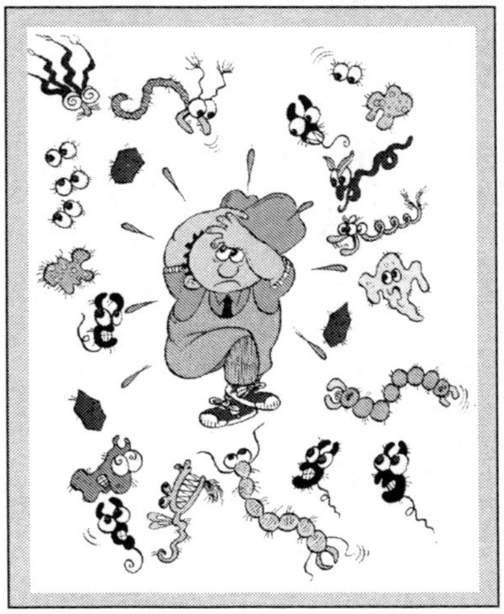

Avoid hazards.

Adapted from ***Retail Best Practices and Guide to Food Safety and Sanitation*** © 2003

RBP. Pg. 338 – 339
RSG. Pg. 256 - 257

3

Slide 3

The retail food industry and food regulatory agencies must deal with:

- New hazards that cause foodborne illness
- A global food supply that carries more risks
- New techniques and equipment for processing, displaying, and serving food
- A change in eating patterns by customers including more ready-to-eat products
- An increase in the number of people who are highly susceptible for foodborne illness due to age and condition.

The FDA identified more than 30,000 food manufacturers and 20,000 food warehouses that process and store food products.

The retail and foodservice portion of the food industry consists of more than one million establishments with a work force of over 15 million people.

Slide 4

Risk – the chance of injury, damage, or loss. Recall from Session 3 the most important factors include:

- Improper holding temperatures
- Poor personal hygiene
- Contaminated equipment
- Inadequate cooking
- Food from an unsafe source.

A food safety management program should protect foods from the factors known to lead to foodborne illness.

- Focus on the following four keys to foodborne illness prevention:
    - Time and temperature control
    - Good personal hygiene practices and procedures
    - Preventing cross contamination
    - An effective cleaning and sanitation program.

Food safety management programs in retail food establishments must be:

- Customized to fit company policies, the size of the establishment, and the number and types of customers served
- The food safety procedures in your management system must be easy for your employees to understand and implement.

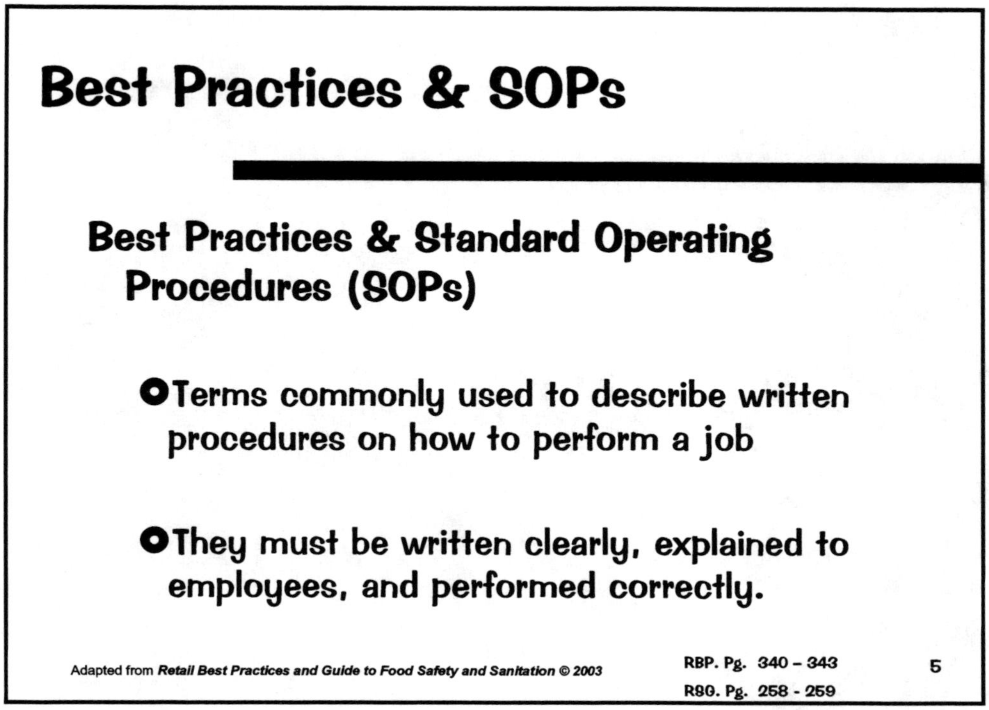

Slide 5

Best Practices & SOPs start with a good set of instructions.

Best Practices and Standard Operating Procedures (SOPs):

- Terms commonly used to describe written procedures on how to perform a job
- They must be written clearly, explained to employees, and performed correctly.

There are many different types of procedures that need to be done correctly to protect the safety of foods.

- A detailed step-by-step form of instruction and training will help the employee do a job properly
- Use Best Practices or SOPs for all functions in a retail food establishment including:
  - Marketing
  - Purchasing
  - Operations
  - Customer services.

Common examples of an SOP:

- Breaking down pieces of equipment for cleaning
- Stocking and rotating products on shelves
- Washing hands properly

- Operating the slicer in the deli or meat department
- Taking care of an emergency.

SOPs include...

Standard Operating Procedures

- Title of task to be completed
- Date the SOP was issued
- Area or department to use the SOP
- Detailed actions to take to complete the task
- Documentation to verify the task was done
- Approval signature.

To see an example of an SOP from a retail food establishment see Figure 10.1 on Page 342 of *Retail Best Practices and Guide to Food Safety and Sanitation.*

Adapted from ***Retail Best Practices and Guide to Food Safety and Sanitation* © 2003**   RBP. Pg. 340 - 343   RSG. Pg. 259 - 261   6

Slide 6

There are many different types of SOPs that can affect the safety and quality of foods during receiving, storage, preparation, and sale. Some examples of SOPs used by retail food establishments include:

- Good personal hygiene programs
- Cross contamination control
- Cleaning and sanitizing programs
- Pest management
- Storage practices of foods and ingredients
- Water and ice handling
- Use of chemicals.

Six important elements should be included in an SOP:

- Title of the task to be completed
- Date the SOP was issued
- Area or department to use the SOP
- Detailed, specific actions that need to be taken to complete the task
- Any documentation that needs to be completed to verify the task was accomplished
- Approval signatures.

To see an example of an SOP from a retail food establishment see Figure 10.1 on Page 342 of the *Retail Best Practices and Guide to Food Safety and Sanitation.*

HACCP – History

Hazard
Analysis
Critical
Control
Point

Inspector Harry HACCP

Adapted from ***Retail Best Practices and Guide to Food Safety and Sanitation* © 2003**

RBP. Pg. 343 – 344
RGG. Pg. 262

7

Slide 7

<u>Hazard Analysis Critical Control Point (HACCP)</u> is a prevention-based program used by many segments of the food industry to ensure food safety.

History of HACCP:

- 1959 – the National Aeronautic and Space Agency (NASA) approached the Pillsbury Company to produce food safe for astronauts to eat in zero gravity conditions in the space capsules
- The project goal was to develop a system that ensured nearly 100% freedom from contamination by microbial, chemical, or physical hazards
- This effort resulted in the beginning of the HACCP concept for food safety management
- HACCP has been recommended for many years in food manufacturing industries.

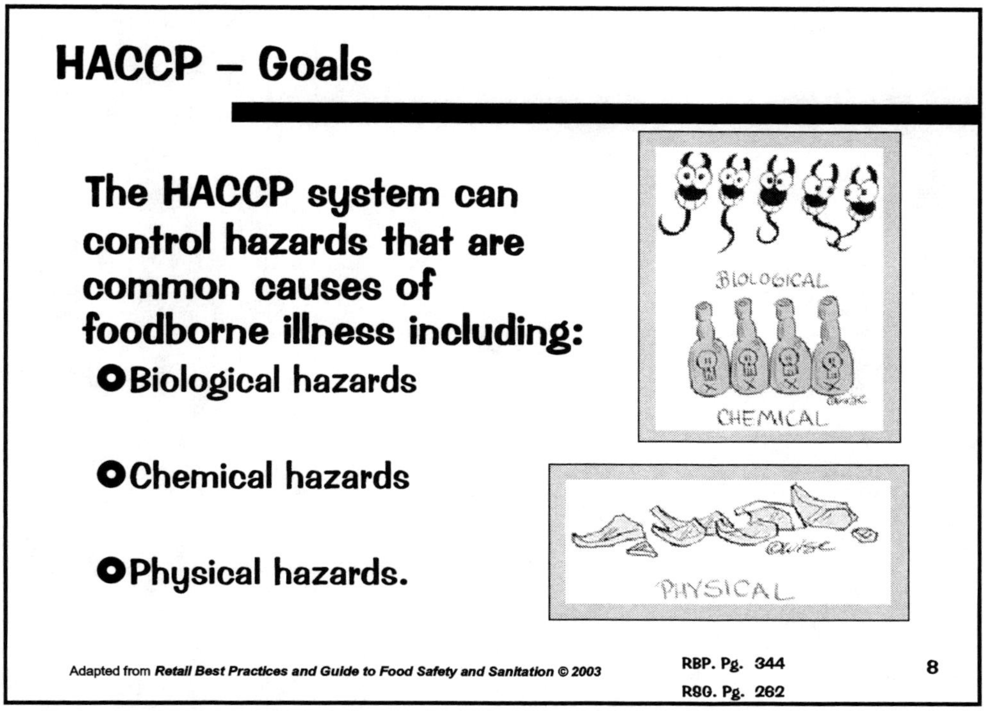

Slide 8

The HACCP system can control hazards such as disease-causing microorganisms, chemical substances and physical objects that are common causes of foodborne illness.

Currently, HACCP is not required for retail food establishments except in a limited number of states and jurisdictions.

However, many retail food establishments have implemented HACCP programs voluntarily to ensure the safety and wholesomeness of the products they produce and sell.

When a HACCP program is developed, a team is chosen to discuss two fundamental questions:

1. Are there any hazards or risks associated with the ingredients, the process, or the finished food?
2. How can I make sure the identified hazards and risks are kept at a safe level?

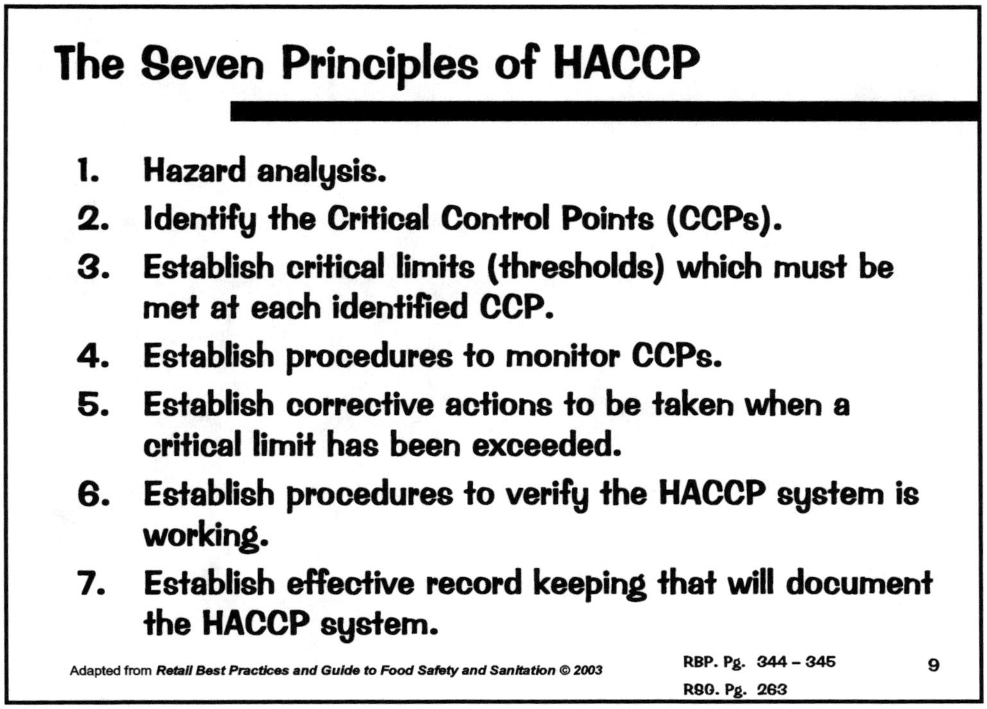

Slide 9

The basic structure of a HACCP system consists of seven principles shown in the slide.

Seven principles in a HACCP system:

1. Hazard analysis.
2. Identify the Critical Control Points (CCPs) in food preparation.
3. Establish critical limits (thresholds) that must be met at each identified CCP.
4. Establish procedures to monitor CCPs.
5. Establish the corrective action to be taken when monitoring indicates a critical limit has been exceeded.
6. Establish procedures to verify the HACCP system is working.
7. Establish effective record keeping that will document the HACCP system.

A HACCP team identifies the potentially hazardous ingredients and foods used and sold at the establishment.

The HACCP team discusses the different processes involved to prepare these items.

After implementation of the seven principles, individuals must be trained and the program must be evaluated to ensure employees understand their roles and responsibilities related to HACCP.

Principle 1: Hazard Analysis
Hazard Identification

The first step is to identify the potential hazards . . . Where are they and what are they?

Hazards can come from anywhere!

Adapted from ***Retail Best Practices and Guide to Food Safety and Sanitation*** © 2003 — RBP. Pg. 345 — RSG. Pg. 263 — 10

Slide 10

Principle 1 – Hazard Analysis – Hazard Identification

What problems could occur?

- Identify hazards associated with:
    - Ingredients that are received
    - Each step in the flow of food
    - The time of purchase or consumption by the customer.

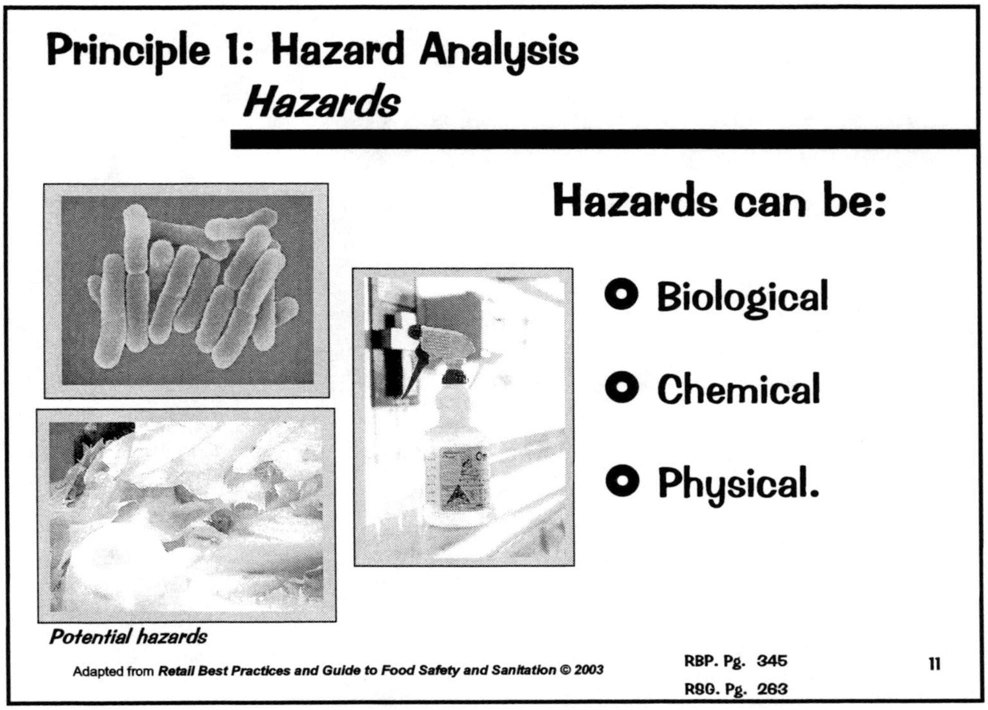

Slide 11

Principle 1 – Hazard Analysis – Hazards

Hazards are biological, chemical or physical agents that may cause an unacceptable consumer health risk.

- Biological
    - Bacteria
    - Viruses
    - Parasites
    - Fungi
- Chemical
    - Man-made
    - Naturally occurring
- Physical
    - Any particle or fragment of an item that is not supposed to be in foods such as:
        - Staples
        - Metal clips
        - Hair
        - Plastic.

Slide 12

Principle 1 – Hazard Analysis – Risk and severity

<u>**Risk**</u> – indicates a level of danger based on how likely the hazard is to occur.
<u>**Severity**</u> – indicates how serious the hazard could be from a public health standpoint.

Assess each hazard for its risk – how likely is this problem to occur?

Risk is influenced by:

- Type of customers served
- Types of foods on menu
- Characteristics of organisms
- Type and frequency of past outbreaks
- Type and size of food establishment
- Extent of employee training.

The hazards most likely to occur or pose the highest risk of disease are identified.

Principle 1: Hazard Analysis
Preventive Measures

Preventing hazards

Preventive measures most often include:

- **Controlling the temperature of foods**
- **Controlling cross contamination**
- **Practicing good personal hygiene**
- **Practicing proper hand washing**
- **Controlling the time foods are held in the temperature danger zone.**

Adapted from ***Retail Best Practices and Guide to Food Safety and Sanitation* © 2003**

RBP. Pg. 345
RSG. Pg. 263

13

Slide 13

Principle 1 – Hazard Analysis – Preventive Measures

Preventive measures should be addressed for those hazards most likely to occur or that pose the highest risk of disease.

Preventive measures may include:

- Controlling the temperature of foods
- Controlling cross contamination
- Practicing good personal hygiene
- Practicing proper hand washing
- Controlling the time foods are held in the temperature danger zone.

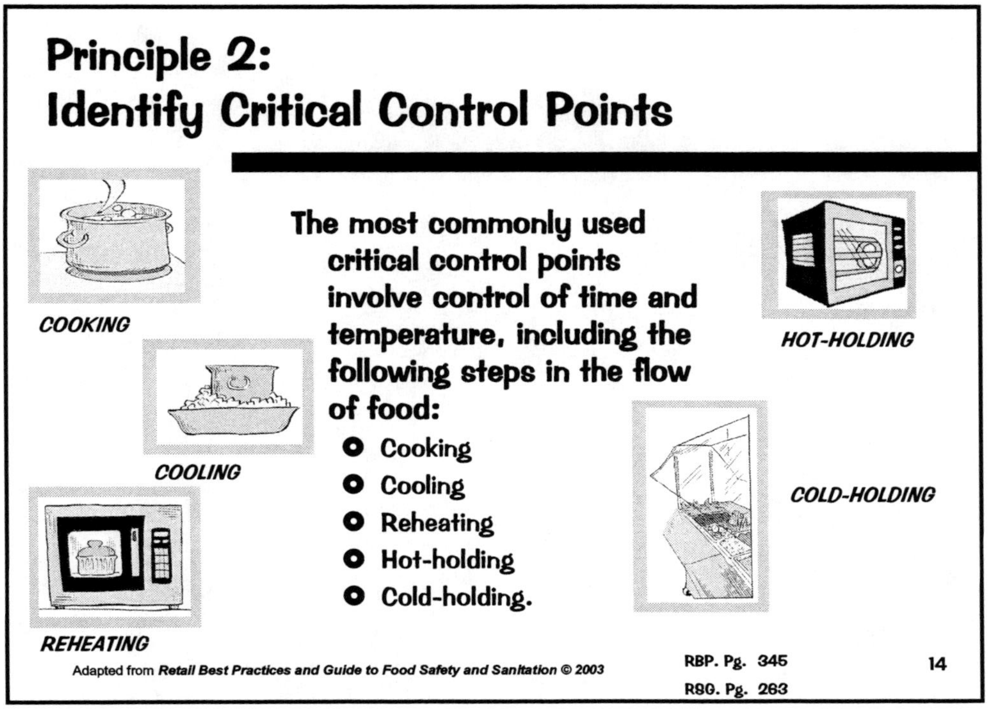

Slide 14

Principle 2 – Identify Critical Control Points (CCPs)

What are the key points in the flow of food where food must be controlled in order to prevent a problem?

- A CCP is a point in which loss of control may result in an unacceptable health risk
- A point in the flow of food where a preventive measure must be put into place to control the identified hazards.

The most commonly used critical control points are:

- Cooking
- Cooling
- Reheating
- Hot-holding
- Cold-holding.

Principle 3: Establish Critical Limits for Each CCP

Critical limits:

- **Are the upper and lower boundaries of food safety**
- **Are set to prevent or eliminate hazards**
- **Usually involve a combination of Time and Temperature controls**
- **Should be very specific [i.e., cook ground beef to 155°F (68°C) for 15 seconds].**

Adapted from ***Retail Best Practices and Guide to Food Safety and Sanitation* © 2003** RBP. Pg. 345 - 346 RSG. Pg. 263 - 264 15

Slide 15

Principle 3 – Establish a critical limit for each CCP

What are the limits of the conditions needed to prevent a problem?

- Critical limits are the upper and lower boundaries of food safety and are set at each CCP to prevent or eliminate the hazards identified during Step 1.
- Critical limits are usually based on time, temperature, pH, and moisture content of a food.
- Critical limits should be very specific. An example would be: "Ground beef must be heated to an internal temperature of 155°F (68°C) or above for at least 15 seconds."
- In retail HACCP plans, critical limits for time and temperature are most often used to keep hazards under control.

Critical Limit – Boundaries of Food Safety

Time
Limit the amount of time a Potentially Hazardous Food (PHF) is in the TDZ to less than 4 hours (cumulative time).

Temperature
Keep potentially hazardous foods below 41°F (5°C) or above 135°F (57°C).

Water Activity
Foods with a water activity (A_W) of .85 or less do not support bacterial growth.

pH (acidity)
Disease-causing bacteria do not grow in foods that have a pH of 4.6 or below.

Principle 4:
Establish Monitoring Procedures

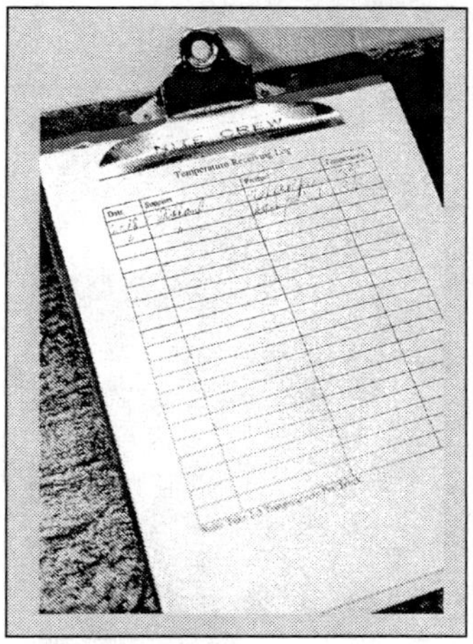

Temperature Log

Monitoring critical limits:

- Should include documentation of findings
- Should only be performed by employees trained on monitoring procedures
- Must be realistic in nature, or "doable."

Adapted from ***Retail Best Practices and Guide to Food Safety and Sanitation* © 2003**

RBP. Pg. 346
RSG. Pg. 264

16

Slide 16

Principle 4 – Monitor Critical Control Points

How will you check to be sure the CCPs are being met?

- Critical control points and critical limits are only effective if they are monitored during the flow of food
- Monitoring includes activities such as measuring temperature and time
- Monitoring helps to ensure the processes involved in the flow of food are done correctly and the hazards are in control.

Monitoring critical limits:

- Should include documentation of findings
- Should only be performed by employees trained on monitoring procedures
- Must be realistic in nature, or "doable."

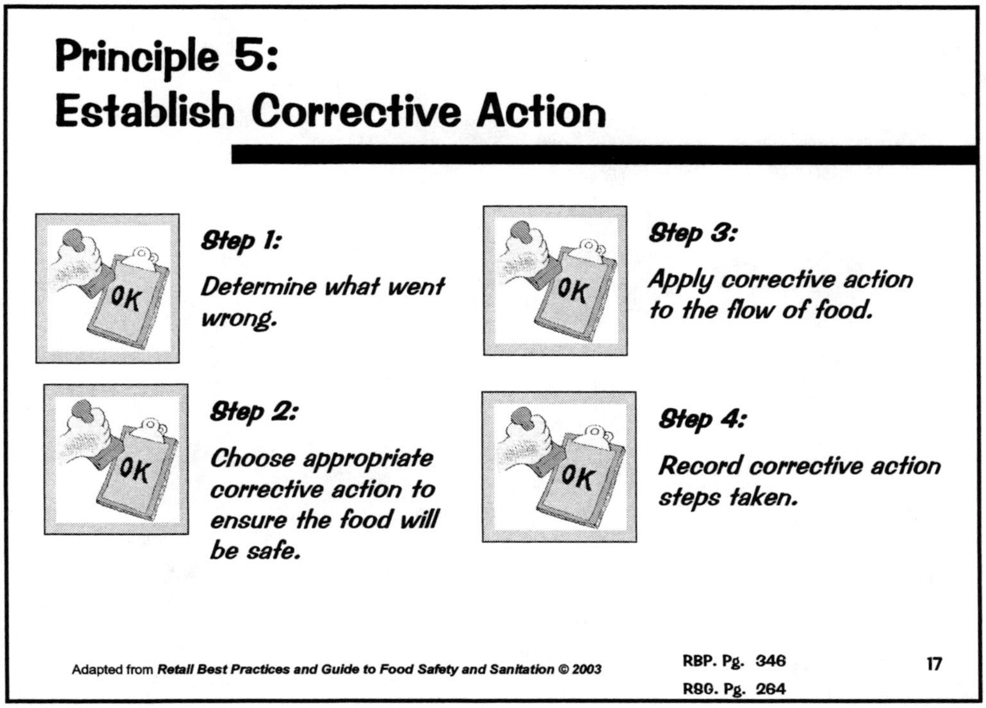

Slide 17

Principle 5 – Take corrective action

What do you do if the CCP is not being met?

- If the critical limit for a CCP is exceeded, something must be done to ensure the food is safe.
- When a critical limit is not met, you must take immediate action to correct the problem. This may involve adjusting product temperature or time, or discarding the product if necessary to ensure the safety of your customers.

Step 1:
Determine what went wrong.

Step 2:
Choose appropriate corrective action to ensure the food will be safe.

Step 3:
Apply corrective action to the flow of food.

Step 4:
Record corrective action steps taken.

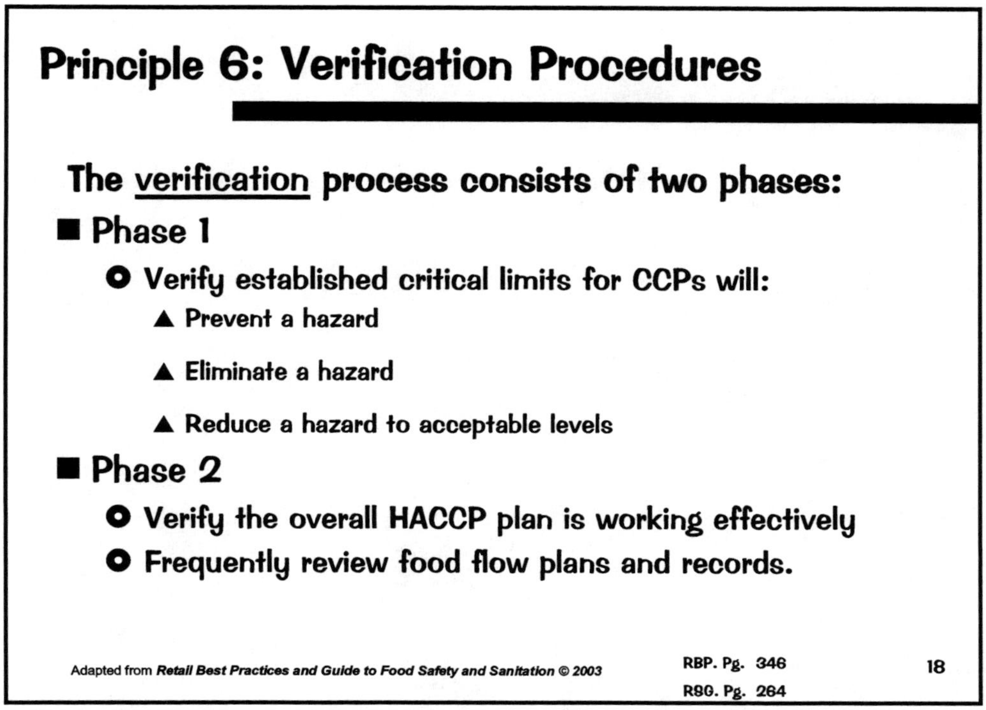

Slide 18

Principle 6 – Establish procedures to verify the HACCP system is working

What must be done to be sure the system is working?

- The verification process consists of two phases:
  - Phase 1
    - Verify established critical limits for CCPs will:
      - Prevent a hazard
      - Eliminate a hazard
      - Reduce a hazard to acceptable levels
  - Phase 2
    - Verify the overall HACCP plan is working effectively
    - Frequently review food flow plans and records.

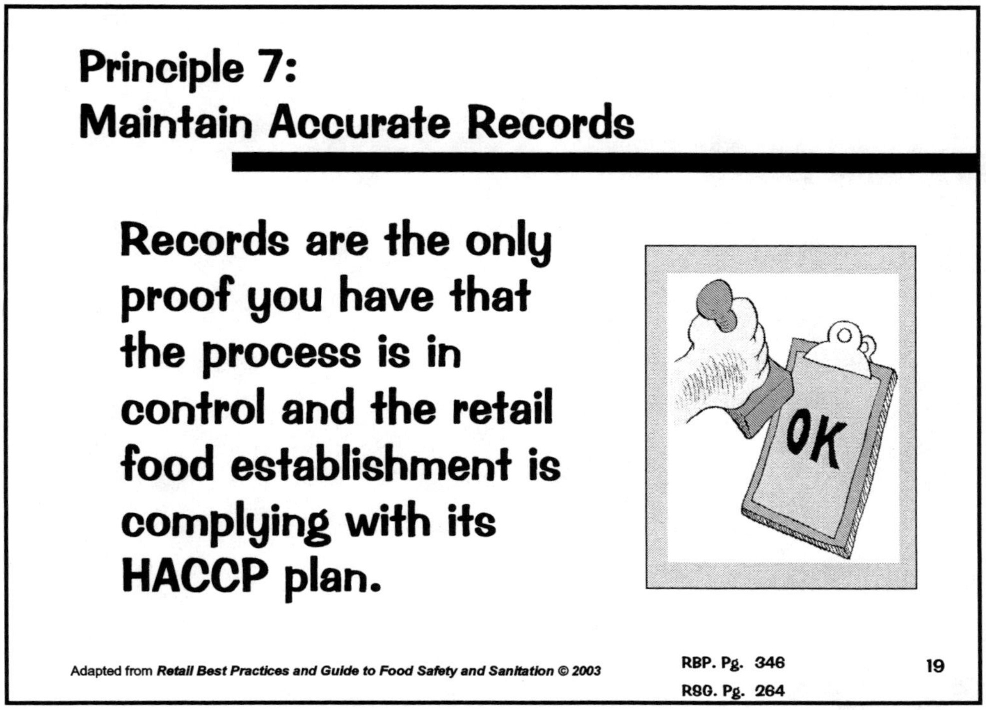

Slide 19

Principle 7 – Establish effective record keeping

What records are needed to show the HACCP plan is working?

- Records are the only proof you have that the process is in control and the retail food establishment is complying with its HACCP plan.
- Record keeping is important for documenting and evaluating the CCPs and CCP limits in the HACCP plan. However, the most important records include the monitoring activities and corrective action procedures since they show the system is working properly.
- Records may be
    - Hard copy (paper)
    - Electronic (spreadsheets, database, temperature logs).

HACCP: A Process Approach

Principle	Process Step Quick Chill	Process Step Reheating
1. Hazard Analysis	Bacteria	Bacteria
2. Identify CCP	Yes	Yes
3. Critical Limit	70°F within 2 hours or less 41°F within 6 hours or less	165°F within 2 hours
4. Monitoring	Temp. every hour	Temp. every 30 min.
5. Corrective Action	Discard if CCP not met	Discard if CCP not met
6. Verification	Food manager	Food manager
7. HACCP Records	Time/temp chart	Time/temp chart

Adapted from *Retail Best Practices and Guide to Food Safety and Sanitation* © 2003

RBP. Pg. 346 - 351
RSG. Pg. 264 - 268

20

Slide 20

Using a Process-specific HACCP Plan:

Process-specific:

- In the retail food industry, numerous types of potentially hazardous foods are produced and sold within a single retail food establishment
- Program is customized to the processes used to produce a particular food or group of similar foods
- It is not manageable to develop a HACCP plan for each one of these food products.

Every process in the flow of food is created from a few basic steps. They may include:

- Received and stored frozen
- Thawed
- Received cold
- Stored cold
- Prepared cold
- Displayed cold
- Cooked
- Hot-held
- Displayed hot
- Cooled
- Reheated
- Service and handling.

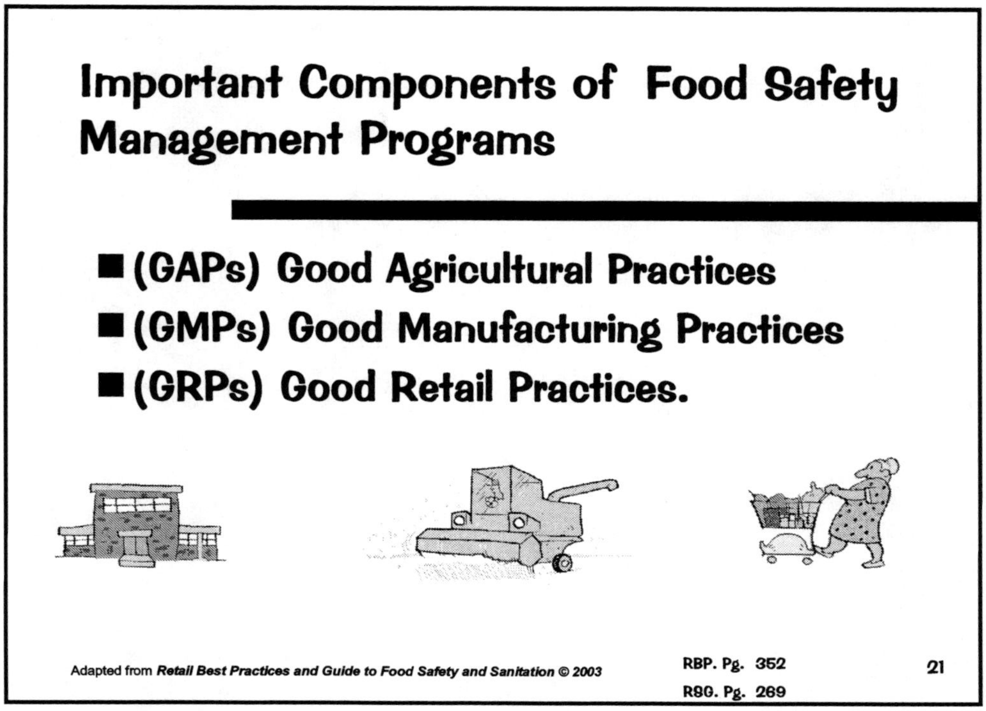

Slide 21

Before the HACCP system can be used, there are several other prerequisite food safety and food quality programs that need to be in place including:

- **Good Agricultural Practices (GAPs)** – used for on-farm production of raw fruit and vegetable products
- **Good Manufacturing Practices (GMPs)** – required for all food processing industries
- **Good Retail Practices (GRPs)** – used in retail food preparation and distribution.

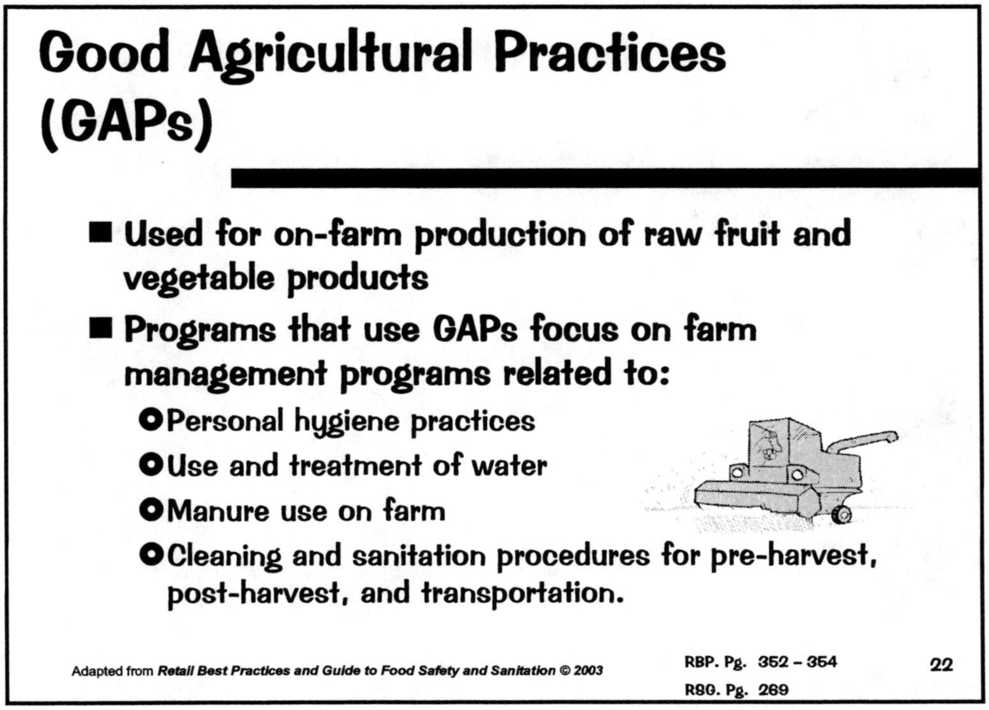

Slide 22

Good Agricultural Practices (GAPs) are used for on-farm production of raw fruit and vegetable products.

Programs that use GAPs focus on farm management programs related to:

- Personal hygiene practices
- Use and treatment of water
- Manure use on farm
- Cleaning and sanitation procedures for pre-harvest, post-harvest, and transportation.

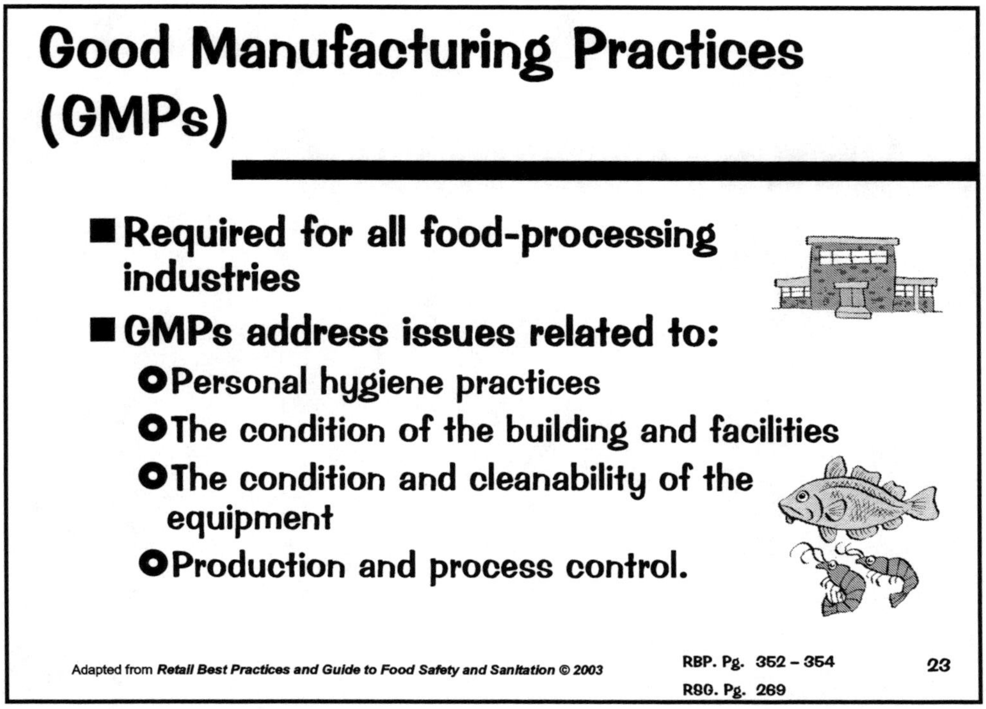

Slide 23

Good Manufacturing Practices (GMPs) are required for all food processing industries.

When implemented properly, GMPs address issues related to:

- Personal hygiene practices
- The condition of the building and facilities
- The condition and cleanability of the equipment
- Production and process control.

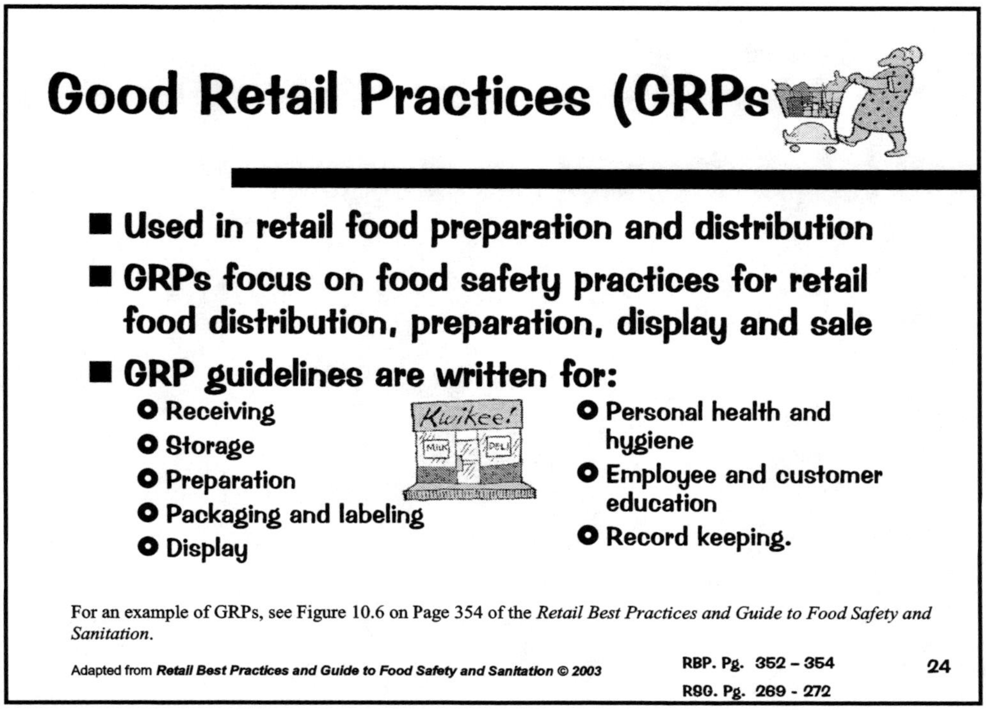

Slide 24

Good Retail Practices (GRPs) are used in retail food preparation and distribution.

The Food Marketing Institute has adopted a program called Good Retail Practices, or GRPs.

- GRPs focus on food safety practices for retail food distribution, preparation, display and sale.
- GRPs can be flexible and written to enhance product quality and shelf life. GRP guidelines are written for:
    - Receiving
    - Storage
    - Preparation
    - Packaging and labeling
    - Display
    - Personal health and hygiene
    - Employee and customer education
    - Record keeping.

For an example of GRPs, see Figure 10.6 on Page 354 of the *Retail Best Practices and Guide to Food Safety and Sanitation.*

Food Recall Procedures

Types of food recalls:

- **Class I**
  - Foods that may cause serious adverse health consequences
- **Class II**
  - Foods that would result in a temporary or reversible health problem
- **Class III**
  - Foods that are not likely to cause danger to health.

Adapted from ***Retail Best Practices and Guide to Food Safety and Sanitation* © 2003**

RBP. Pg. 355

RSG. Pg. 272 - 273

25

Slide 25

Food Recall Procedures

On occasion, contaminated products may be received from food manufacturers, slaughterhouses, or farms that can present a hazard to your customers.

In these situations, the company that supplied you with contaminated product may initiate a "food recall."

Food recall procedure:

- A food recall is the system used to protect customers from using a product that may be unsafe
- Most of the time, a recall involves removing the product from retail shelves so it cannot be sold
- It may also involve retrieving product already sold or providing information to customers that have already purchased the product
- In the event of a food safety emergency or other emergencies involving security and loss prevention issues, FMI members can use the 24-hour Crisis Notification System to contact FMI staff after regular working hours.

Types of food recalls:

Class I
Foods that may cause serious adverse health consequences
Class II
Foods that would result in a temporary or reversible health problem
Class III
Foods that are not likely to cause danger to health.

Product Coding

Use product coding to track:

- **Stock rotation**
- **Shelf life**

Product coding is customized to the operation.

Adapted from ***Retail Best Practices and Guide to Food Safety and Sanitation* © 2003**

RBP. Pg. 355 – 356

RSG. Pg. 273

26

Slide 26

Coding

- Date and source codes are printed on an item to indicate its shelf life
- Date codes simplify rotation, one of the more common uses of product coding is to ensure foods are displayed and used on a First-in,First-out (FIFO) basis and products with an expired shelf life should be discarded
- Product coding will be different in almost every retail food establishment because it needs to be customized to your in-house operations.

Slide 27

Safety Stop! 10.1

Sample Test Questions

1. HACCP consists of:
   A. 4 steps.
   B. 5 steps.
   C. 7 steps.
   D. 3 steps.

2. The chances that a harmful effect will occur due to ingestion of a hazard in food is an appropriate definition for:
   A. Risk.
   B. Physical hazard.
   C. HACCP.
   D. GRP.

3. Risk is influenced by:
   A. Types of customers served.
   B. Size of the retail food establishment.
   C. Type and frequency of past outbreaks.
   D. All of the above.

4. A good food safety management program should focus on
   A. Time and temperature control.
   B. Good personal hygiene practices and procedures.
   C. Elimination of cross contamination.
   D. All of the above.

5. SOP stands for:
   A. Standing Operation Process.
   B. Standard Operating Procedure.
   C. Safe Operation Procedure.
   D. Safety Operating Process.

6. This HACCP concept originated from:
   A. USDA.
   B. FDA.
   C. NASA.
   D. FMI.

7. Biological, chemical or physical agents that may cause an unacceptable consumer health risk are called:
   A. Potentially hazardous foods.
   B. Hazards.
   C. Bacteria.
   D. Viruses.

8. Which of the following is an example of a common critical control point in a HACCP plan?
   A. Washing fruits and vegetables.
   B. Hand washing.
   C. Inspecting procedures for quality.
   D. Cooking foods to required temperature.

9. CCP stands for:
   A. Criticism Conference Place.
   B. Critical Correction Point.
   C. Critical Control Point.
   D. Crisis Control Point.

10. The upper and lower boundaries of food safety which are set at each CCP to prevent or eliminate the hazards are called:
    A. Correction points.
    B. Critical limits.
    C. Checkpoints.
    D. Critical checkpoints.

Real-life

Slide 13
Critical Control Point Violation
Josh was heating a bag of precooked Red Chili Con Carne (pork chili) for the hot soup station. He heated the chili to155°F (68°C). He placed the chili into the soup well with the appropriate clean label and placed a lid on it. What critical control point did Josh violate?

Josh violated the time and temperature critical control point for reheating. When reheating a pre-cooked food, according to the *FDA Food Code*, the food must be reheated to 165°F (74°C) within 2 hours in order to kill all biological hazards associated with it.

Slide 14
Critical Control Point Violation
Employees of Kwik Convenient Food Mart made most of its sandwiches during the 11 to 6 night shift. Maggie was working a double shift so she was happy to start making sandwiches early. She pulled the cut meats and sandwich spreads around 10 p.m., working at a prep table at room temperature [approximately 75°F (24°C)] just after she dropped the evening deposit. As it turned out it was a really busy night at the gas pumps. It was early morning, about 3:00 a.m. when she got to wrap the last sandwich for the case. Maggie had logged when she started but her production was scattered over a four- to five-hour period. What critical control points did Maggie violate? What should have been done to control for this?

Cold-holding was violated because she left some food out for more than four hours. Maggie should have taken out smaller portions and set a timer to remind her of the time windows for production, keeping production in short intervals and significantly less than four hours.

Slide 25
Food Recall
Grocery Groceries, Inc. received notice from the manufacturer that during a routine plant inspection some of their chocolates may have been contaminated with peanut particles, a known allergen. The manufacturer has issued a recall for all their chocolates with the code PDW0608. What class recall is this considered and what steps should Grocery Groceries, Inc. take to ensure the safety of their customers?

This food recall is considered to be a Class I recall because people with peanut allergies could potentially go into anaphylactic shock and die if they eat peanuts or peanut particles. They should begin by immediately pulling the chocolate products coded PDW0608 from their shelves. They should attempt to contact any customers that have purchased the product and inform them of the recall. They should follow all additional procedures outlined in their policies and procedures manuals including reimbursement of funds to affected customers, additional food safety procedure, and future purchasing plans from the manufacturer recalling the product.

Making it Real

Standard Operating Procedures – Slides 5 – 6

- Standard operating procedures often have very specific guidelines that should be followed. In fact, even if you believe you are following the procedure correctly, you may be missing a step.
  - Supplies
    - Standard Operating Procedure for
      - Hand washing
      - Cleaning a piece of equipment
      - Another SOP that would be suitable for your training
    - Handwashing sink or appropriate materials for the procedure you have chosen
  - Procedure
    - Have the learners:
      - Perform the task as they have been trained or believe the SOP is written
      - Read the actual SOP
      - Re-perform the task as it is written

- Teaching Tip – You may be surprised how many people believe they are performing a task properly, but are actually missing important steps
- This can be done for GRPs as well.

HACCP Plan – Slides 7 – 19

- HACCP plans can be difficult to interpret, especially if you have never developed one.
  - Supplies
    - A simple food preparation or procedure such as
      - Slicing cheese on a meat slicer
      - Halving chicken
  - Procedure
    - Have the workers write a HACCP plan for the procedure. Be sure to include each of the following principles:
      1. Hazard analysis.
      2. Identify the Critical Control Points (CCPs) in food preparation.
      3. Establish critical limits (thresholds) that must be met at each identified CCP.
      4. Establish procedures to monitor CCPs.
      5. Establish the corrective action to be taken when monitoring indicates a critical limit has been exceeded.
      6. Establish procedures to verify the HACCP system is working.
      7. Establish effective record keeping that will document the HACCP system.
  - Teaching Tip – This can be a group exercise having each group of learners develop a HACCP plan for different food preparation procedures.
  - This can be done for SOPs and GRPs as well.

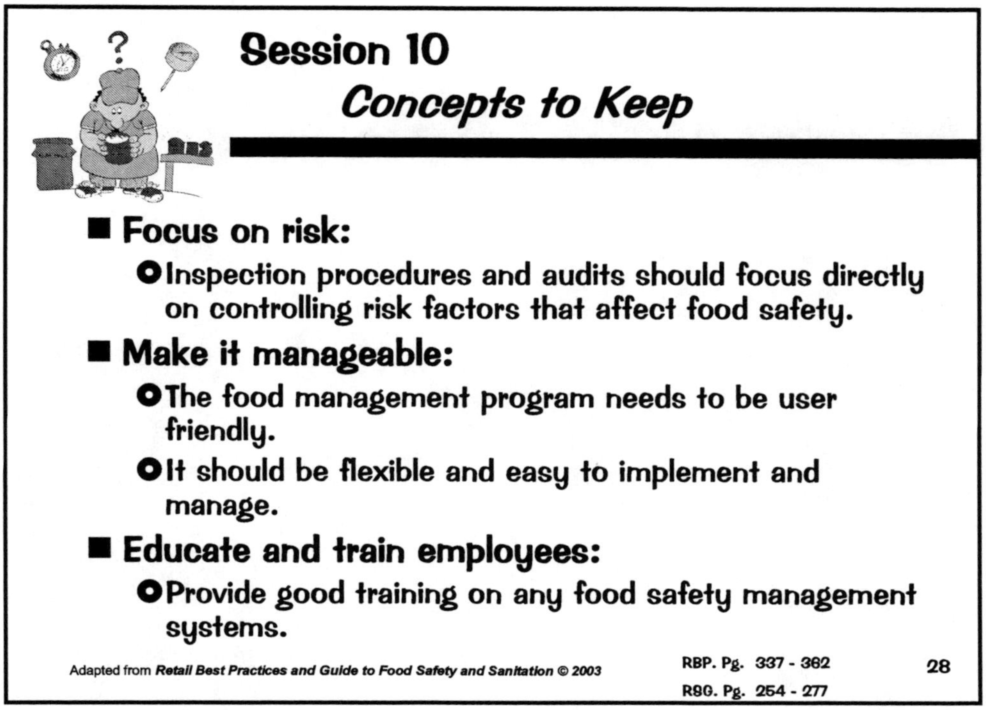

Slide 28

Concepts to Keep

As you develop a food safety management program for your establishment, use the following guidelines:

- **Focus on risk:**
  - Inspection procedures and audits for food safety management programs should focus directly on controlling risk factors that affect food safety.
- **Make it manageable:**
  - The food management program needs to be user friendly.
  - It should be easy to implement and manage.
  - Food safety management programs need to be flexible and related to specific operational procedures.
- **Strengthen communication:**
  - The food industry and regulatory professionals should partner and work together.
  - Education and training need to be encouraged for members of the retail food industry and for regulators.

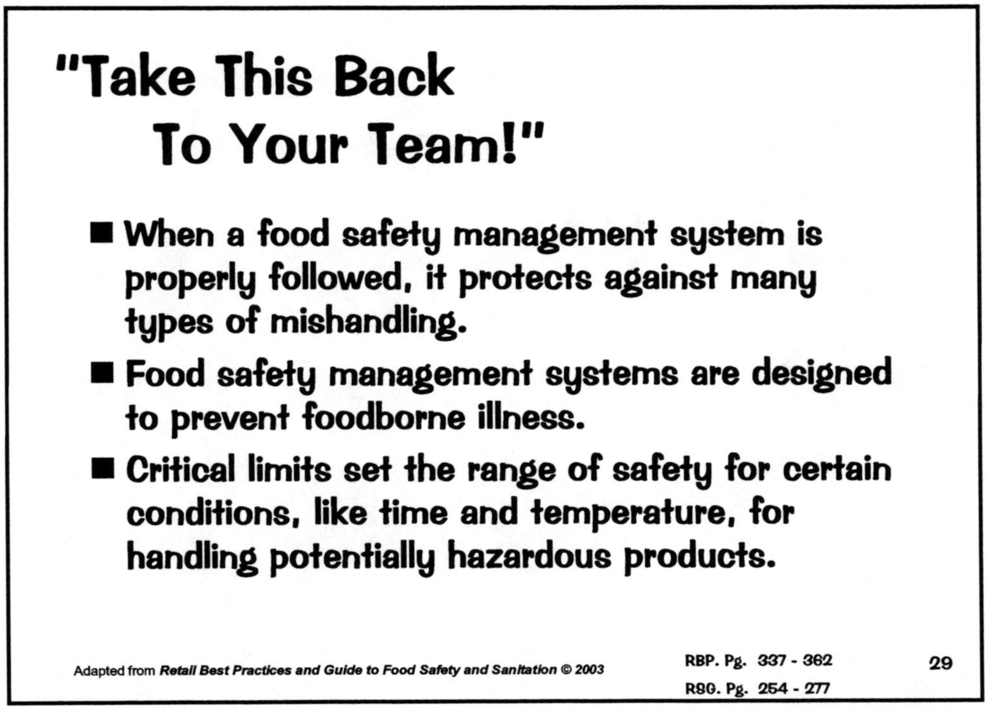

Slide 29

Take This Back to Your Team!

- When a food safety management system is properly followed, it protects against many types of mishandling.
- Food safety management systems are designed to prevent foodborne illness.
- Critical limits set the range of safety for certain conditions, like time and temperature, for handling potentially hazardous products.

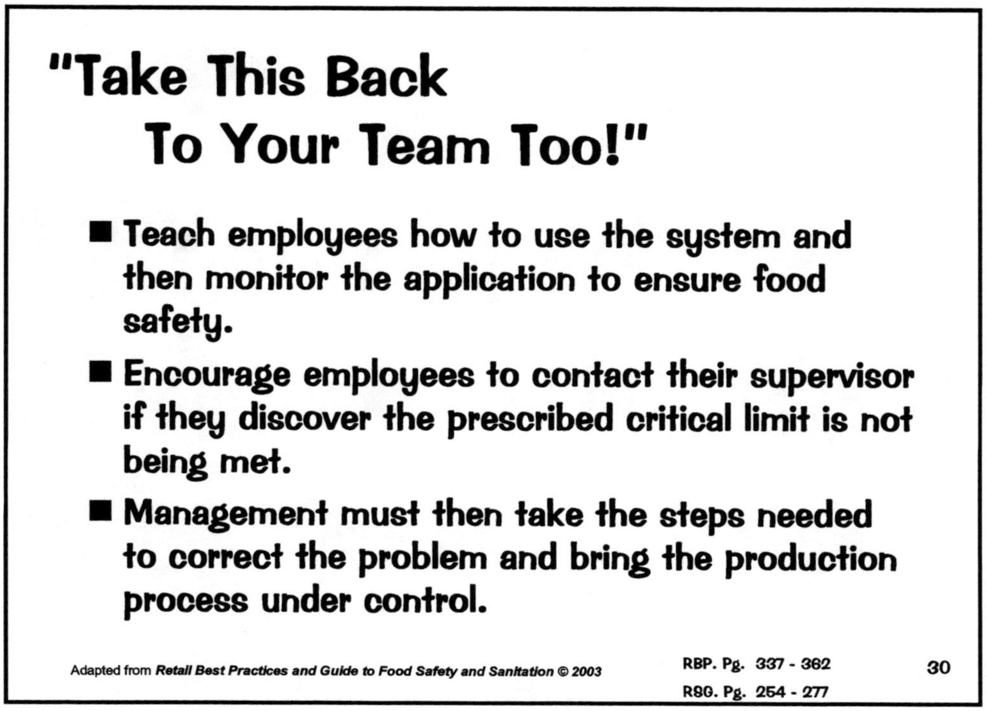

Slide 30

Take This Back to Your Team Too!

- Teach employees how to use the system and then how to effectively monitor the application to ensure food safety.
- Encourage employees to contact their supervisor if they discover the prescribed critical limit is not being met.
- Management must then take the steps needed to correct the problem and bring the production process under control.

Session 11

Food Safety Regulations – Session 11 Key Points

Slide 1

Training Tip:

The content of each slide provides the most essential information in the session. The written text below each slide is provided to help you understand and better teach the essential information. The written text may also be used to answer questions that may be posed during the session. Depending on the length of your training session, you may not be able to address all the information contained beneath each slide.

Overview:

Session 11 reviews the current government agencies and regulations governing food safety. From handling local health inspections to dealing with nutritional labeling and food recalls, this chapter looks at the roles of state, local and national governments in providing safe food.

Suggested timeline:

8-hour course: 15 minutes using the PowerPoint© slides only.

- 11 ***Slides.***

16-hour course: 30 minutes using the 18 PowerPoint© slides and the 1 ***Safety Stop!*** slide containing:

- 7 ***Sample Test Questions***
- 2 ***Real-life*** scenarios
- 2 ***Making it Real***

Video suggestions and supplemental resources:

Videos:

Food Safety Success: Supermarket Application of the Food Code, Food Marketing Institute, 1-301-843-3084

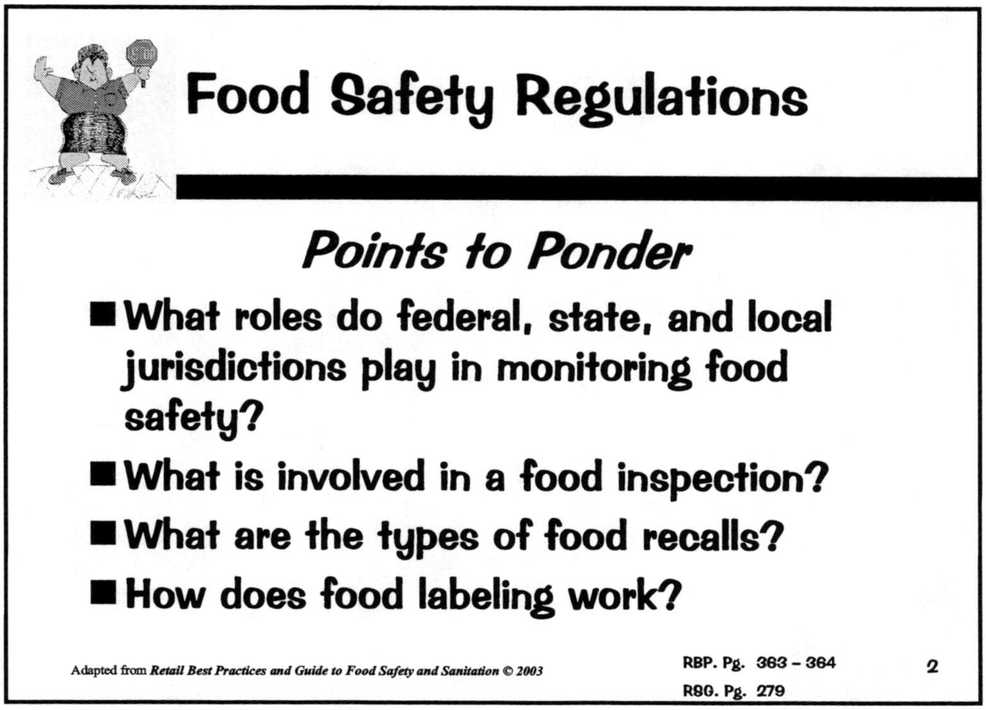

Slide 2

Points to Ponder

- What roles do federal, state, and local jurisdictions play in monitoring food safety?
- What is involved in a food inspection?
- What are the types of food recalls?
- How does food-labeling work?

State and Local Regulations

You must have the proper permits displayed in order to operate.

Post licenses and permits in a prominent location.

Adapted from *Retail Best Practices and Guide to Food Safety and Sanitation* © 2003

RBP. Pg. 364 – 365
RSG. Pg. 280 - 281

3

Slide 3

You must have the proper permits to be able to operate.

- It is unlawful to operate a retail food establishment unless you have a valid permit issued by the regulatory authority in your area.
- Post the permit in a prominent location. Permits are generally not transferable from one person to another, from one retail food establishment to another, or from one food operation to another.
- If you add a partner or incorporate, these actions are considered ownership changes and may require you to apply for a new permit to operate.

Inspections

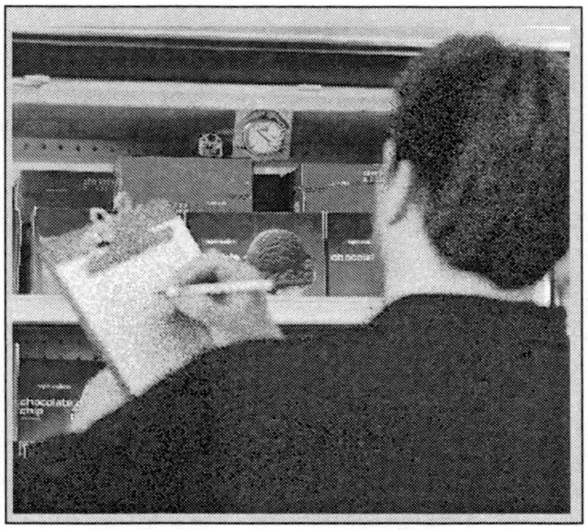

Inspection

If a critical violation such as temperature abuse or cross contamination is discovered during an inspection, corrective action must be taken immediately.

Adapted from ***Retail Best Practices and Guide to Food Safety and Sanitation* © 2003**

RBP. Pg. 365 - 367
RGG. Pg. 281 - 282

4

Slide 4

How frequently will your organization be inspected?

The risk a retail food establishment poses is calculated using such factors as:

- The sanitation history of the establishment
- Number of home meal replacements sold
- The amount of food preparation and handling conducted on-site
- Condition of building
- Critical violations observed.

A routine inspection or audit normally consists of three phases:

- Pre-inspection
    - Always cooperate with inspecting personnel. Greet the inspector when he or she arrives and ask to see their official identification.
    - Discussion between inspector and manager about previous inspections.
- Inspection is conducted
    - Current inspection.
    - If possible, the manager should accompany the inspector during the inspection to answer any questions and initiate immediate corrective action when necessary.

- Post-inspection
    - Discussion between inspector and manager about the results of the current inspection.

Most inspections will focus on:

- Foods and supplies
- Personal hygiene and employee health
- Temperatures of food and food-holding equipment
- Cleaning and sanitizing procedures
- Equipment and utensils
- Water supply and waste disposal
- Pest control and other aspects of the operation that might compromise food safety.

If you have a critical violation such as temperature abuse or cross contamination discovered during an inspection, corrective action must be taken immediately.

- Failure to do so could result in fines and possible closure of the establishment
- Lesser violations should be corrected as soon as possible, but always before the next inspection or audit.

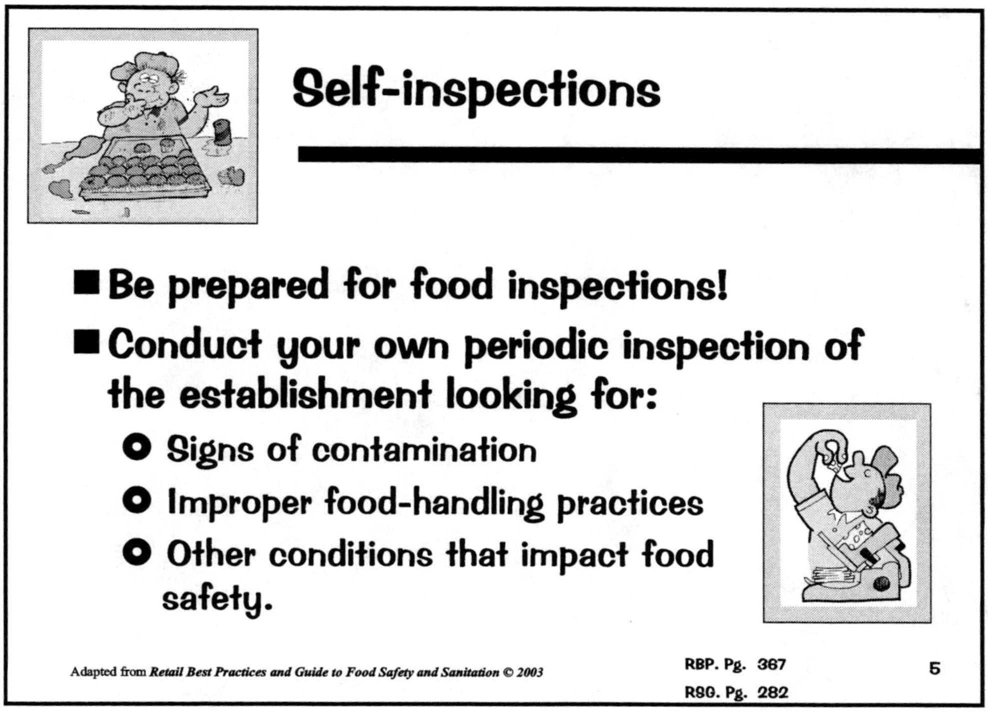

Slide 5

Food managers should periodically conduct self-inspections of their establishment to ensure proper sanitation and food safety.

During a self-inspection, the manager should look for:

- Signs of contamination
- Improper food-handling practices
- Other conditions that may put food safety in jeopardy
- Personal hygiene and health.

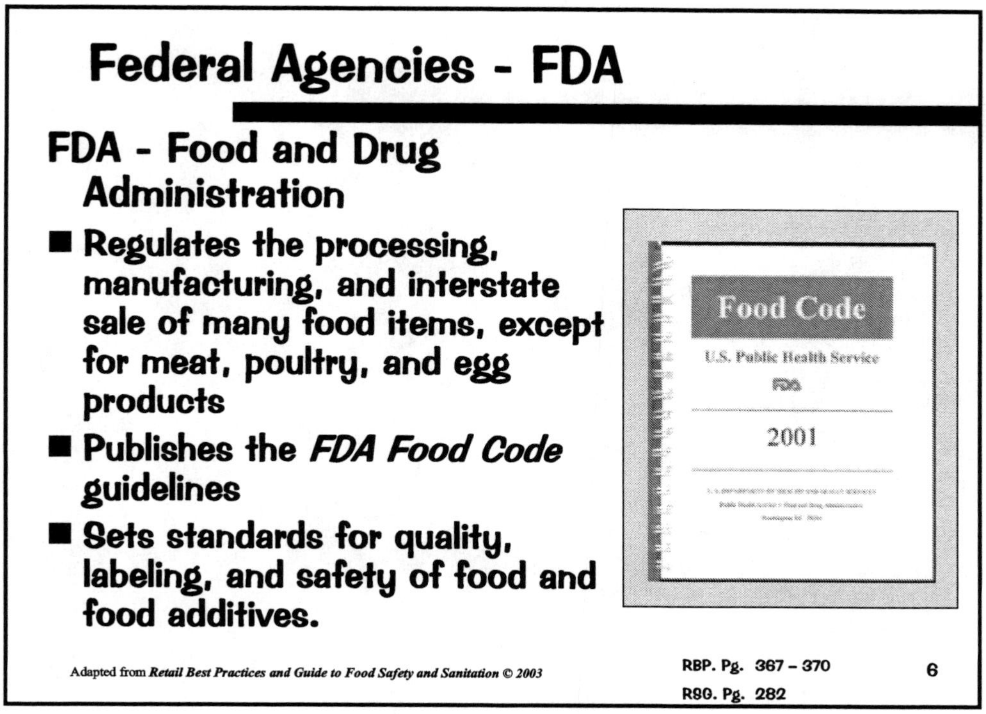

Slide 6

FDA – Food and Drug Administration

Primary function or duties:

- Regulates the processing, manufacturing, and interstate sale of many food items, except for meat, poultry, and egg products
- Sets standards with respect to composition, quality, labeling, and safety of foods and food additives
- Publishes the *FDA Food Code* which serves as a model for retail food programs
- Protects the public's health by preventing the adulteration and misbranding of food
- Ensures food shipped in interstate commerce is safe, pure, wholesome, sanitary, and honestly packaged and labeled
- Maintains a list of interstate certified shellfish shippers and a list of interstate milk shippers.

Federal Agencies - USDA

U. S. Department of Agriculture

- Responsible for inspection of
  - Meats
  - Poultry
  - Processed meat and poultry products
  - Meat and poultry processing facilities
- Voluntary grading services for meat, poultry, eggs, dairy products, fruits and vegetables.

Adapted from *Retail Best Practices and Guide to Food Safety and Sanitation* © 2003

RBP. Pg. 370 – 371
RSG. Pg. 283

7

Slide 7

USDA – U. S. Department of Agriculture

- Inspects domestic and imported meats, poultry, and processed meat and poultry products. The USDA maintains a list of approved facilities for meat and poultry processing.
- Conducts voluntary grading services for red meats, poultry and eggs, dairy products, and fruits and vegetables.

Federal Agencies - USDC

U.S. Department of Commerce

- Voluntary grading standards for fishery products
- National Marine Fisheries Service (NMFS) voluntary inspection of
  - Fishery products
  - Fishery processors.

Adapted from *Retail Best Practices and Guide to Food Safety and Sanitation* © 2003

RBP. Pg. 371 - 372
RSG. Pg. 283

8

Slide 8

USDC – U.S. Department of Commerce

- Develops grade standards for processed fishery products.

NMFS – National Marine Fisheries Service

- Provides voluntary inspection service for processed fishery products
- Maintains a list of approved fish processors and fishery products on a permanent basis.

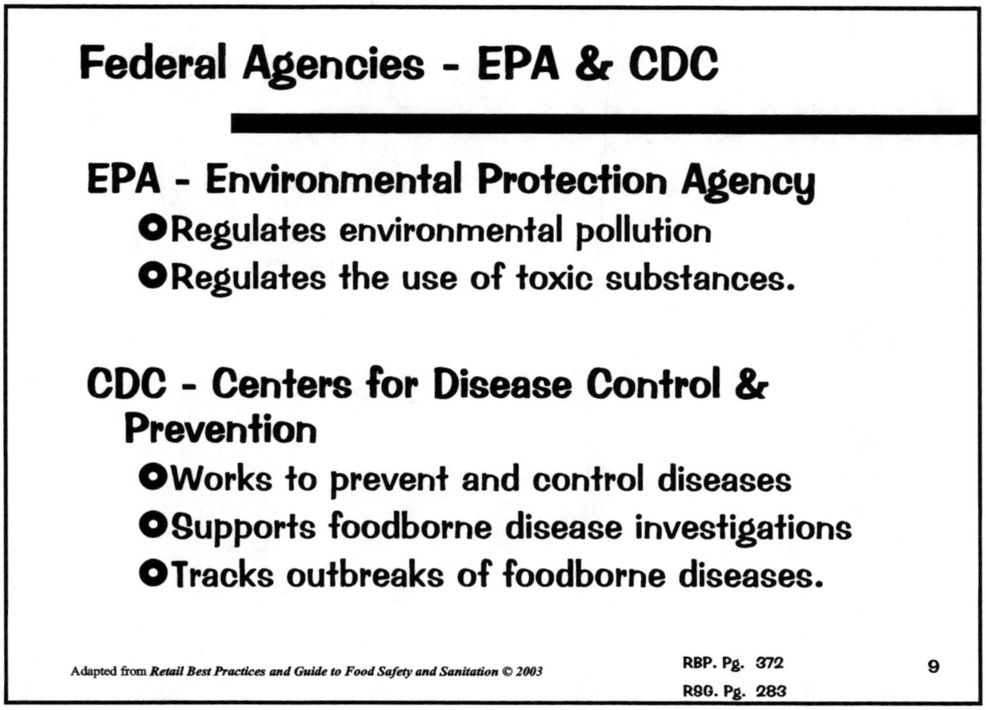

Slide 9

EPA – Environmental Protection Agency

- Created to prevent, control, and reduce air, land, and water pollution
- Regulates the use of toxic substances (pesticides, sanitizers, and other chemicals), monitors compliance, and provides technical assistance to states.

CDC – Centers for Disease Control and Prevention

- Responsible for protecting public health through the prevention and control of diseases
- Supports foodborne disease investigations and prepares annual summaries and statistics on outbreaks of foodborne diseases transmitted through food and water.

Federal Agencies - OSHA & FTC

OSHA - Occupational Safety & Health Administration

- Enforces health and safety standards in the workplace.

FTC - Federal Trade Commission

- Enforces laws regarding marketing practices and national advertising of food and other products.

Adapted from *Retail Best Practices and Guide to Food Safety and Sanitation* © 2003

RBP. Pg. 373
RSG. Pg. 284

10

Slide 10

OSHA – Occupational Safety and Health Administration

- Created in the U.S. Department of Labor to enforce the Occupational Safety and Health Act. It enforces health standards and regulations on safety, noise, and other workplace-related hazards.

FTC – Federal Trade Commission

- Involved in enforcing various laws regarding marketing practices and national advertising of foods and other products.

Food Recalls

Types of Food Recalls	
Class I	Foods that may cause serious adverse health consequences
Class II	Foods that would result in a temporary or reversible health problem
Class III	Foods that are not likely to cause danger to health.

Adapted from *Retail Best Practices and Guide to Food Safety and Sanitation* © 2003

RBP. Pg. 373 – 376
RSG. Pg. 286

11

Slide 11

The FDA can request a food recall if it has determined a potential hazard exists.

- Food recalls are "voluntary" with the possible exception of a recall involving infant formula
- In the vast majority of cases, food processors voluntarily remove products from the marketplace to keep consumers safe
- If the company doesn't agree with the potential hazard, the FDA may issue public press releases and seize the product in order to remove it from commerce
- Recalled foods should be promptly removed from retail shelves and cannot be sold.

Types of Food Recalls	
Class I	**Foods that may cause serious adverse health consequences (i.e., foodborne disease outbreak caused by *Listeria Monocytogenes* in ready to eat luncheon meats).**
Class II	**Foods that would result in a temporary or reversible health problem (i.e., foodborne disease outbreak caused by *Salmonella* in refrigerated pasta).**
Class III	**Foods that are not likely to cause danger to health (i.e., foodborne disease outbreak caused by pesticide residue in butter).**

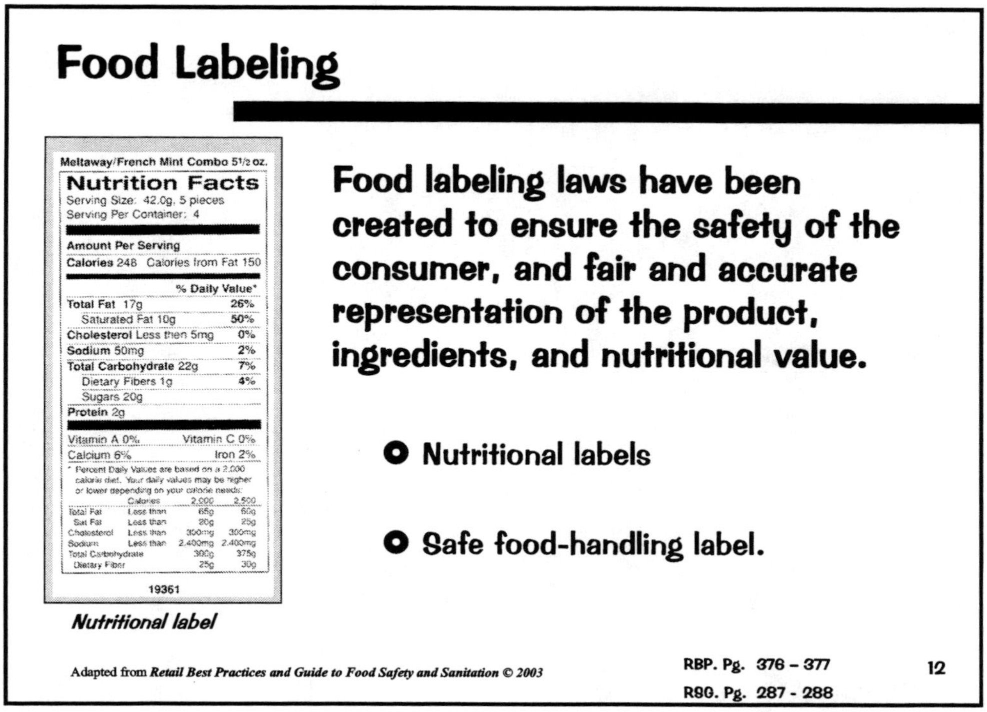

Slide 12

Nutrition and ingredient labeling:

- The USDA and the FDA Nutrition enforce Labeling and Education Act jointly.
    - Law provides information on specific nutritional guidelines and defines what needs to be on a nutrition label
- Labels provide information for processed foods on protein, fat, carbohydrate, and mineral content
    - Label helps customers make informed choices when selecting foods
- Although not mandatory at this time, the USDA has published guidelines for demonstrating voluntary compliance to nutritional labeling requirements.
- Supermarkets are encouraged to provide consumers accurate nutritional information using "point-of-purchase" product labels, brochures or signage
- Products prepared and packaged in the retail establishment must include a label for the consumer:
    - The name of the food
    - A list of ingredients
    - The quantity of ingredients.

Safe food-handling label

- Provides helpful food-handling information for the consumer on all raw or partially pre-cooked (not ready-to-eat) meat and poultry
- Designed to educate the consumer for storage, preparation, and handling of raw meat and poultry products in the home.

Slide 13

Product Dating Labels

- A date stamped on a food package to help a retail food establishment know how long to display the product for sale
- Can help customers understand the time limit to purchase or use the product at its best quality
- Date labels are on almost all food products
- Date labels are usually found on perishable, potentially hazardous foods such as meat, poultry, eggs, and dairy items
- Date labels are voluntary and not required by law.

Two types of dating labels:

- **"Sell by" date** – tells the retail food establishment how long to display the product for sale
    - Food products should be removed from the retail shelves immediately after the sell by date
    - Customers should buy the product before the date expires
- **"Best if used by (or before)" date** – is recommended for the best flavor or quality
    - If handled properly the product quality will last two to five days past the "sell by" date

- **"Use by" date** – the last date recommended for the use of the product while at peak quality
    - The manufacturer of the product usually determines the date by using quality control and shelf-life tests
    - Federal laws regulate date marking of infant formula products.

Slide 14

Recent Initiatives in Food Safety

National Food Safety Initiative (federal government):

- An initiative from the federal government that works to reduce the incidence of foodborne illness by strengthening and improving food safety practices and policies. The initiative seeks to:
    - Expand education efforts aimed at improving food handling in homes and retail outlets.
    - Improve inspections and expand monitoring efforts.
    - Increase research to develop new and more rapid methods to detect foodborne pathogens and to develop preventive techniques.
    - Improve intergovernmental communications and coordination of responses to foodborne outbreaks. The FDA, CDC, USDA, and EPA will form a new intergovernmental group to improve federal, state, and local responses to outbreaks of foodborne illnesses.
    - Determine the frequency and severity of foodborne disease.
    - Determine the proportion of common foodborne diseases that results from eating specific foods.
    - Describe the epidemiology of new and emerging bacterial, parasitic, and viral foodborne pathogens.

The Fight BAC! TM Campaign:

- An educational program developed by industry, government, and consumer groups to teach customers about safe food handling.
- The focal point of the campaign is BAC (short for bacteria), a character used to make customers aware of the invisible germs that cause foodborne illness. Even though customers cannot see, smell, or taste BAC, it and millions more germs like it are in and on food and food-contact surfaces.
- Fight Bac! ™ Campaign uses media and community outreach programs to teach customers how to keep food safe from harmful bacteria through four simple steps: wash hands and surfaces often, prevent cross contamination, cook foods to proper temperature, and refrigerate foods quickly.
- The partners in the Fight Bac! ™ Campaign use newspaper articles, public service announcements, printed brochures, and school curricula to teach customers how to lower their risk of foodborne illness.

Consumer Advisory in Food Safety

Consumer advisory

- Food establishments that sell or serve raw or undercooked animal foods or ingredients for human consumption must inform their customers about the increased risk associated with eating these kinds of foods.

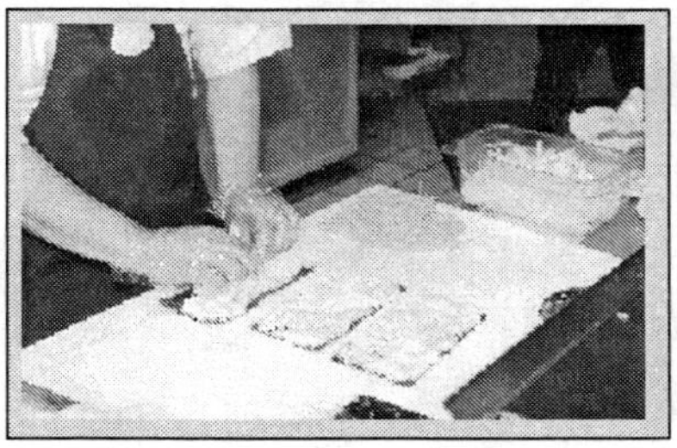

Sushi preparation

Adapted from *Retail Best Practices and Guide to Food Safety and Sanitation © 2003*

RBP. Pg. 380 – 381
RSG. Pg. 290 - 291

15

Slide 15

Consumer Advisory

- Food establishments that sell or serve raw or undercooked animal foods or ingredients for human consumption must inform their customers about the increased risk associated with eating these kinds of foods
    - The disclosure is used to make customers aware of the dangers of raw or undercooked foods and of items that either contain or may contain raw or undercooked ingredients
    - The reminder is a notice about the relationship between thorough cooking and food safety.

Disclosure is satisfied when items are described such as oysters on the half-shell (raw oysters), and raw-egg Caesar salad; or items are asterisked to a footnote that states the items:

- Are served raw, or undercooked, or
- Contain (or may contain) raw or undercooked ingredients.

The reminder is satisfied when the items requiring disclosure are asterisked to a footnote that states information about:

- The safety of these items, written information is available upon request.
- Eating raw or undercooked meats, poultry, seafood, shellfish, or eggs, which may increase your risk of foodborne illness.

- Increased risk of foodborne illness when the above products are eaten if you have certain medical conditions.
- Disclosure should appear at the point where the customer selects the food.
- Both the disclosure and the reminder need to accompany the information from which the customer makes a selection. That information could appear in many forms such as display signage, package labels, and consumer brochures.
- Consumer advisories may be tailored to be product specific if a food establishment either has a limited menu or offers only certain animal-derived foods in a raw or undercooked, ready-to-eat form.

Slide 16

Safety Stop! 11.1

Sample Test Questions

1. Which government agency is responsible for ensuring the safety of all foods except meat, poultry and egg products?
   A. The Consumer Advisory.
   B. The Centers for Disease Control and Prevention.
   C. The Food and Drug Administration.
   D. The U. S. Department of Agriculture.

2. Which of the following factors are considered when setting inspection frequencies for the risk-based approach to inspections?
   A. The sanitation history of the establishment.
   B. Number of home meal replacements sold.
   C. Number of critical violations observed.
   D. All of the above.

3. Which federal agency conducts voluntary inspections for processed fishery products?

   A. The Centers for Disease Control and Prevention.

   B. The Food and Drug Administration.

   C. U.S. Department of Commerce.

   D. The U. S. Department of Agriculture.

4. A food recall that would result in a temporary or reversible health problem would be a

   A. Class I recall.

   B. Class II recall.

   C. Class III recall.

   D. Class IV recall.

5. Nutritional labeling for raw meats

   A. Is required on all foods.

   B. Indicates how long the product is for sale.

   C. Is voluntary.

   D. Provides food-handling information for raw foods.

6. During an routine health inspection of an establishment, the manager should:

   A. Ask to see credentials.

   B. Accompany the inspector when possible.

   C. Discuss any violations.

   D. All of the above.

7. Food managers should periodically conduct self-inspections of their establishment looking for:

   A. Signs of contamination.

   B. Improper food-handling practices.

   C. Conditions that may put food safety in jeopardy.

   D. All of the above.

Real-life

Slide 4

Routine Health Inspection

Marissa was the manager on duty for the morning shift. At approximately 11:30 a.m. at the height of their business in the deli and self-service hot and cold food lines, the local health inspector came in for a routine visit. Marissa felt she was too busy to accompany him on his visit, so she had an hourly employee walk with him. She failed the inspection. What could Marissa have done differently?

You are never too busy to accompany the health inspector on a routine visit. By accompanying the inspector, it provides you the opportunity to discuss any potential violations and describe any violations they may find you are already aware of. They may be receptive and understanding if you have viable work orders to prove you are fixing a piece of equipment or if they see a sense of urgency toward fixing minor infractions. You can also coach and teach your employees on anything they may be doing improperly – this will allow the inspector to see your comprehension of food laws.

Slide 11

Food Recall

Carlos is a manager of a superstore. He received a call from his supervisor that there has been a food recall on ground beef with the code A0616. What steps should Carlos take?

Answer

- Inspect storage and display coolers for ground beef and the code for the product being recalled
- If product is found, it should be labeled and placed in a secure area away from other foods
- If it is suspected the product may have been sold to customers the crisis management team should contact customers using procedures outline in the company recall procedures.

Making it Real

Product Labels – Slide 12

Bring in product dating labels, safe food-handling labels and nutritional labels from various products. Take the time to review the labels for content. Ask the learners questions about the labels such as what information is required or optional. Have the learners describe important information on the labels; you never know when a customer may ask them a question about a product label!

- Supplies
    - Various labels
- If this is a multi-day session, have the students bring labels with them from home and ask them questions about the labels.

Fight BAC! – Slide 14

Contact the Fight BAC! Campaign for complementary materials (http://www.fightbac.org/main.cfm). Bring the materials to class and disseminate them. Many of their materials will allow you to review pertinent information that has been discussed throughout this training.

- Supplies
    - Fight BAC! Materials
        - Time and temperature information
        - Cooking temperatures
        - Cleaning procedures
        - Handwashing procedures.

Session Eleven:
Concepts to Keep

- Keeping food safe takes the efforts of many agencies as well as the individual food establishments.
- Expect local health departments to inspect your establishment on a regular basis.
- Understand and comply with food regulations and laws.

Adapted from *Retail Best Practices and Guide to Food Safety and Sanitation © 2003*

RBP. Pg. 363 – 385
RSG. Pg. 278 - 292

17

Slide 17

Concepts to Keep

- Keeping food safe takes the efforts of many agencies as well as the individual food establishments.
- Expect local health departments to inspect your establishment on a regular basis.
- Understand and comply with food regulations and laws.

"Take This Back to Your Team!"

- Always join inspectors while they survey your establishment so you can answer their questions.
- Immediately take corrective action on any violations an inspector finds.
- Accurately labeling foods is an important practice.
- Protect your customers with consumer advisories.

Adapted from *Retail Best Practices and Guide to Food Safety and Sanitation* © 2003

RBP. Pg. 363 – 385
RSG. Pg. 278 - 292

18

Slide 18

Take This Back to Your Team!

- Always join inspectors while they survey your establishment so you can answer their questions.
- Take immediate action on any violations.
- Accurately labeling foods is an important practice.
- Protect your customers with consumer advisories.

Appendices

Appendix A

8-hour Training Session – Hidden Slides

In order to incorporate all of the necessary information into an 8-hour session, you will need to:

1. Reduce or eliminate the Safety Stop! slides and interactive exercises.
2. Eliminate some content slides.

On the enclosed CD, you will find an 8-hour training folder. This folder contains a set of slides that we suggest you use for your 8-hour training sessions. Below you will find a list of the slides from the 16-hour session that will not be reviewed in the 8-hour session.

These slides are "hidden," not deleted from each session. The slide number at the bottom of each slide will still match the slide number listed on the lesson pages in the manual. You can customize your training by "hiding" additional slides, or removing the "Hide" command on the hidden slides. See the Microsoft PowerPoint Owners Manual for more information on using the "Hide Slide" command.

Hidden Slides in the 8-hour training sessions:

Session One: 5, 11, 12, 15, and 17

Session Two: 5, 7 – 9, 12, 20, 23, 35, 39, 41 – 46, 52, 55

Session Three: 3, 5, 8, 10, 13 – 14, 16, 19, 29, 39, 45, 46

Session Four: 7, 15, 23, 30, 47

Session Five: 6, 12, 15 – 17, 20, 22 – 23, 26 – 30, 33

Session Six: 7 – 9, 14, 17, 20, 23

Session Seven: 4 – 5, 13, 21

Session Eight: 3, 6 – 11, 13 – 14, 19

Session Nine: 3, 4, 8, 13

Session Ten: 3, 11 – 12, 21 – 23, 27

Session Eleven: 5, 8 – 10, 12, 14, 16

Appendix B

SuperSafeMark Food Safety Training Program

16-hour Sample Agenda

Day 1

8:00 a.m. – 8:15 a.m.:	Welcome, Introductions and Training Goals Setting the Ground Rules for the Training Program Meeting Participant Expectations
8:15 a.m. – 8:45 a.m.:	Session 1 – Retail Food Safety (20 slides) ***Safety Stop!*** 1.1 review
8:45 a.m. – 9:15 a.m.:	Session 2 – Hazards to Food Safety – Part I (23 slides [1 – 23]) ***Safety Stop!*** 2.1 and 2.2 reviews
9:15 a.m. – 10:00 a.m.:	Session 2 – Hazards to Food Safety – Part II (23 slides [24 – 46]) ***Safety Stop!*** 2.3, 2.4 and 2.5 reviews
10:00 a.m. – 10:15 a.m.:	Break
10:15 a.m. – 10:45 a.m.:	Session 2 – Hazards to Food Safety – Part III (13 slides [47 - 59]) ***Safety Stop!*** 2.6 review
10:45 a.m. – 11:45 a.m.:	Session 3 – Preventing Foodborne Illness – Part I (19 slides [1 –19]) ***Safety Stop!*** 3.1 and 3.2 reviews
11:45 a.m. –12:45 p.m.:	Lunch
12:45 p.m. – 1:45 p.m.:	Session 3 – Preventing Foodborne Illness – Part II (30 slides [20 – 49]) ***Safety Stop!*** 3.3, 3.4, and 3.5 reviews
1:45p.m. – 3:00 p.m.:	Session 4 – Following the Food Product Flow (30 slides [1 – 30]) ***Safety Stop!***! 4.1, 4.2, and 4.3 reviews
3:00 p.m. – 3:15 p.m.:	Break
3:15 p.m. – 4:00 p.m.:	Session 4 – Following the Food Product Flow (21 slides [31 – 51]) ***Safety Stop!*** 4.4 and 4.5 reviews
4:00 p.m. – 5:00 p.m.:	Session 5 – Facilities, Equipment and Utensils (36 slides) ***Safety Stop!*** 5.1, 5.2, and 5.3 reviews
5:00 p.m.:	Adjourn and Review Session

Day 2

8:00 a.m. – 8:15 a.m.	Review of Day 1
8:15 a.m. – 9:15 a.m.	Session 6 – Cleaning and Sanitizing Operations (26 slides) ***Safety Stop!*** 6.1, 6.2, and 6.3 reviews
9:15 a.m. – 10:00 a.m.:	Session 7 – Environmental Sanitation and Maintenance (24 slides) ***Safety Stop!*** 7.1 and 7.2 reviews
10:00 a.m. – 10:15 a.m.:	Break
10:15 a.m. – 11:00 a.m.:	Session 8 – Accident Prevention and Crisis Management (21 slides) ***Safety Stop!*** 8.1 review
11:00 a.m. – 11:30 a.m.:	Session 9 – Education and Training (15 slides) ***Safety Stop!*** 9.1 review
11:30 a.m. – 12:30 p.m.:	Lunch
12:30 p m. – 1:45 p.m.:	Session 10 – Food Safety Management Programs (30 slides) ***Safety Stop!*** 10.1 review
1:45 p.m. – 2:15 p.m.:	Session 11 – Food Safety Regulations (18 slides) ***Safety Stop!*** 11.1 review
2:15 p.m. – 3:00 p.m.:	Final Review Session
3:00 p.m. – 3:15 p.m.:	Break
3:15 p.m. – 5:00 p.m.:	Food Safety Certification Exam.

SuperSafeMark Food Safety Training Program

8-hour Sample Agenda

8:00 a.m. – 8:10 a.m.:	Welcome, Introductions and Training Goals Setting the Ground Rules for the Training Program Meeting Participant Expectations
8:10 a.m. – 8:30 a.m.:	Session 1 – Retail Food Safety (15 slides)
8:30 a.m. – 9:30 a.m.:	Session 2 – Hazards to Food Safety (42 slides)
9:30 a.m. – 10:15 a.m.:	Session 3 – Preventing Foodborne Illness (37 slides)
10:15 a.m. – 10:30 a.m.:	Break
10:30 a.m. – 10:45 a.m.:	Session 3 ***Making it Real*** exercises • ***Safety Stop!*** 3.2 ***Making it Real*** Calibration of Temperature-measuring Devices exercise • ***Safety Stop!*** 3.4 ***Making it Real*** Proper Hand Washing Procedures exercise
10:45 a.m. – 12:00 noon:	Session 4 – Following the Food Product Flow (46 slides)
12:00 noon –12:30 p.m.:	Lunch
12:30 p.m. – 1:00 p.m.:	Session 5 – Facilities, Equipment and Utensils (22 slides)
1:00 p.m. – 1:30 p.m.:	Session 6 – Cleaning and Sanitizing Operations (19 slides)
1:30 p.m. – 2:00 p.m.:	Session 7 – Environmental Sanitation and Maintenance (20 slides)
2:00 p.m. – 2:15 p.m.:	Session 8 – Accident Prevention and Crisis Management (11 slides)
2:15 p.m. – 2:30 p.m.:	Break
2:30 p.m. – 2:45 p.m.:	Session 9 – Education and Training (12 slides)
2:45 p.m. – 3:30 p.m.:	Session 10 – Food Safety Management Programs (23 slides)
3:30 p.m. – 3:45 p.m.:	Session 11 – Food Safety Regulations (11 slides)
3:45 p.m. – 4:00 p.m.:	Final Review Session
4:00 p.m. – 5:30 p.m.:	Food Safety Certification Exam

Appendix C

SuperSafeMark 16-hour Session – Final Review Questions

Session 1
***Retail Food Safety* – (10 points)**

1. Identify three negative outcomes of a foodborne illness (2).
2. Define foodborne disease outbreak (1).
3. Most cases of foodborne illness are caused by foods that have been mishandled. Identify 3 ways foods can be mishandled in retail food establishments (2).
4. What 4 groups of people are highly susceptible to foodborne illness (2)?
5. Define contamination (1).
6. Identify four possible sources of contamination (2).

Session 2
***Hazards to Food Safety* – (20 points)**

1. What is the difference between foodborne infection and foodborne intoxication (1)?
2. What is a toxin-mediated infection (1)?
3. What is onset time (1)?
4. What biological agents are the leading causes of foodborne illness (1)?
5. List the three categories of hazards that are known to cause foodborne illness (1).
6. Why do bacteria form spores (1)?
7. What 6 factors do bacteria need for growth (2)?

8. What are the two most critical factors affecting the growth of bacteria in foods (1)?

9. Disease-causing bacteria grow best in the temperature danger zone. What temperature range is considered the temperature danger zone (1)?

10. What are potentially hazardous foods? Give four examples (2).

11. Give two examples of a foodborne intoxication (2).

12. Give two examples of a foodborne infection (2).

13. Give two examples of a toxin-mediated infection (2).

14. Name two diseases caused by parasites (2).

Session 3

***Preventing Foodborne Illness* – (13 points)**

1. Almost all foodborne illnesses are linked to 4 general food-handling errors. What are they (2)?

2. Identify three things food handlers can do in retail food establishments to demonstrate good personal hygiene (2).

3. Identify three situations when food handlers should wash their hands (2).

4. What four diseases must employees report to the person in charge when they have been diagnosed with one (2)?

5. What is cross contamination (1)?

6. Identify three ways to prevent cross contamination (2).

7. Describe the correct procedure for calibrating a bi-metallic stem thermometer (2).

Session 4

Following the Food Product Flow – (25 points)

1. Foods prepared in a private home may not be sold in a retail food establishment. Why (1)?

2. What is sensory evaluation of food (1)?

3. Describe the proper technique for measuring the temperature of unpackaged food (1).

4. Describe the proper technique(s) for measuring the temperature of packaged foods (1).

5. Identify three types of defects that can result in the rejection of canned foods in a retail food establishment (2).

6. What is the difference between inspection for wholesomeness and grading (2)?

7. What are three characteristics of fresh red meat (2)?

8. What are three characteristics of fresh poultry (2)?

9. What are three characteristics of fresh fish (2)?

10. What is FIFO (1)?

11. How far off the floor should foods be stored in storage areas? Why (1)?

12. What is the final cooking temperature for a) pork roasts, b) ground beef and ground pork, and c) poultry and stuffed meats (2)?

13. What is the final cooking temperature required for raw animal foods cooked in a microwave oven (1)?

14. Identify four ways to reduce the amount of time required for cooling foods (2).

15. Identify four ways to safely thaw foods (2).

16. What are the temperature and time requirements for re-heating foods (1)?

17. Identify two ways to prevent contamination and cross contamination at a self-service food operation (1).

Session 5

***Facilities, Equipment, and Utensils* (11 points)**

1. Identify two benefits of a properly designed and equipped retail food establishment (1).

2. Identify four features of equipment that should be taken into account when deciding which type of equipment to purchase (2).

3. Identify two organizations that evaluate the design and construction of food processing equipment for health and safety (2).

4. Floor mounted equipment must be how many inches off the floor or sealed to a masonry base? Why (1)?

5. List 5 characteristics of materials that can be used for food production equipment (2).

6. Why is wood not a recommended material as a construction material for food equipment (1)?

7. Why are ventilation systems installed in areas where food is cooked and equipment and utensils are washed (1)?

8. Why is proper lighting important in food production areas (1)?

Session 6

***Cleaning and Sanitizing Operations* (13 points)**

1. What is the purpose of cleaning (1)?

2. Identify three factors that affect how easily soil is removed from the surfaces of equipment and utensils (2).

3. What is the purpose of sanitizing (1)?

4. Identify the basic steps in the manual dishwashing procedure (2).

5. Identify three types of chemical sanitizers (2).

6. How long must equipment and utensils be immersed in chemical sanitizers to assure proper reduction of microbes (1)?

7. How hot must water be when it is used to sanitize equipment and utensils (1)?

8. What procedure should be used to dry equipment and utensils that have been properly cleaned and sanitized (1)?

9. What is the purpose of a materials safety data sheet (MSDS)? What are some pieces of information the MSDS contains (2)?

Session 7

***Environmental Sanitation and Maintenance* (15 points)**

1. Identify two features that should be considered with selecting materials for floors, walls and ceilings in retail food establishments (1).

2. What do you call the molding that is used to make the cleaning of the areas where the floor and walls intersect in a food production area (1)? (NOTE: The molding is usually ceramic tiling)

3. What is an air gap and what does it do (1)?

4. Identify the equipment and supplies that are considered standard features of a properly equipped handwashing station (2).

5. Identify three features of a properly designed and equipped toilet facility (2).

6. Identify three ways to control flies and flying insects in a food establishment (2).

7. Identify two ways to control cockroaches in a food establishment (1).

8. What are some common signs of rodent infestation (1)?

9. Identify three ways to control rodents in a food establishment (1).

10. What is Integrated Pest Management? What are the advantages of this method for pest control (2)?

11. How does proper storage and disposal of garbage and refuse help reduce pest problems in retail food establishments (1)?

Session 8

***Accident Prevention and Crisis Management* (8 points)**

1. Identify two ways retail food establishments can prevent accidents (1).

2. According to the *FDA Food Code*, what are three rules for properly handling employee medications in a retail food establishment (2)?

3. Identify three types of injury that are common in retail food establishments. How can they be best eliminated or controlled (2)?

4. Identify three situations where a crisis management plan should be employed (2)?

5. What type of fire extinguisher is recommended for use when fighting grease and electrical fires (1)?

Session 9

***Education and Training* (6 points)**

1. What are three benefits to training (2)?

2. What are three barriers to effective training (2)?

3. What are some factors that make an environment conducive to learning (1)?

4. What are some ways to evaluate the effectiveness of training (1)?

Session 10

***Food Safety Management Programs* (17 points)**

1. A food safety management system should protect foods from factors that can lead to foodborne illness. Name 3 risk factors that are known to contribute to foodborne disease outbreaks (2).

2. A total food safety management system in retail food establishments consists of Standard Operating Procedures (SOPs), Good Retail Practices (GRPs), and a Hazard Analysis Critical Control Point system. What is meant by the term SOP? Give two examples of SOPs (2).

3. What is meant by the term GRP? Give two examples of GRPs (2).

4. What is a hazard (1)?

5. What is hazard analysis (1)?

6. What is a critical control point (1)?

7. Identify four critical control points (2).

8. What is a critical limit (1)?

9. What purpose does monitoring serve in HACCP (1)?

10. When should a HACCP plan be modified (1)?

11. What is a food recall (1)?

12. Identify two uses for product coding (2).

Session 11

Food Safety Regulations (6 points)

1. Which governmental agencies have the most influence on the day-to-day operation of retail food establishments (1)?

2. What are the basic phases of a retail food inspection (1)?

3. What are some of the activities of the FDA (1)?

4. What are some of the activities of the USDA (1)?

5. What is the *FDA Food Code* (1)?

6. What is the purpose of a consumer advisory as referenced in the *FDA Model Food Code* (1)?

Appendix D

SuperSafeMark 16-hour Session – Final Review Questions and Answers

Chapter 1
***Retail Food Safety* – (10 points)**

1. Identify three negative outcomes of a foodborne illness (2).
   - Loss of business
   - Loss of reputation
   - Lawsuits and legal fees
   - Possible closure by regulatory agencies
   - Injury to or loss of life
   - Expenses to victims.

2. Define foodborne disease outbreak (1).
   - When two or more unrelated people experience a similar illness after consuming a common food.

3. Most cases of foodborne illness are caused by foods that have been mishandled. Identify 3 ways foods can be mishandled in retail food establishments. (2 points)
   - Improperly cooked
   - Held at improper temperatures
   - Exposed to disease-causing agents by cross contamination
   - In contact with contaminated equipment that has not been properly cleaned and sanitized.
   - Obtained from unsafe sources.

4. What 4 groups of people are highly susceptible to foodborne illness (2)?
   - Elderly
   - Infants and children
   - Pregnant and lactating women
   - People with weakened immune system.

5. Define contamination (1).
   - The presence of harmful substances in foods that can lead to foodborne illness.

6. Identify four possible sources of contamination (2).
   - Raw foods
   - Contaminated water
   - Air
   - Food workers
   - Packages
   - Rodents and insects
   - Contaminated equipment.

Chapter 2

Hazards to Food Safety – (20 points)

1. What is the difference between foodborne infection and foodborne intoxication (1)?
   - In an infection the live cells leads to the illness and in an intoxication a preformed toxin leads to the illness.

2. What is a toxin-mediated infection (1)?
   - In a toxin mediated infection the live cell must be consumed with the food and then the live cell produces a toxin in the human gut.

3. What is onset time (1)?
   - The amount of time that elapses between eating a contaminated food and the first signs of the illness occur.

4. What biological agents are the leading causes of foodborne illness (1)?
   - Bacteria
   - Viruses.

5. List the three categories of hazards that are known to cause foodborne illness (1).
   - Biological
   - Chemical
   - Physical.

6. Why do bacteria form spores (1)?
   - To survive in adverse environmental conditions. Spores are very resistant to heat.

7. What 6 factors do bacteria need for growth (2)?
   - Food
   - Acid >4.6
   - Temperature 41°F (5°C) to 135°F (57°C) (Known as the Temperature Danger Zone)
   - Time >4 hours
   - Oxygen (conditions vary) (i.e., aerobic, anaerobic, facultative)
   - Moisture A_w>0.85.

8. What are the two most critical factors affecting the growth of bacteria in foods (1)?
   - Time and Temperature.

9. Disease-causing bacteria grow best in the temperature danger zone. What temperature range is considered the temperature danger zone (1)?
   - 41°F (5°C) to 135°F (57°C).

10. What are potentially hazardous foods? Give four examples (2).
    - Foods that are capable of supporting the rapid and continuous growth of disease-causing bacteria including:
      - Fish
      - Poultry
      - Red meat
      - Dairy products
      - Eggs
      - Re-fried beans
      - Cooked rice.

11. Give two examples of a foodborne intoxication (2).
    - *Clostridium botulinum*
    - *Staphylococcus aureus*
    - Seafood toxins
    - *Bacillus cereus* can also be an intoxication, however it is usually associated with a toxin-mediated infection.

12. Give two examples of a foodborne infection (2).
    - *Salmonella*
    - *Campylobacter*
    - Viruses
    - Parasites.

13. Give two examples of a toxin-mediated infection (2).
    - *Clostridium perfringens*
    - Shiga toxin-producing *E. coli*
    - *Bacillus cereus.*

14. Name two diseases caused by parasites (2).
    - Trichinosis
    - Cryptosporidisis
    - Anisakasis.

Chapter 3

Preventing Foodborne Illness – (13 points)

1. Almost all foodborne illnesses are linked to 4 general food-handling errors. What are they (2)?
   - Time and temperature abuse
   - Poor personal hygiene
   - Cross contamination
   - Improper cleaning and sanitizing.

2. Identify three things food handlers can do in retail food establishments to demonstrate good personal hygiene (2).
   - Wash their hands properly
   - Wear clean clothing
   - Wear proper hair restraints
   - No smoking or eating in food production and warewashing areas.
   - Keep their fingernails short and not wear false fingernails
   - Not go to work when they are ill.

3. Identify three situations when food handlers should wash their hands (2).
    - Before food preparation
    - After touching bare human body parts, except clean hands and clean exposed arms
    - After using the toilet
    - When returning to a work station
    - After coughing, sneezing, using a handkerchief or tissue, using tobacco, eating, or drinking
    - During food preparation as often as necessary to remove soil and contamination and to prevent cross contamination.
    - Before putting on gloves for working with food
    - After engaging in any activities that may contaminate hands (taking out the garbage, wiping counters or tables, handling cleaning chemicals, picking up dropped items, etc.)
    - After caring for or touching service animals or aquatic animals.

4. What four diseases must employees report to the person in charge when they have been diagnosed with one (2)?
    - *Salmonella* Typhi
    - *Shigella*
    - Hepatitis A virus
    - Shiga toxin-producing *E. coli*
    - Note: all other symptoms of diarrheal illness and flu-like symptoms should be reported.

5. What is cross contamination (1)?
    - Transfer of hazardous substances from one food to another (typically from raw to ready-to-eat foods).

6. Identify three ways to prevent cross contamination (2).
    - Store ready-to-eat foods above raw foods
    - Clean and sanitize equipment and utensils after raw then ready-to-eat foods
    - Use separate equipment (i.e., cutting boards) for raw and ready-to-eat foods
    - Good personal hygiene.

7. Describe the correct procedure for calibrating a bi-metallic stem thermometer (2).

   - Insert the probe into a cup of crushed ice. Add enough cold water to the cup to remove any air pockets that might remain. Wait until the temperature stabilizes in the ice and water mixture and adjust the needle to 32°F (0°C). -Immerse the sensing portion of the temperature measuring device in an ice water bath (50% crushed ice, 50% water) calibrated to freezing temperatures or in boiling water calibrated to boiling temperatures.

Chapter 4

Following the Food Product Flow – (25 points)

1. Foods prepared in a private home may not be sold in a retail food establishment. Why (1)?

   - The processes used for making home-prepared foods may not ensure the destruction of the deadly bacteria *Clostridium botulinum*.

2. What is sensory evaluation of food (1)?

   - Use of taste, smell, touch, and site to determine the culinary quality of food.

3. Describe the proper technique for measuring the temperature of unpackaged food (1).

   - Place the sensing part of the temperature-measuring probe in the thickest part of the food (typically the geometric center of the food).

4. Describe the proper technique(s) for measuring the temperature of packaged foods (1).

   - Place the sensing part of the temperature measuring probe in-between two packages or fold the package around the probe.

5. Identify three types of defects that can result in the rejection of canned foods in a retail food establishment (2).

   - Bulging cans
   - Excessive rust especially around the side seams or can lid
   - Dents on seams or can lids.

6. What is the difference between inspection for wholesomeness and grading (2)?

   - Inspection for wholesomeness is used to identify conditions that would indicate that the food is contaminated.
   - Grading is the evaluation to characterize food quality.

7. What are three characteristics of fresh red meat (2)?
   - Received at the proper temperature
   - Bright red color
   - Firm and elastic to the touch
   - No off odors and aroma.

8. What are three characteristics of fresh poultry (2)?
   - Received at the proper temperature
   - No darkened wing tips or mucus build-up under the wings
   - Firm and elastic to the touch
   - No off odors and aroma.

9. What are three characteristics of fresh fish (2)?
   - Received at the proper temperature
   - Bright red gills and bulging clear eyes
   - Firm and elastic to the touch
   - No off odors and aroma.

10. What is FIFO (1)?
    - First-in, First-out – an inventory and storage system for assuring the proper rotation of foods during storage.

11. How far off the floor should foods be stored in storage areas? Why (1)?
    - Six-inches
    - To facilitate cleaning and discourage pest harborage.

12. What is the final cooking temperature for a) pork roasts, b) ground beef and ground pork, and c) poultry and stuffed meats (2)?
    - a) 145°F (62°C)
    - b) 155°F (68°C)
    - c) 165°F (74°C).

13. What is the final cooking temperature required for raw animal foods cooked in a microwave oven (1)?
    - 165°F (74°C).

14. Identify four ways to reduce the amount of time required for cooling foods (2).
    - Break bulk foods down into smaller portions
    - Use ice as an ingredient for concentrated foods
    - Use an ice water bath
    - Distribute into shallow pans
    - Frequent stirring of liquid foods
    - Use quick chill units.

15. Identify four ways to safely thaw foods (2).
    - In a refrigerator
    - Under cool, running water
    - In a microwave oven and immediately transferred to conventional cooking equipment
    - As part of the cooking process.

16. What are the temperature and time requirements for re-heating foods (1)?
    - Reheat foods to 165°F (74°C) within two hours.

17. Identify two ways to prevent contamination and cross contamination at a self-service food operation (1).
    - Have customers get a clean plate each time the get food.
    - Use properly designed utensils and use only one utensil for each food item.
    - Use sneeze shields to protect food from contamination by customers.
    - Do not put raw animal foods on a self-service bar.
    - Keep foods out of the temperature danger zone.

Chapter 5

Facilities, Equipment, and Utensils (11 points)

1. Identify two benefits of a properly designed and equipped retail food establishment (1).
    - Workers may perform their tasks more efficiently
    - A safer place for employees to work in
    - A more sanitary environment for food handling activities.

2. Identify four features of equipment that should be taken into account when deciding which type of equipment to purchase (2).
   - Design
   - Construction
   - Durability
   - Ability to clean easily
   - Size
   - Cost
   - Safety
   - Overall ability to do the job.

3. Identify two organizations that evaluate the design and construction of food processing equipment for health and safety (2).
   - NSF International
   - Underwriter's Laboratories.

4. Floor mounted equipment must be how many inches off the floor or sealed to a masonry base? Why (1)?
   - Six-inches off the floor
   - To prevent trash from accumulating and pest harborage.

5. List 5 characteristics of materials that can be used for food production equipment (2).
   - Non toxic
   - Non absorbent
   - Smooth
   - Easily cleanable
   - Durable under normal use
   - Will not chip or pit.

6. Why is wood not a recommended material as a construction material for food equipment (1)?
   - Wood is absorbent and can harbor the attachment and growth of microorganisms.

7. Why are ventilation systems installed in areas where food is cooked and equipment and utensils are washed (1)?
   - To remove potential contaminants that may be present in the air such as grease, condensate, smoke, fumes, and toxic gases.

8. Why is proper lighting important in food production areas (1)?
    - It provides a better working environment for employees and it shows when work areas are dirty and if they have been properly cleaned.

Chapter 6

***Cleaning and Sanitizing Operations* (13 points)**

1. What is the purpose of cleaning (1)?
    - To remove soil and visible dirt from surfaces.

2. Identify three factors that affect how easily soil is removed from the surfaces of equipment and utensils (2).
    - Type of soil to be removed
    - Water quality
    - The detergent or cleaner to be used.
    - Water temperature.
    - Water velocity or force.
    - Time detergent remains in contact with the surface.
    - The concentration of cleaner.

3. What is the purpose of sanitizing (1)?
    - To reduce food contaminants on a properly cleaned surface to safe and acceptable levels.

4. Identify the basic steps in the manual dishwashing procedure (2).
    - Pre-flush
    - Wash
    - Rinse
    - Sanitize
    - Air dry.

5. Identify three types of chemical sanitizers (2).
    - Chlorine
    - Iodophors
    - Quaternary ammonia compounds.

6. How long must equipment and utensils be immersed in chemical sanitizers to assure proper reduction of microbes (1)?
    - At least 10 seconds for chlorine and at least 30 seconds for iodophors and quaternary ammonia compounds.

7. How hot must water be when it is used to sanitize equipment and utensils (1)?
    - 171°F (77°C).

8. What procedure should be used to dry equipment and utensils that have been properly cleaned and sanitized (1)?
    - Air dry.

9. What is the purpose of a materials safety data sheet (MSDS)? What are some pieces of information the MSDS contains (2)?
    - It provides information about the safety and toxicity of hazardous substances.
    - An MSDS will contain information on the following:

      i) Chemical composition

      ii) Physical characteristics

      iii) Suggested personal protective equipment

      iv) Recommended first aid procedures from exposure.

Chapter 7

Environmental Sanitation and Maintenance (9 points)

1. Identify two features that should be considered with selecting materials for floors, walls and ceilings in retail food establishments (1).
   - Sanitation
   - Safety
   - Durability
   - Comfort
   - Cost.

2. What do you call the molding that is used to make the cleaning of the areas where the floor and walls intersect in a food production area (1)? (NOTE: The molding is usually ceramic tiling)
   - Coving.

3. What is an air gap and what does it do (1)?
   - The physical space between the opening of a faucet and the flood rim of a basin.
   - It prevents back siphonage.

4. Identify the equipment and supplies that are considered standard features of a properly equipped handwashing station (2).
   - Hot and cold running water under pressure through a mixing faucet.
   - A supply of soap
   - A means to dry hands.
   - A waste container to collect disposable towels when they are used.

5. Identify three features of a properly designed and equipped toilet facility (2).
    - Employee restrooms must be conveniently located and accessible to employees during all hours of operation.
    - Toilet rooms must be completely enclosed and provided with tight-fitting and self-closing doors.
    - Toilet fixtures must be durable and easily cleanable.
    - Handwashing stations must be provided.
    - Floors, walls and fixtures in toilet areas must be clean and well maintained.
    - Toilet tissue must be supplied at each toilet.
    - Easy-to-clean containers for waste materials and at least one covered container in toilet rooms used by women.

6. Identify three ways to control flies and flying insects in a food establishment (2).
    - Air curtains
    - Screens
    - Self closing doors
    - Electronic zappers.

7. Identify two ways to control cockroaches in a food establishment (1).
    - Implementing an integrated pest management program (IPM)
    - Inspect incoming shipments
    - Tight threshold under doors
    - A good cleaning and sanitation program
    - Eliminate food and shelter.

8. What are some common signs of rodent infestation (1)?
    - Rodent droppings
    - Rub marks
    - Burrows
    - Gnawing
    - Footprints.

9. Identify three ways to control rodents in a food establishment (1).
   - Implementing an integrated pest management program (IPM)
   - Eliminate food and shelter
   - A good cleaning and sanitation program
   - Prevent entry with proper construction and facility design.

10. What is Integrated Pest Management? What are the advantages of this method for pest control (2)?
    - A system that uses a combination of biological, mechanical, and chemical controls to reduce pest population to acceptable levels.
    - A good IPM program reduces the risk of contamination due to pests and pesticides.

11. How does proper storage and disposal of garbage and refuse help reduce pest problems in retail food establishments (1)?
    - Eliminates a potential food supply for the pests.

Chapter 8

Accident Prevention and Crisis Management (7 points)

1. Identify two ways retail food establishments can prevent accidents (1).
   - Eliminating hazards in the environment
   - Providing personal protective equipment for employees
   - Training employees about potential hazards and how to prevent or avoid them.

2. According to the *FDA Food Code*, what are three rules for properly handling employee medications in a retail food establishment (2)?
   - Only those medications necessary for the employee's health are allowed in a retail food establishment.
   - Employee medications must be clearly labeled and stored in an area away from food, equipment, utensils, linens, and single-use items like straws, eating utensils, and napkins.
   - Medications stored in a designated refrigerator must be kept in a clearly labeled and covered container located on the lowest shelf of the refrigerator.
   - Medications cannot be stored in display or storage cases, or walk-in units.

3. Identify three types of injury that are common in retail food establishments. How can they be best eliminated or controlled (2)?
   - Cuts – use sharp utensils
   - Burns – use personal protective equipment
   - Falls – anti skid mats, proper use of ladders, and clean up spills.

4. Identify three situations where a crisis management plan should be employed (2)?
   - Foodborne illness
   - Fire
   - Loss of utility services
   - Natural disasters.

5. What type of fire extinguisher is recommended for use when fighting grease and electrical fires (1)?
   - Carbon dioxide (CO_2), or type B/C extinguishers are recommended because it is safe and leaves no damaging residue on machinery or food-contact surfaces.

Chapter 9

Education and Training (6 points)

1. What are three benefits to training (2)?
   - Customer satisfaction
   - Lower employee turnover
   - Lower costs
   - Fewer accidents
   - Better quality products.

2. What are three barriers to effective training (2)?
   - Lack of time
   - Lack of commitment
   - Lack of knowledge
   - Lack of ability to train
   - The culture of the organization.

3. What are some factors that make an environment conducive to learning (1)?
   - Well lit
   - Quiet
   - Comfortable temperature.

4. What are some ways to evaluate the effectiveness of training (1)?
   - Observe if good or improved food handling practices are taking place.
   - Written or oral exams.
   - Ask the learner to demonstrate the techniques they have been taught.

Chapter 10

Food Safety Management Programs (17 points)

1. A food safety management system should protect foods from factors that can lead to foodborne illness. Name 3 risk factors that are known to contribute to foodborne disease outbreaks (2).
   - Improper holding temperatures
   - Poor personal hygiene practices
   - Cross contamination
   - Inadequate cooking
   - Food from unsafe sources
   - Contaminated equipment

2. A total food safety management system in retail food establishments consists of Standard Operating Procedures (SOPs), Good Retail Practices (GRPs), and a Hazard Analysis Critical Control Point system. What is meant by the term SOP? Give two examples of SOPs (2).
   - Standard Operating Procedures are written procedures on how to perform a job properly. Examples may include good personal hygiene programs, cleaning and sanitizing programs, pest management, storage practices, water and ice handling, and use of chemicals.

3. What is meant by the term GRP? Give two examples of GRPs (2).
   - Good Retail Practices are minimum sanitary and processing requirements necessary to ensure the production of safe and wholesome food.
   - Examples may include:
     i) Guidelines for receiving, storing, preparing, packaging, and display
     ii) Personal health
     iii) Employee and customer education
     iv) Record keeping.

4. What is a hazard (1)?

   - Hazards are biological, chemical, or physical agents that may be introduced to food by certain food production practices or the intended use of the product. Hazards are commonly introduced into food by poor food handling practices and contaminated equipment. Some foods may be naturally contaminated as in the case of raw meats, poultry, fish, and unwashed fruits and vegetables.

5. What is hazard analysis (1)?

   - Hazard Analysis involves identifying *hazards* that might be introduced to food by certain food production practices or the intended use of the product. Hazard analysis starts with a review of the menu or product list to identify all of the potentially hazardous foods served to the customers.

6. What is a critical control point (1)?

   - A CCP is an operation (practice, preparation, step or procedure) in the flow of food that will prevent, eliminate, or reduce hazards to acceptable levels.

7. Identify four critical control points (2).

   - The four most commonly used CCP's are:

     i) Cooking

     ii) Cooling

     iii) Reheating

     iv) Hot/cold holding.

8. What is a critical limit (1)?

   - Critical limits are the upper and lower boundaries of food safety. When these boundaries are exceeded, a hazard may exist or could develop. Critical limits must be set to make certain that each critical control point effectively blocks a biological, chemical, or physical hazard.

9. What purpose does monitoring serve in HACCP (1)?

   - Monitoring involves making observations and measurements to determine whether a critical control point is under control. Monitoring also shows when a critical limit has been exceeded at a critical control point. Monitoring is a critical part of a HACCP system and provides written documentation that can be used to verify that the HACCP is working properly.

10. When should a HACCP plan be modified (1)?

    - The HACCP system should be reviewed and, if necessary, modified to accommodate changes in your clientele (i.e. more high risk clients), the items on your menu (addition of potentially hazardous foods or substitution of low-risk foods for high-risk foods), or the processes used to prepare HACCP products.

11. What is a food recall (1)?
    - A food recall is the system used to prevent customers from using a product that may be unsafe. The purpose of a recall is to remove a food product from commerce when there is reason to believe it may be adulterated (injurious to health or unfit for human consumption) or misbranded (false or misleading labeling and/or packaging).

12. Identify two uses for product coding (2).
    - To assure that foods are displayed and used on a FIFO basis.
    - To assure that food products with an expired shelf life are discarded properly.
    - To assure that potentially hazardous foods are held for the proper amount of time at the required storage and display temperatures.

Chapter 11

Food Safety Regulations (6 points)

1. Which governmental agencies have the most influence on the day-to-day operation of retail food establishments (1)?
    - State and local regulatory agencies.

2. What are the basic phases of a retail food inspection (1)?
    - Pre-inspection conference
    - Inspection
    - Post-inspection conference.

3. What are some of the activities of the FDA (1)?
    - Inspection of all foods except for meat and poultry
    - Facilitate the development of the Food Code
    - Work with states to effectively implement the recommendations of the Food Code.

4. What are some of the activities of the USDA (1)?
    - Inspect meat and poultry products
    - Grading of food products, such as meats, poultry, pork, eggs and produce.

5. What is the *FDA Food Code* (1)?

   - A document published by the FDA that serves as a model for safe food handling practices. The code is a set of recommendations, and is not federal law or federal regulation.

6. What is the purpose of a consumer advisory as referenced in the *FDA Model Food Code* (1)?

   - To notify customers about the increased risk associated with eating raw or undercooked animal foods or ingredients.
   - The FDA additionally issues consumer advisories for products, such as methyl mercury in seafood and consumption of deli products by high-risk populations.

Appendix E

Posters

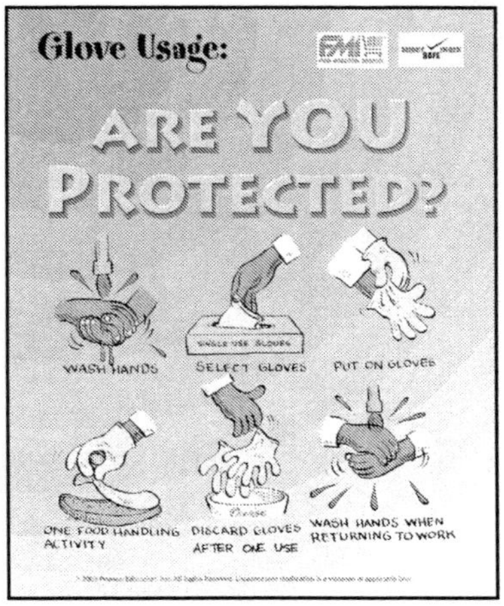

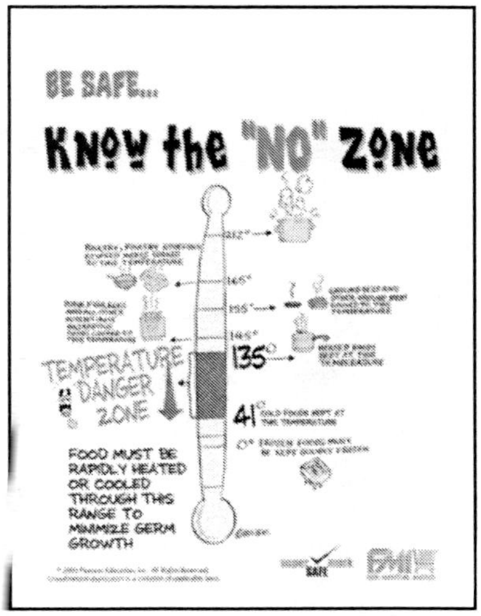
BE SAFE...
Know the "NO" Zone
TEMPERATURE DANGER ZONE
135°
41°
FOOD MUST BE RAPIDLY HEATED OR COOLED THROUGH THIS RANGE TO MINIMIZE GERM GROWTH

Don't Mix...
Raw and Ready-to-Eat Foods!

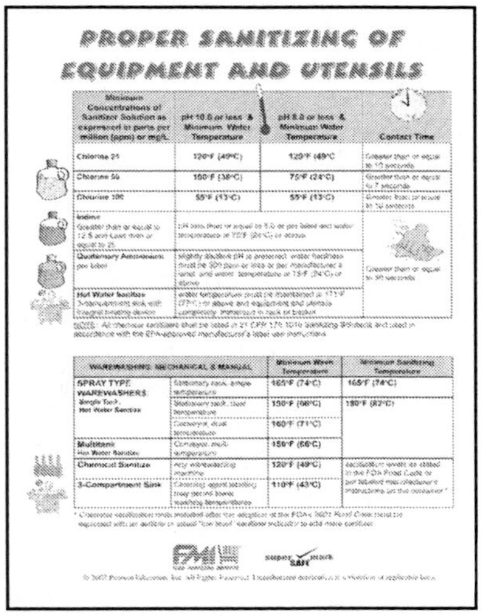
PROPER SANITIZING OF EQUIPMENT AND UTENSILS

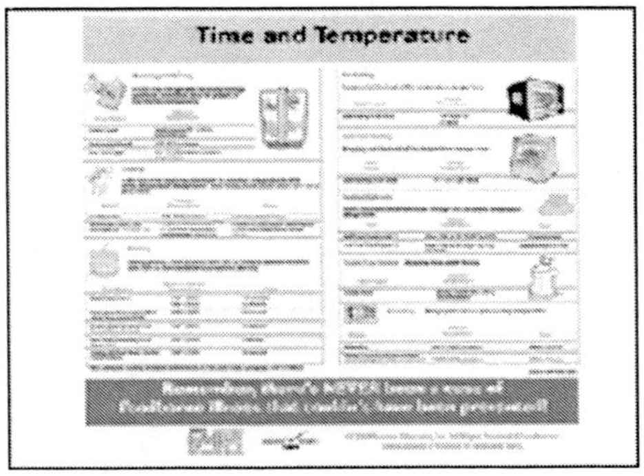
Time and Temperature

Wash Your Hands!
SOAP

Appendix F

Microsoft PowerPoint® Slideshows: Viewing, Installation & Customization

The PowerPoint® Training Sessions can be run from the enclosed CD (CD No. 1 – 16-hour, or CD No. 2 – 8-hour) or downloaded onto your computers hard drive and run from there. In order to run the Training Sessions on your computer you need to have Microsoft PowerPoint® installed on your computer.

Running PowerPoint® Training Sessions from the enclosed CD

1) Insert Compact Disk (16-hour or 8-hour) into your computers CDROM drive.
2) Click on the "Start" button in the bottom left corner of the screen.
3) Click on "Run"
4) Type the name of the drive you are using (usually the "D" drive) and a colon, i.e. "D:"
5) A screen will appear that shows all of the folders on the CD
   - Session 1
   - Session 2
   - Session 3
   - Session 4
   - Session 5
   - Session 6
   - Session 7
   - Session 8
   - Session 9
   - Session 10
   - Session 11
   - A "Read Me" file
6) Double-click on the training session you want to open.
7) To run the training session as a slideshow:
   - Click on View
   - Click on Slideshow
   - Or press F5

Installing the Training Sessions onto Your Computer

Before beginning installation, it is suggested all other programs be closed. Before you attempt to install the Training Sessions on you computer, it is suggested you carefully read all of the instructions first. Then follow the instructions as you install your new PowerPoint® Training Sessions onto your computer.

1) Insert Compact Disk (16-hour or 8-hour) into your computers CDROM drive.
2) Click on the "Start" button in the bottom left corner of the screen.
3) Click on "Run"
4) Type the name of the drive you are using (usually the "D" drive) and a colon, i.e. "D:"
5) A screen will appear that shows all of the folders on the CD
   - Session 1
   - Session 2
   - Session 3
   - Session 4
   - Session 5
   - Session 6
   - Session 7
   - Session 8
   - Session 9
   - Session 10
   - Session 11
   - A "Read Me" file
6) Double-click on the training session you want to save to your hard drive. This will open the file.
7) Click on File
8) Click on Save As
9) Open the directory (or create a directory) where you want to save the file
10) Click Save.

You may want to create a directory to house you new PowerPoint® Training Sessions on. To do this, follow these instructions:

Creating a Directory (Optional)

It is suggested you create a directory on your computer to house your new PowerPoint® Training Sessions. To do this, follow these instructions:

1) Follow the above instructions (PowerPoint® Training Session Installation) to Step 9
2) Click on the down arrow to the left of the box called "Save In"
3) Click on the "C" drive (or the drive where you want to house your training sessions).
4) Click on the "Create New Folder" icon to the right of the "Save In" box (looks like a file folder with an asterisk [*] above it

5) Type in a name for the folder. The folder name should be no longer than 8 letters and not contain spaces. We suggest "FoodSafe"
6) Click OK
7) Click Save.

Customizing Your Slideshows

In order to customize your Microsoft PowerPoint®© Slideshows, you will need to have Microsoft PowerPoint®© installed on your computer.

All of your new Microsoft PowerPoint®© Trainers Sessions are fully customizable as long as you own, and have installed on your computer, the Microsoft PowerPoint®© program. In order to customize your slides, you will need to open the file you want to edit in Microsoft PowerPoint®©. To open your session in PowerPoint®© follow these instructions:

1. Open your previously installed Microsoft PowerPoint®© program
2. Open the file folder where you saved your presentations (we suggested "C:\Foodsafe") during the "Installing Training Sessions."
3. Double click on the Training Session you want to edit. This will open the file in Microsoft PowerPoint®©.

Once the Training Session you want to edit is open, follow these instructions (also listed in the ("Image Bank CD – CD No. 4 Image bank with illustrations and photos used for slide customization") section of the "How to Use the Trainer's Kit" introduction. That section will guide you through inserting and deleting photos and text from the supplied "Image Bank" CD.

1. Open the presentation you wish to edit.
2. Scroll to the slide you wish to edit.
3. Click on "Insert."
4. Place your mouse over "Picture."
5. Click on "From File."
6. From the "Insert Picture" box, choose your compact disk drive.
7. Double click on the picture you want to insert.

To delete a photo, illustration, or text:

1. Click on the image you want to delete.
2. Press delete.

